ECOLOGICAL INTERACTIONS IN SOIL:

PLANTS, MICROBES AND ANIMALS

ECOLOGICAL INTERACTIONS
IN SOIL

PLANTS, MICROBES AND ANIMALS

SPECIAL PUBLICATION NUMBER 4 OF THE
BRITISH ECOLOGICAL SOCIETY

EDITED BY

A. H. FITTER

Department of Biology,
University of York,
York YO1 5DD

WITH

D. ATKINSON

East Malling Research Station,
East Malling,
Maidstone,
Kent ME19 6BJ

D. J. READ

Department of Botany,
University of Sheffield,
Sheffield S10 2TN

M. B. USHER

Department of Biology,
University of York,
York YO1 5DD

BLACKWELL SCIENTIFIC PUBLICATIONS

OXFORD   LONDON   EDINBURGH

BOSTON   PALO ALTO   MELBOURNE

1985

First published 1985
Reprinted 1988

Set by Spire Print Services Ltd, Salisbury,
Wilts.
Printed in Great Britain.

DISTRIBUTORS

USA
    Blackwell Scientific Publications Inc
    P O Box 50009, Palo Alto
    California 94303

Canada
    Oxford University Press
    70 Wynford Drive
    Don Mills
    Ontario M3C 1J9

Australia
    Blackwell Scientific Publications
    (Australia) Pty Ltd
    107 Barry Street, Carlton
    Victoria 3053

British Library
Cataloguing in Publication Data

Ecological interactions in soil: plants,
    microbes and animals.—(Special
    publication of the British Ecological
    Society, ISSN 0262–7027; no.4)
1. Soil ecology
I. Fitter, A.H. II. Atkinson, D.
1944-   III. Series
574.5′26404   QH541.5S6

ISBN 0-632-01386-9

CONTENTS

Preface   vii

D. C. Coleman. Through a ped darkly: an ecological assessment of root-soil-microbial-faunal interactions   1

R. Fogel. Roots as primary producers in below-ground ecosystems   23

R. I. Fairley & I. J. Alexander. Methods of calculating fine root production in forests   37

D. Atkinson. Spatial and temporal aspects of root distribution as indicated by the use of a root observation laboratory   43

A. Carpenter, J. M. Cherrett, J. B. Ford, M. Thomas & E. Evans. An inexpensive rhizotron for research on soil and litter-living organisms   67

D. J. Gibson, I. A. Colquhoun & P. Greig-Smith. A new method for measuring nutrient supply rates in soils using ion-exchange resins   73

R. I. Fairley. Grass root production in restored soil following open-cast mining.   81

A. H. Fitter. Functional significance of root morphology and root system architecture.   87

E. I. Newman. The rhizosphere: carbon sources and microbial populations   107

P. C. Brooks, D. S. Powlson & D. S. Jenkinson. The microbial biomass in soil   123

K. E. Giller & J. M. Day. Nitrogen fixation in the rhizosphere: significance in natural and agricultural systems   127

L. A. Kapustka, P. T. Arnold & P. T. Lattimore. Interactive responses of associative diazotrophs from a Nebraska Sand Hills grassland   149

K. Paustian. Influence of fungal growth pattern on decomposition and nitrogen mineralization in a model system   159

R. C. Ferrier & I. J. Alexander. The persistence under field conditions of excised fine roots and mycorrhizas of spruce   175

J. M. Lynch. Microbial saprophytic activity on straw and other residues: consequences for the plant   181

D. J. Read, R. Francis & R. D. Finlay. Mycorrhizal mycelia and nutrient cycling in plant communities   193

v

G. Wang, D. P. Stribley, P. B. Tinker & C. Walker. Soil pH and vesicular-arbuscular mycorrhizas — 219

K. Killham. VA mycorrhizal mediation of trace and minor element uptake in perennial grasses: relation to livestock herbage — 225

H. Petersen, R. V. O'Neill & R. H. Gardner. Use of an ecosystem model for testing ecosystem responses to inaccuracies of root and microflora productivity estimates — 233

M. B. Usher. Population and community dynamics in the soil ecosystem — 243

R. Harmer & I. J. Alexander. Effects of root exclusion on nitrogen transformations and decomposition processes in spruce humus — 267

R. G. Booth & M. B. Usher. Relationships between Collembola and their environment in a maritime Antarctic moss-turf habitat — 279

R. A. Brown. Effects of some root-grazing arthropods on the growth of sugar-beet — 285

S. Visser. Role of the soil invertebrates in determining the composition of soil microbial communities — 297

R. D. Finlay. Interactions between soil micro-arthropods and endomycorrhizal associations of higher plants — 319

P. J. A. Shaw. Grazing preferences of *Onychiurus armatus* (Insecta: Collembola) for mycorrhizal and saprophytic fungi in pine plantations — 333

O. W. Heal & J. Dighton. Resource quality and trophic structure in the soil system — 339

M. Clarholm. Possible roles for roots, bacteria, protozoa and fungi in supplying nitrogen to plants — 355

H. A. Verhoef & R. G. M. De Goede. Effects of collembolan grazing on nitrogen dynamics in a coniferous forest — 367

J. M. Anderson, S. A. Huish, P. Ineson, M. A. Leonard & P. R. Splatt. Interactions of invertebrates, micro-organisms and tree roots in nitrogen and mineral element fluxes in deciduous woodland soils — 377

J. P. Curry, M. Kelly & T. Bolger. Role of invertebrates in the decomposition of *Salix* litter in reclaimed cutover peat — 393

J. A. Springett. Effect of introducing *Allolobophora longa* Ude on root distribution and some soil properties in New Zealand pastures — 399

J. Miles. Soil in the ecosystem — 407

Author index — 429

Subject index — 444

Preface

This volume contains papers presented at a British Ecological Society meeting held in York from 16 to 18 April 1984. The meeting brought together researchers on all aspects of the soil biota, from the roots through the various microbial groups to the soil fauna. There have been several meetings recently on interactions between parts of this complex, for example on root–fungus interactions (particularly mycorrhizas) and on animal–microbial interactions. Our aim was to link together all these studies so that, for example, groups working on decomposition could talk to those interested in root production or in grazing interactions; in particular we hoped to provide an opportunity to discuss higher-order interactions, in which producer, grazer and predator, for example, interact in the soil environment. We also wished to provide an opportunity to bring together workers in agricultural and natural ecosystems.

These thoughts dictated the order and selection of papers, and the arrangement in this volume progresses from studies on roots alone, to root–microbial, animal–microbial and finally more complex interactions. In both the planning and delivery stages the meeting served to highlight several areas where little information was available or where, excitingly, the data are now beginning to appear.

One impression that stands out strongly is of the need to forge links between above-ground and below-ground ecological studies. Both in quantification and in the development of theory, 'above-ground' ecologists seem to have outstripped their 'subterranean' colleagues. It is clear, though, that we are at last beginning to put numbers to our diagrams. In the opening paper, Coleman points to the complexity of below-ground food webs, with up to seven possible links in a chain. Does this imply a greater complexity than is found in above-ground webs or simply that there are more alternative pathways? Without numbers it is difficult to answer such questions, and it may be that the application of above-ground theory to below-ground events is not a simple matter, as suggested by Heal & Dighton and Usher. It is certain, though, that invertebrates have a profound effect on microbial populations (Visser) and on microbially-mediated processes (Anderson *et al.*). The consequences of such interactions may be unexpected, as for example, the 'priming' effect suggested by Clarholm, and greatly influence plant growth, above- and below-ground. The essential connectedness of the above- and below-ground subsystems is emphasized by Miles.

Production estimates are another area where quantification seems now to be possible. Fogel shows that in many natural communities, the greater part of net primary production goes underground, in striking contrast to the agricultural communities that have been the models for much theoretical development. The idea that above-ground food chains are mainly grazer-initiated, whereas those below-ground are based on decomposers, is widely held, and highlights our lack of knowledge of below-ground herbivory, discussed here by Brown. Of course, much of this production may go straight into mycorrhizal hyphae, and we need to know much more of the consequences of mycorrhizal grazing (Finlay and Shaw) in parallel with studies of grazing on soil fungi in general.

It is increasingly realized that the mycorrhizal condition is the norm in natural vegetation, and the implications of this, particularly in terms of nutrient transport through interplant connections, discussed here by Read, must be profound. Since colonization of roots by hyphae already attached to other roots seems widespread, and since plant uptake of P and, at least for ectomycorrhizas, of N and even water occurs through hyphae, we must re-examine concepts of competition between roots based on the soil model, which are themselves recent in origin. It could even be that the resource competed for is the mycorrhizal hyphae.

The growth, distribution and activity of roots in soil has been strangely neglected in studies of the soil ecosystem. They are considered here at two levels (soil profile and individual root system) by Atkinson and Fitter. Until recently the rhizosphere was viewed as a site of intense microbial (and hence, presumably but not demonstrably, faunal) activity. One strong theme to emerge from the meeting was of the relative inactivity of the rhizosphere populations, and even the significance of rhizosphere dinitrogen fixation, a process recently felt to have exciting agricultural potential, is questioned in a critical review by Giller & Day. The importance of the dead root as an energy source for soil microbes was stressed by Newman, leaving open the question of the importance of grazing on living roots. The relative quantitative significance of above- and below-ground grazing, particularly in the light of the new production estimates, remains unclear. It may be that it is the root–fungal complex that should be viewed as the grazing resource.

As any good meeting should, the symposium on which this volume is based raised new questions in the minds of the participants. As editors, we hope the volume serves the same purpose to its readers. In putting it together we have tried to represent the flavour of the meeting by including both the spoken papers and many of the posters. The latter have been written on a more economical scale, explaining the great range of lengths of paper presented here. We have nevertheless mixed them in with the other papers, both invited and offered, to create what we hope is a coherent order. All the papers were refereed and in most cases these comments were given to authors before the meeting, the final versions being accepted a few weeks later, giving authors a chance to take discussions and comments from participants into account.

Finally we would like to thank the anonymous referees who read and commented on manuscripts, both rapidly and effectively; Roger Booth, Peter Shaw and Maureen Chapman who helped with the organization at the meeting; and Avril Harrison, who, single-handed and with unfailing good humour, handled all the bookings and correspondence.

ALASTAIR FITTER
DAVID ATKINSON
DAVID READ
MICHAEL B. USHER

Through a ped darkly: an ecological assessment of root–soil–microbial–faunal interactions

D. C. COLEMAN

Department of Zoology/Entomology and Natural Resource Ecology Laboratory, Colorado State University, Fort Collins, Colorado 80523, USA

SUMMARY

1 Studies of ecological interactions in the soil environment require skilled usage of concepts from physics, chemistry, mathematics and biology. This volume offers numerous examples of the multi-disciplinary nature of our work.

2 In each instance—roots as primary producers, root/microbe interactions, interactions of roots, microbes and fauna, and interactions of all three groups with the sand–silt–clay–organic matter milieu (soil)—significant new information has appeared over the last 5 years.

3 Viewing total-system aspects, there have been important findings on resource quality, total-system manipulations, including usage of biocides, in-the-field 'meso-cosms', agro-ecosystems, land reclamation and other systems.

4 The added complexity of theoretical aspects of food web design, stability in soil systems, and the short-term physiological and long-term evolutionary implications give us a cornucopia of research topics to study in the next several years.

INTRODUCTION

Overview of the symposium

Our topic, 'Ecological Interactions in the Soil Environment,' is broad and could be more extensively covered in a week than in the two and a half days of the meeting that this volume reports. We will address several major topics in root–soil–microbe–fauna interactions. I shall present some aspects on which we have new information and areas that, if concentrated upon over the next few years, should yield large dividends in information and new insights into soil ecological interactions.

Ecology has been termed 'physiology under the worst possible conditions' (Brock 1966). Interpretation of this remark depends on whether it is observer oriented or soil-organism oriented. Certainly, the organisms living in this opaque gaseous–liquid–solid milieu are growing, if not optimally, then most ingeniously in ways that are only now becoming apparent to us. Perhaps a more venturesome approach to our topic is given by the American poet Wallace Stevens (1954).

1

Throw away the lights, the definitions,
And say of what you see in the dark
That it is this, or that it is that,
But do not use the rotted names.
How should you walk in that space and know
Nothing of the madness of space,
Nothing of its jocular procreations?
Throw the lights away. Nothing must stand
Between you and the shapes you take
When the crust of shape has been destroyed.

Let us take the latter course, which, I suggest, using scientific licence, is an invitation to use a holistic, functional approach, i.e. to travel through a ped darkly.

International nature of our work

Work in soil ecology has been truly international in orientation from its inception. I can touch only briefly on these aspects, details of which are well known to many ecologists.

The area of our work on which I would like to concentrate first is the soil itself, its formation, and its development. There have been some true giants in the field, including Lawes and Gilbert in Britain, Berthelot and Pasteur in France, Liebig in Germany, Müller in Denmark, Hilgard and Jenny in America, and Dokuchaev in Russia. A synoptic overview of the field of soil chemistry and biology was presented by Sir E. J. Russell (1923) in a concise and scholarly review entitled 'The development of the idea of a soil population.' He noted the early interest of Georgius Agricola in nitrification four centuries ago, which then, as now, spurred production and industry for warlike pursuits, i.e. production of nitrates for gunpowder.

The history of the controversy between Lawes & Gilbert on the one hand and Liebig on the other is well known. Liebig helped emphasize a dichotomy between organic chemistry (leading to 'humus' formation) and inorganic chemistry, which persists to this day and remains firmly rooted, particularly in North America.

There are a number of historical and psychological reasons for this dichotomy, and we will explore some avenues for removing these barriers. As will become apparent, the interconnected nature of living and dead organic matter above, on, and below the ground surface can best be considered in the context of the spatial milieu in which it exists.

Being primarily biologists, we think of the myriad of biological interactions that occur in the soil system. However, as shown in the detailed diagram of a soil microsite (Fig. 1), there is an immensely variegated array of pores, particles, surfaces, and interfaces. In the last case, entire books have been written about interfaces, and their role in microbial ecology and ecology in general (Marshall 1976; Berkeley *et al.* 1980).

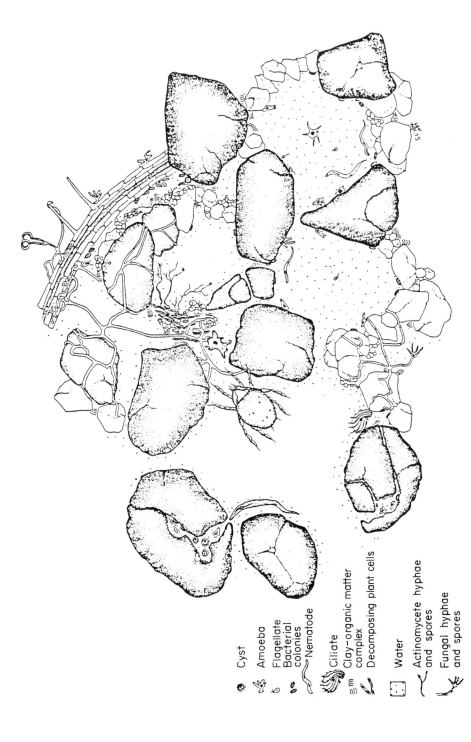

FIG. 1. Trophic relationships among different groups of soil organisms are controlled by accessibility to their resources in this illustration of approximately 1 cm² of a highly structured microzone in the surface horizon of a grassland soil (Rose & Elliott, unpubl.).

Cyst
Amoeba
Flagellate
Bacterial colonies
Nematode
Ciliate
Clay-organic matter complex
Decomposing plant cells
Water
Actinomycete hyphae and spores
Fungal hyphae and spores

Soil development: a 'small world'

One of the new themes of the 1980s for ecological work will undoubtedly be miniaturization and integration. Just as many computers have been developed for home and personal use around powerful microchips, we will see more ecological work developing around an integrative approach to, and study of, microaggregates. This work was pioneered by Kubiëna (1938), but has had few adherents until recently. The ways in which particles of sand, silt, and clay are brought together, often via microbial and plant activity (Griffiths 1965; Foster 1981; Tisdall & Oades

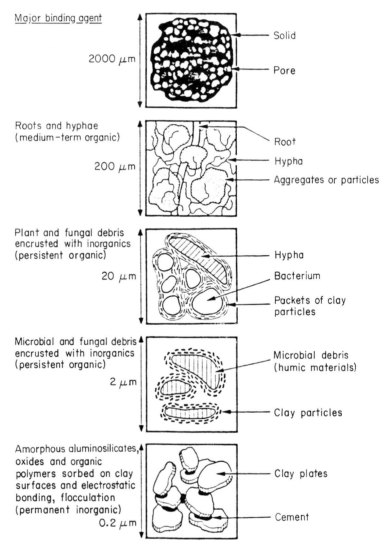

FIG. 2. Soil microaggregates, across five orders of magnitude, beginning at the level of clay particles, through plant and fungal debris, up to a 2 mm diameter soil crumb. (From Tisdall & Oades 1982.)

1982) and the ways they affect subsequent organic matter and nutrient dynamics are fascinating. An integrative scheme for levels of aggregation from a fine clay particle up to a 2 mm microaggregate is shown in Fig. 2. A series of these form a macroaggregate, or crumb. In a native, undisturbed soil, the arrangement of crumbs in an aggregate is called a *ped*, which is a diagnostic soil structure, reflecting the parent material and the various factors that affected the soil formation process.

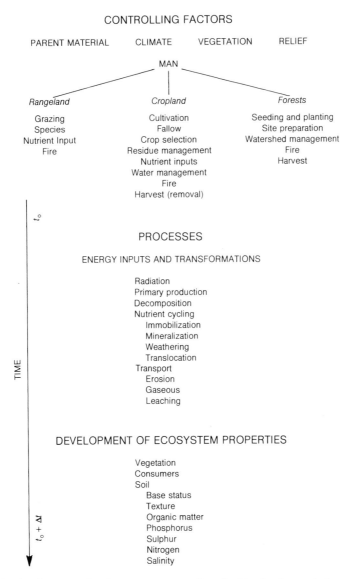

FIG. 3. Soil-forming factors, and major processes, over time, lead to development of ecosystems properties. (From Coleman *et al.* 1983.)

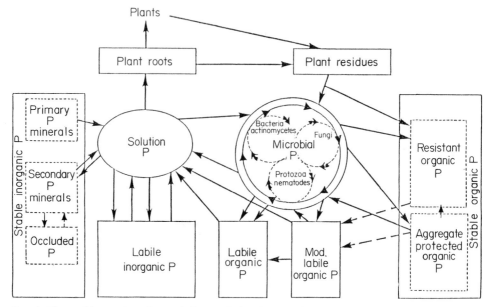

FIG. 4. Flows of inorganic and organic forms of phosphorus. Microbial and faunal interactions affect flows of labile inorganic and organic P. (From Stewart & McKercher 1982.)

Jenny (1941) succinctly described an equation for soil formation:

$$\text{soil formation} = f(cl, o, p, r, t),$$

where cl = climate, o = organism, p = parent material, r = relief, and t = time. It may be of greater use if we envisage soil-forming factors, with several key processes, including immobilization, mineralization, leaching, etc., which lead over time $(t + \Delta t)$ to soil properties (Fig. 3; Coleman, Reid, & Cole 1983). This viewpoint was first noted, in a different form, by Stephens (1947); cited by Burges (1960).

Soil-forming processes occur over thousands of years in chronosequences. They have been characterized by Crocker & Dickson (1957) and Dickson & Crocker (1953) for nitrogen dynamics and Walker (1965) for phosphorus. Because P is a fundamental limiting nutrient in both aquatic (Hutchinson 1957) and terrestrial (Cole & Heil 1981) milieux, the availability of labile inorganic forms of this element (as phosphate ions) may be crucial for subsequent inputs and fluxes of nitrogen and carbon in the ecosystem. Some of the biotic and chemical transformations are summarized (Fig. 4) by Stewart & McKercher (1982). As I shall emphasize repeatedly, the turnover rate of a relatively small pool (i.e. labile inorganic) is far more important than pool size. For example, the larger but less dynamic non-labile inorganic and organic pools may contribute only a small fraction to annual nutrient fluxes, compared with the small highly labile pool.

One of the major findings of groups in our laboratory and elsewhere in North America and Europe has been the central role microbial production and turnover plays in these processes. This interplay between microbes as primary decomposers and organic matter as a substrate to be decomposed has been incorporated in a

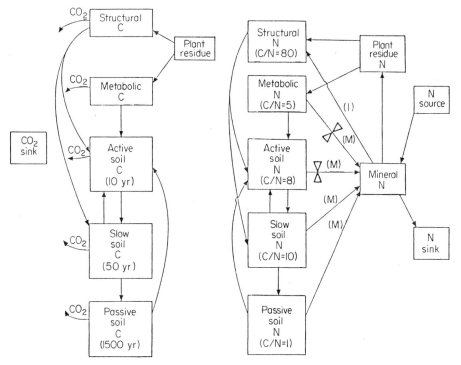

FIG. 5. Flows of structural, metabolic, and soil N and C in a soil organic matter model. Note faunal–microbial interactions, represented by ⋈ from 'metabolic' and 'active' N boxes. 'I' = immobilization and 'M' = mineralization processes. (Modified from Parton *et al*. 1984.)

mathematical simulation model for an agro-ecosystem by Parton *et al.* (1984). The model (Fig. 5), developed to simulate organic matter dynamics over decades and centuries, does not explicitly separate bacteria and fungi or include activities of grazing fauna. However, by accounting for inputs of organic matter of various quality (e.g. C:N ratio, fraction that is labile), the model adequately mimics observed changes in organic matter content on continuously cropped versus alternate wheat–fallow rotations in dryland wheat fields of the northern Great Plains. Interestingly, another version of the model, developed in conjunction with J. Persson and colleagues in the Swedish arable lands project, accurately reflects changes in organic matter inputs in regularly burned fields versus fields that have stubble incorporated into them.

MAJOR TOPICS TO BE COVERED

Our topics run the gamut of subjects concerning the soil environment: the root as primary producer; interactions with soil microbes; effects of grazing animals; interactions in soil and ecosystem function. These constitute some twenty papers plus posters on the same topics. Rather than attempt to give a synopsis of each presentation, I will make a few comments on each topic, discuss some new developments, and then attempt a synthesis of concepts and information.

The root as primary producer

Drs Fogel, Atkinson and Fitter present information on roots as substrates but also 'outposts', as it were, in the continuous struggle required for nutrient uptake by plants and growth and maintenance.

Roots are amazingly labile and highly varied in their growth and reactions to environmental stimuli, both abiotic and biotic. A complex array of techniques has been developed for determining distribution and activity of functional (meta-bolically active) roots (Singh & Coleman 1977), which has been summarized by Böhm (1979).

The amount of detailed information required for population, community, or ecosystem level research will vary but, as noted by all three speakers, we must have better measurement of changes in root production and intra-season and annual turnover. The crude practice of sequential harvesting of fibrous roots may give erroneous estimates of below-ground production. Biomass estimates depend heavily on identification of maxima and minima in standing crops of roots. If these are the result of sampling bias, rather than real trends, either overestimates or under-estimates of true production will result (Singh *et al.* 1984). With from 30 to 50% or more of high-quality material coming from rhizodeposition (Shamoot, McDonald & Bartholomew 1968; Sauerbeck & Johnen 1976; other references in Fogel, p. 23), we obviously require suitable integrative techniques to measure these inputs; isotopic tracers, including stable isotopes, and other functional measurements will be useful tools for many of us.

Interactions with soil microbes

Here we arrive at the crux of several issues before us as soil ecologists. Whether dealing with rhizosphere, soil, or litter; inputs or outputs; co-operative or dis-operative (Allee *et al.* 1949) interactions, surely a careful study of rhizosphere phenomena (Newman; p. 107; Giller & Day, p. 127), litter decomposition (Paustian, p. 159), mycorrhizas and nutrient cycling (Read, Francis & Finley, p. 193), and root-microbial (including protozoa) associations (Clarholm, p. 355) will be infor-mative areas of ecosystem functional studies.

One of the major challenges in soil ecology, as in all science, is to keep an observant eye out for important patterns, processes, and their implications. Some examples of these are: (i) if annual and perennial plant rhizospheres are fundamen-tally different, what does this mean for the ways in which we study root-microbial interactions? (ii) Does the root–microbe–fauna complex significantly affect sym-biotic and nonsymbiotic N_2 fixation? (iii) Do the intra- and interspecific bridges between plants via mycorrhizas (Chiariello, Hickman & Mooney 1982; Read, Francis & Finlay, p. 193) form an important mechanism for both N (Ames *et al.* 1983) and P uptake, and does this extend to water uptake and other often limiting factors as well?

The effects of root–microbe interactions on organic matter transformations are an increasing area of inquiry. Both in terms of negative effects of roots on aggregating properties of native soil polysaccharides (Reid & Goss 1981, 1982a) and as net inhibitors of mineralization of labile ^{14}C organic matter (Sparling, Cheshire & Mundie 1982; Reid & Goss 1982b, 1983), it seems apparent that simple hetero-trophic soil incubations are missing some important interactions provided by a growing root system.

Effects of grazing animals

Studies of mesofauna and macrofauna and their effects on other biota, including bacteria, fungi, roots, and root–mycorrhizal associations continue as a burgeoning area of research (Anderson, Ineson & Huish 1983; Seastedt 1984). Several years ago, my colleagues and I espoused a somewhat simplistic approach to micro-bial–faunal interactions (Coleman 1976), emphasizing the positive, system-level benefits of enhanced nutrient return (Coleman et al. 1977). With the benefit of hindsight, I shall emphasize the multifarious nature of these interactions. In study-ing plant–animal and animal–microbial interactions, we should know the physiological state of the grazer and grazed populations. Thus, under certain condi-tions, with actively growing tissues, a low to moderate amount of grazing (Fig. 6) may enhance plant or microbial growth (Dyer & Bokhari 1976; Dyer et al. 1982; Hunt et al. 1977). Fungal growth response to collembolan grazing is closely depen-dent on the nutritional state (nutrient content) of the fungal hyphae being grazed. Do fungus-grazing animals preferentially feed on the young, actively growing parts of hyphae? Recent work by Ingham et al. (1985) in our laboratory indicates that fungal-feeding nematodes consume amounts of fungal cytoplasm greater than the amounts that stain with FDA (Söderström 1977).

Perhaps more important for field-related phenomena, what is the importance of the repeatedly demonstrated preference of certain fungal-feeding fauna for particu-lar species of fungi (Coleman & Macfadyen 1966; Parkinson, Visser & Whittaker

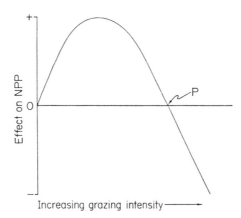

FIG. 6. Optimization of production as a function of grazing. Note peak at low → moderate grazing rates. (From Dyer et al. 1982).

1979; Newell 1984) or bacteria (Führer 1961) in subsequent ecosystem function? Do these interactions extend more generally to 'disease suppressive' soils (Schroth & Hancock 1981)? Recent work from Australia shows some intriguing amoebal–pathogenic fungi interactions which bear further study (Chakraborty, Old & Warcup 1983; Chakraborty & Warcup 1983). Recent reviews of microarthropod feeding relationships are given by Petersen & Luxton (1982) and Cancela da Fonseca & Poinsot-Balaguer (1983), and by Visser (p. 297).

An additional topic of great interest is mycorrhiza–fauna interactions. Extensive research in crop plants (e.g. *Gossypium*, *Citrus*) has shown little, if any, effect of fungal-feeding nematodes on subsequent plant growth (Hussey & Roncadori 1981, 1982)). Therefore, it will be interesting to see if microarthropod feeding on VAM external hyphae is generally deleterious or not; an approach to this is discussed by Finlay (p. 319).

Other aspects of plant–animal–microbial interactions will be addressed in the next section, on soil and ecosystem function.

Interactions in soil and ecosystem function

An extensive and intriguing set of topics are presented within the general rubric of soil interactions and ecosystem function. We need to be as percipient and integrative as possible. One approach that may be of considerable utility now and in the future is the 'unit community' concept of Swift (1976). He noted that many organisms in the decomposer community play similar roles and that dynamics of various cellulolytic or ligninolytic microflora may differ between neighbouring branches or logs in a forest floor, but the overall processes and rates at which they proceed will be largely independent of the species composition in the substrate itself. This is paralleled by earlier observations of Fager (1968), who showed the stochastic nature of meso- and macrofaunal colonization and exploitation of experimental oak logs in Wytham Wood.

The problem of resource quality and how best to quantify it arises in several papers in our meeting. A holistic approach should include study of the mineral, as well as the biotic, components. The work of Kilbertus & Vannier (1981), and Touchot, Kilbertus & Vannier (1983) is exemplary in this regard. They observed that eu-edaphic species of *Tomocerus* or, in laboratory conditions, *Folsomia candida* tended to ingest small amounts of argillic (clay) material, which helped detoxify various phenolic compounds in oak (*Quercus*) leaves. Even more interesting, bacterial development in the aggregates formed was promoted during gut passage. Thus, the bacterial colonies were included in a mucus layer surrounded, in turn, by a layer of clay minerals.

The long-term effect of these processes on faecal-material decomposition requires further study. Jongerius (1964), Zachariae (1965), Rusek (1975), Kühnelt (1976), and others have commented on the nature, extent, and persistence (over several years) of faecal pellets of microarthropods in a wide variety of ecosystems.

SYSTEM-LEVEL MANIPULATION OF SOIL ORGANISMS

In recent years, there has been a renewed interest in manipulation of system components in soils. This developed from some of the uses of atomic energy in ecology and was summarized in various publications sponsored by the US Atomic Energy Commission (e.g. Odum & Pigeon 1970). More specific studies of manipulation of soil arthropods were noted by Edwards (1967) and soil biota in general by Coleman & Cowley (1973). The latter study showed marked changes in soil respiration; but, as one might expect, only when primary decomposers (bacteria and fungi) were drastically reduced or totally removed by several Megarad (10^4Gray)-level doses of gamma radiation. Use of high-energy gamma-emitting sources is restricted in its utility worldwide, and other, more general approaches were desired.

In our studies at CSU, we decided on a 'bottom-up' approach, assembling oligo-specific groups of microflora and fauna, with and without seedlings growing in them, in gnotobiotic microcosms. Our main conclusions have been reported in Coleman *et al.* (1983, 1984) and are discussed only briefly here.

In either heterotrophic microcosms or with heterotrophs and autotrophs combined, we observed significant mineralization of organic N and P (Fig. 7), or nitrogen alone (Fig. 8). Because of inherently greater net mineralization potential by saprophytic fungi than bacteria in some of our experiments, the holophagic nature of the protozoa or bacterial-feeding nematodes seems to have a greater impact on the nutrient return than the fungus/fungivore association (Ingham *et al.* 1985).

It became apparent that, to approach a true system-level species richness more readily, we would need to begin studies using a 'top-down' approach, with selective elimination of target groups of organisms. The rationale for this approach, i.e. using 'perturbation experiments', has been ably summarized by Usher, Booth & Sparkes

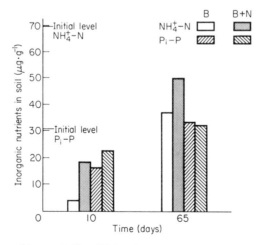

FIG. 7. Mineralization of inorganic N and P in soil microcosms with (B + N), and without (B) bacterial-feeding nematodes. Significant differences ($P < 0.05$) were observed for N and P on day 10. (From Anderson *et al.* 1981).

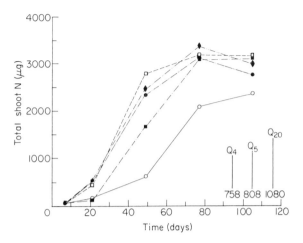

F<small>IG</small>. 8. Mineralization of N, and facilitated uptake, in the presence of bacterial or fungal grazers (----) versus plant alone (——). (From Ingham *et al.* 1985).

(1982), and assessed for real and idealized communities by Bender, Case & Gilpin (1984), who observed various responses, depending on pulsatile ('pulse') versus continued ('press') types of perturbations. One of the major difficulties of performing whole-system perturbation experiments is adequately to distinguish between direct and indirect effects of organism removal.

Drawing on some earlier concepts reported before a British Ecological Society audience (Coleman 1976), we tried manipulating soil 'functional groups' (e.g. bacterial-feeders, fungal-feeders, etc.) where possible. We devised a major field experiment, using six biocides specific for the following groups: bacteria, saprophytic fungi, phycomycetous fungi (including vesicular-arbuscular mycorrhizae), nematodes, acari, and all arthropods. This study ran for 6 months and generated large amounts of interesting but as yet unpublished data.

Unfortunately, most of the available biocides are specific for most or all organisms in a major taxonomic group, such as arthropods or nematodes. These transcend, of course, the desired functional constructs, e.g. bacteriophages, fungiphages. Therefore, we have proceeded with an alternative approach, namely, a mathematical simulation model. We set up a conceptual model of nitrogen flow consisting of shoots, roots, two categories of substrates, and some eleven groups of fauna that feed directly or indirectly on bacteria (including actinomycetes), fungi (including VAM), or other fauna.

In developing this model, we were impressed by two things: the many interconnections occurring, particularly at the higher trophic levels, and the long chain length of some of the pathways. For example, going from substrate to microflora to amoebae, via intermediate predators, to predatory mites entails going through seven trophic levels. This is longer than any food chains studied in the classic paper of F. E. Smith (1969) or the recent book of Pimm (1982).

As noted by May (1973a,b) and Pimm (1982), these systems, containing a plethora of omnivores, should be inherently unstable, both in terms of populational responses and associated nutrient-cycling responses (DeAngelis 1980). This is an area where much new research is currently being conducted in theoretical ecology. Drawing on extensive data summarized by Briand (1983), Yodzis (1984) addressed the issue of commonness or rarity of omnivory. He concluded that omnivory tends to occur more prevalently between organisms in similar trophic positions than in those far apart. So far as I can ascertain, there was no input of information from litter-soil ecosystems in the array of forty data sets, from terrestrial and aquatic systems. We have an excellent opportunity to assess whether omnivory, particularly feeding on prey more than one trophic step away, is unusually common in litter-soil systems, and, if so, the implications for ecosystem resilience.

However, as noted by Coleman *et al.* (1983), Yeates & Coleman (1982), and others, the 'real system' in soil is truly resilient, contains many cryptobiotic forms, and refuses to function in as simplistic a manner as we have so far managed to portray it. The main problem is how to represent considerable detail, yet keep the model to manageable proportions. It is apparent that we must adequately represent the extent to which much of the soil biota enter into, and exit from, quiescent, dormant, or cryptobiotic stages (Demeure *et al.* 1979; Stout *et al.* 1982).

Some initial simulation runs of the biocide model (Fig. 9) showed that predatory forms of soil arthropods and nematodes were very sensitive to changes in abundance of the bacterial or fungal populations (H. W. Hunt *et al.*, unpublished data).

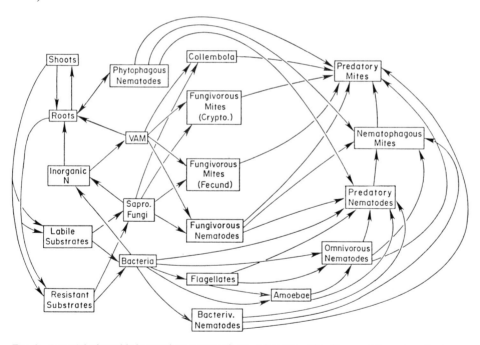

FIG. 9. A model of trophic interactions among plants, substrates, microflora, and fauna in a shortgrass prairie. (From Hunt *et al.* unpubl.)

New research in the next decade

One of our major concerns should be how best to cope with various levels of resolution in the soil–root–biota system. It is fruitful to review what others have said about the 'cardinal entities', as it were, in ecology. Watt (1968) claimed that all of ecology considers matter, energy, space, time, and diversity. Having dealt with all of these to an extent, I would like to pursue some further ideas on time and size. In an address to this society, Burges (1960) discussed with great insight the manners in which time and size affect ecological processes, both successional and longer-term, including soil formation and the chronosequences which I mentioned briefly earlier.

I shall address some aspects of size. Burges expressed concerns about methods to sample communities which are small in relation to ourselves, concerns which were further amplified by Macfadyen (1969). Organisms within the community play markedly different roles as a function of size. Thus, organisms (microbes, protozoa, small nematodes, rotifers) which exist in water films, in various micropores within soil, are probably operating in a quite different 'world' from the organisms which move into and out of pores, independently of water films. A third, and perhaps even more distinctive category are those members of the macrofauna which move soil around by external means (fossorial mammals) or internal means (earthworms) (Hole 1981).

Of course, Charles Darwin (1837, 1881) played a pioneering role in investigating activities of earthworms. Numerous other books and papers refer to earthworms' function in soil, summarized by Edwards & Lofty (1978), Bouché (1983) and Lavelle, Zaidi & Schaefer (1983). Recent work in New Zealand (Sharpley *et al.* 1979; Syers *et al.* 1979) gives further indication of how these organisms affect nitrogen and phosphorus dynamics in pastures. The paper by Springett (p. 399)

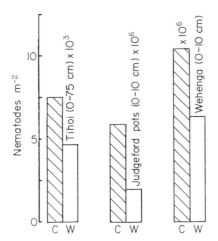

Fig. 10. Nematodes per m^2 in field sites in New Zealand (Tihoi, Wehenga) or in pot trials (Judgeford) with (W) or without (C) earthworms present. (From Yeates 1981).

develops some themes in this area, such as impacts on root growth. My comments are meant to be complementary to hers.

In addition to having an effect on plant debris through trituration and comminution, earthworms in New Zealand were found markedly to depress soil nematode populations in both laboratory (pot experiments) and field sites. Nematode standing crops were only 40–50% of those in treatments (Fig. 10) which had no earthworms (Yeates 1981). Similar decreases, in laboratory studies, have been observed after earthworms' ingestion of protozoa (Piearce & Phillips 1980). We have no information yet on effects on population growth rates of these 'prey' populations but, with our concern for measurement of turnover rates mentioned earlier, this is an important factor to consider in future ecological field studies.

Perhaps geophagy should be more generally considered with other members of the soil fauna. I assume that a certain amount of soil is ingested by millipedes and isopods. The structural and biochemical impacts (e.g. P mineralization) of the little-studied soil-eating termites has also been noted (Wood *et al.* 1983).

A further aspect of ecosystem function is the time at which activity occurs. Dr Verhoef discusses interesting aspects of collembolan activity in pine forests in mid-winter (p. 367). The cumulative effect of soil organisms living at or near 0°C, whether cryophilic or not, is an important area for further work in ecosystem function.

Distributions of organisms, relative to organic matter in soil

One of the interesting findings to emerge from our soil ecology studies over the last few years has been the commonality of patterns of distribution and abundance of soil organisms. Both roots and VA mycorrhizal hyphae proliferate locally when they reach a zone of organic matter enrichment (St John, Coleman & Reid, 1983a,b). The implications of proliferation of many types of organisms, such as roots, mycorrhizal fungi, microbes, or fauna with associated organic matter may well be an area of interest which will continue to develop in soil ecology over the next several years.

Studies in agro-ecosystems

Another aspect, in the basic/applied ecology realm should be considered by all of us. It is apparent that, with greatly increasing amounts of land area going into minimum- or zero-tillage, we should be alert to developments and opportunities in this area. The following is some recent information from studies we have conducted in Eastern Colorado, in dry-land wheat (Elliott *et al.* 1984a,b). The results discussed here are from a fallow, no-till treatment in a silt-loam soil (Argiustoll).

The site was sampled over five times in summer 1982 (5 June–13 September). The season was quite moist, with adequate water available for microbial and faunal activity the entire summer. We observed marked dynamics in faunal activity throughout the summer. Acari were greatest in early June, and continued to decrease through to September (Fig. 11). Holophagous nematodes (principally

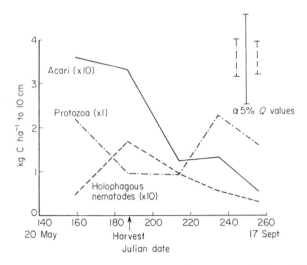

FIG. 11. Soil faunal dynamics in the fallow side of no-till plots at Akron, Colorado (1982). Note early tenfold greater biomass of protozoa, mostly amoebae and flagellates. Means between dates farther apart than confidence intervals are significantly different at $P < 0.05$. (From Elliott *et al.* 1984).

bacterial feeders) reached a peak in early July, while protozoa had early and late summer peaks of biomass. Microbial C and P were low in early and late summer, corresponding with peaks of protozoan biomass (Fig. 12). We suggest that these results correspond well with the classical results of Cutler *et al.* (1922), and the microcosm studies of Coleman *et al.* (1978) and Cole *et al.* (1978). Perhaps these interactions can be considered along with the aspects of phytotoxicity, nitrogen fixation, etc., covered by Lynch in this symposium (p. 181).

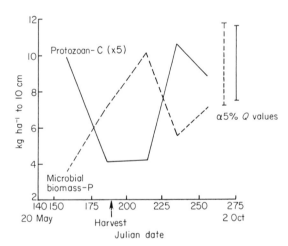

FIG. 12. Dynamics of microbial biomass P and protozoan C in fallow no-till and stubble-mulch plots, Akron, Colorado, showing inverse relationship. Sample dates are the same as in Fig. 11. Confidence intervals calculated as in Fig. 11 (From Elliott *et al.* 1984b).

Seastedt, T.R. (1984). The role of microarthropods in decomposition and mineralization processes. *Annual Review of Entomology*, **29**, 25–46.

Shamoot, S., McDonald, L. & Bartholomew, W.V. (1968). Rhizo-deposition of organic debris in soil. *Soil Science Society of America, Proceedings*, **32**, 817–820.

Sharpley, A.N., Syers, J.K. & Springett, J.A. (1979). Effects of surface-casting earthworms on the transport of phosphorus and nitrogen in surface runoff from pasture. *Soil Biology and Biochemistry*, **11**, 459–462.

Singh, J.S. & Coleman, D.C. (1977). Evaluation of functional root biomass and translocation of photo–assimilated carbon–14 in a shortgrass prairie ecosystem. In *A Synthesis of Plant-associated Processes* (Ed. J.K. Marshall), pp. 123–131. Range Science Department Science Series, No. 26. Colorado State University, Fort Collins.

Singh, J.S., Lauenroth, W.K., Hunt H.W. & Swift, D.M. (1984). Bias and random errors in estimators of net root production: a simulation approach. *Ecology*, **65**, 1760–1764.

Smith, F.E. (1969). Effects of enrichment in mathematical models. In *Eutrophication: Causes, Consequences and Correctives,* pp. 631–645. National Academy of Sciences, Washington, D.C.

Sparling, G.P., Cheshire, M.V. & Mundie, C.M. (1982). Effect of barley plants on the decomposition of ^{14}C labelled soil organic matter. *Journal of Soil Science*, **33**, 89–100.

Söderstrom, B.E. (1977). Vital staining of fungi in pure culture and in soil with fluorescein diacetate. *Soil Biology and Biochemistry*, **9**, 59–63.

Stevens, W. (1954). *The Collected Poems of Wallace Stevens.* Alfred Knopf, New York.

Stewart, J.W.B. & McKercher, R.B. (1982). Phosphorus cycle. In *Experimental Microbial Ecology* (Eds R. G. Burns & J. H. Slater) pp. 221–238. Blackwell Scientific Publications, Oxford.

Stout, J.D., Bamforth, S.S. & Louiser, J.D. (1982). Protozoa. In *Methods of Soil Analysis, Part 2. Chemical and Microbiological Properties—Agronomy Monograph No. 9* (2nd edn), pp. 1103–1120. ASA–SSSA, Madison, Wisconsin.

Swain, T. (1978). Plant–animal coevolution: A synoptic view of the Paleozoic and Mesozoic. In *Biochemical Aspects of Plant and Animal Coevolution* (Ed. J. B. Harborne), pp. 3–19. Academic Press, London.

Swift, M.J. (1976). Species diversity and the structure of microbial communities in terrestrial habitats. In *The Role of Terrestrial and Aquatic Organisms in Decomposition Processes* (Ed. J. M. Anderson & A. Macfadyen), pp. 185–222. Blackwell Scientific Publications, Oxford.

Syers, J.K., Sharpley, A.N. & Keeney, D.R. (1979). Cycling of nitrogen by surface–casting earthworms in a pasture ecosystem. *Soil Biology and Biochemistry*, **11**, 181–185.

Tisdall, J.M. & Oades, J.M. (1982). Organic matter and water-stable aggregates in soils. *Journal of Soil Science*, **33**, 141–163.

Touchot, F., Kilbertus, G. & Vannier, G. (1983). Role d'un collembole (*Folsomia candida*) au cours de la dégradation des litiéres de charme et de chene, en presence ou en absence d'argile. In *New Trends in Soil Biology* (Eds P. Lebrun, H.M. André, A. Demedts, C. Grégoire–Wibo & G. Wauthy), pp. 269–380. Dieu–Brichart, Ottignies-Louvain-la-Neuve, Belgium.

Usher, M.B., Booth, R.G. & & Sparkes, K.E. (1982). A review of progress in understanding the organization of communities of soil arthropods. *Pedobiologia*, **23**, 126–144.

Walker, T.W. (1965). The significance of phosphorus in pedogenesis. In *Experimental Pedology.* (Eds E.G. Hallsworth & D.G. Crawford), pp. 185–222. Butterworths, London.

Watt, K.E.F. (1968). *Ecology and Resource Management.* McGraw-Hill, New York.

Wilson, D.S. (1980). *The Natural Selection of Populations and Communities.* Benjamin Cummings, Menlo Park, California.

Wood, T.G., Johnson, R.A. & Anderson, J.M. (1983). Modification of soils in Nigerian savanna by soil-feeding *Cubitermes* (Isoptera, Termitidae). *Soil Biology and Biochemistry*, **15**, 575–579.

Wynne-Edwards, V.C. (1962). *Animal Dispersion in Relation to Social Behaviour.* Oliver and Boyd, Edinburgh.

Yeates, G.W. (1981). Soil nematode populations depressed in the presence of earthworms. *Pedobiologia*, **22**, 191–195.

Yeates, G.W. & Coleman D.C. (1982). Role of nematodes in decomposition. In *Nematodes in Soil Ecosystems* (Ed. D.W. Freckman), pp. 55–80. University of Texas Press, Austin.

Yodzis, P. (1984). How rare is omnivory? *Ecology*, **65**, 321–323.

Zachariae, G. (1965). Spuren tierischer Tätigkeit im Boden des Buchenwaldes. *Forstwissenschaftliche Forschungen*, **20**, 1–68. Paul Parey, Hamburg.

Roots as primary producers in below-ground ecosystems

ROBERT FOGEL

University of Michigan Herbarium, Ann Arbor, MI 48109, USA

SUMMARY

1 Evidence from a variety of forest, shrub, and grassland ecosystems suggests that roots and mycorrhizas constitute 40–85% of total net primary production. The estimated fluxes represent a sizeable drain on the carbohydrate reserves of plants and comrpise a major route for the return of associated nutrients to the soil.

2 Very often these productivity estimates are rough approximations based on sequential biomass sampling, and fail to account for the simultaneous production and decomposition of roots during active periods of net biomass increase.

3 Inclusion of losses due to detritus production, exudation, and herbivory would improve these estimates.

INTRODUCTION

Primary production by vascular plants provides most of the organic materials fueling decomposition and other processes supporting soil organisms. Net primary production is the rate of storage of organic matter in plant tissues in excess of respiratory utilization during a given period (Odum 1971). Two general methods of determining annual net production exist. One, the gas-exchange approach, is based on measuring the rate of photosynthesis throughout the year and then subtracting concomitant respiration, and the other, the harvest approach, on measuring increases in biomass during a year. The latter method provides information of more direct interest to soil biologists, e.g. amounts of organic matter entering the soil and rates for decomposition and other processes. One variant of a commonly used expression summarizes the parameters used to calculate net primary production:

$$\text{net production} = \Delta B + L$$

where ΔB is the annual increment in biomass, L is the loss due to detritus production (e.g. leaves, branches, bark, inflorescences, seeds, and roots), exudation, volatilization, leaching and herbivory.

Accurate estimation of root parameters in the production equation has proven one of the most intractable tasks for soil biologists due to non-uniformities in the size and distribution of roots, the physical problems in separating roots from soil, and difficulties in collecting exudates and volatiles. As a consequence of these problems and the effort needed to measure root biomass, one approach for estimating net primary production has been to ignore below-ground parameters entirely or to estimate them indirectly. Newbold stated in 1967 that no accurate estimates of below-ground production in forests had been obtained and as recently as 1975 Whittaker & Marks, stated that 'the amounts of annual loss of root hairs and roots from plants in the field are almost unknown.' Leaf litterfall provided an obvious, easily measured process for the input of organic matter and associated nutrients to the soil and was believed to constitute the major route for the return of nutrients to the soil (Gray & Williams 1971; Heilman & Gessel 1963). If below-ground production was included it was estimated indirectly, based on the assumption that the ratio of production to mass must be similar for the root system and the shoot system (Andersson 1970; Kira & Ogawa 1968; Newbold 1967; Nihlgard 1972; Whittaker *et al.* 1979). The ratio of root production to mass was generally assumed to range from 20 to 100% of the above-ground ratio (Harris, Santantonio & McGinty 1980; Whittaker & Marks 1975). The use of indirect estimates has persisted. As recently as 1981, Pastor & Bockheim assumed the ratio of root production to mass to be equal to the ratio for above-ground biomass in the study of an aspen–mixed hardwood ecosystem.

   This review examines the results of an alternative approach, the harvest method, for measuring root parameters and determining their significance in net primary production. Methods and assumptions in the measurements are examined; information available on production and loss parameters evaluated; and finally net primary production estimates for different forest and herbaceous ecosystems compared.

ESTIMATION OF PRIMARY PRODUCTION PARAMETERS

Biomass increment

Measurement of initial standing crops

The harvest approach for estimating root production employs measurements of changes in standing crops to determine the significance of root parameters in the net primary production equation. The first problem in this approach is devising a sampling scheme for measuring root biomass. Three basic factors have to be considered in sampling roots: (i) diameter of roots, (ii) spatial distribution, and (iii) temporal distribution. The roots of monocotyledenous plants such as barley are small and fairly uniform, ranging in diameter from less than $0 \cdot 1$ to $0 \cdot 7$ mm (Russell 1977). After germination, the root system develops by repeated branching, producing seminal axes, nodal axes, first order laterals, and second order laterals. Ultimately, the root system occupies a soil volume determined in part by genetics, environmental factors, and plant density (Nye & Tinker 1977; Russell 1977). An acceptable sampling scheme must account for the variation in root density resulting from the interaction of these factors and temporal differences in the spread of roots through the soil. Distribution of the root system of the entire crop is of greater interest in ecosystems research than that of a single plant. Classification of roots by size is only of interest if the size classes reflect functional or spatial differences.

Classification or description of root system architecture is of more utility in forest ecosystem studies, due to the large size differences among different orders of tree roots and their associated functional differences. One commonly employed classification scheme for conifers divides roots functionally into those providing structural support for the tree, termed long lateral roots, and those active in the absorption of water and nutrients, termed short roots; the latter are often ectomycorrhizal (Fogel 1983). Lateral roots are generally subdivided into large and fine root categories on the basis of diameter; either 2 or 5 mm is often used as the upper diameter limit for fine roots. Complicating the functional classification of roots is the lack of short roots in the Cupressaceae, Taxodiaceae, and most angiosperms. Further complicating a functional classification is the inclusion of suberized, unsuberized, and endomycorrhizal roots in the fine root category for those species lacking short roots and inclusion of ectomycorrhizal and non-mycorrhizal roots in this category for species having short roots. From the standpoint of sampling, root systems can be considered to consist of two parts: (i) fine roots less than 2 mm in diameter, and (ii) structural roots, larger than 2 mm in diameter, which provide the framework supporting the fine roots.

Up to a dozen or more large lateral roots may arise irregularly from the stem base in conifers (Eis 1974; Fayle 1975), especially on steeply sloping ground, or approximately equidistantly around the stem (Deans & Ford 1983; Eis 1974; McMinn 1963). The horizontal spread of lateral roots varies with the size of crown, stocking density, and age of tree (McMinn 1963). Root density, diameter, and average depth of main lateral roots all decrease with distance from the stem (Hermann 1977; McMinn 1963). From 80 to 90% of structural root biomass is found within 1–1·2 m of the stem base in young red pine *Pinus resinosa* and sitka spruces *Picea sitchensis* (Deans 1981; Fayle 1975). Full occupation of the soil may take 10 years in Douglas fir *Pseudotsuga menziesii* (McMinn 1963) or up to 20 years in red pine (Fayle 1975). Most of the lateral roots are found in the upper 50 cm of soil with sinker roots reaching far greater depths (Fayle 1975; Deans & Ford 1983). The difficulty in severing large-diameter lateral roots precludes the use of coring techniques for biomass sampling. An additional problem results from the cross-sectional area of cores (generally less than 75 mm in diameter) being inadequate to sample roots present in low densities. As a consequence, biomass of large structural roots is best determined by excavation of whole trees or soil monoliths–laborious techniques very often resulting in only a small number of root systems being sampled. The data are often expressed as top to root ratios (Bray 1963) or as allometric equations relating root mass to stem diameter (Dice 1970; Ovington & Madgwick 1959).

Soil coring has been used extensively to measure fine root and mycorrhiza biomass since these roots are easily obtained by this method. Fine root density has been reported to be independent of stem proximity (Moir & Bachelard 1969), decreasing with distance from stem (Ford & Deans 1977; Roberts 1976), or reaching a maximum at 0·5–1·0 m from the stem (Persson 1980). Localized concentrations of roots may occur due to discrete patches of decaying organic matter or

obstructions such as fragipans or cobbles (Fogel 1983). Most fine roots and mycorrhizas are found in the top 20 cm of the soil, depending on soil aeration and
fertility (Fogel 1983). The discontinuous distribution of fine roots and mycorrhizas
may require a large number of soil cores and/or batching of soil samples to reduce
the variance in biomass to an acceptable level. Once cores have been collected,
considerable time is spent separating roots from the soil by hand-sorting or by
wet-sieving. The choice of separation technique affects the efficiency of recovery;
wet-sieving yields estimates *c*. 30% larger than hand-sorting (Fogel 1983).

Available estimates suggest that tree root biomass comprises 15–25% of the
total tree biomass, although individual estimates range from 9 to 44% (Fogel
1983). In grasslands and tundra, below-ground components of the plant may reach
75–98% of plant biomass (Dahlman 1968; Shaver & Billings 1975). Root biomass
estimates for conifer stands less than 200 years old range from 3 to 85 Mg ha^{-1};
an estimated 209 Mg ha^{-1} has been reported for an old-growth Douglas fir stand
(Fogel 1983). Root mass in stands of deciduous trees less than 200 years old ranges
from 17 to 95 Mg ha^{-1} (Santantonio, Hermann & Overton 1977). Fine root estimates range from 1 to 12·6 Mg ha^{-1} with a mean of 5 (Fogel 1983). It is highly
probable that mycorrhizas have not been included in the fine root estimates despite
their importance in nutrient absorption (Fogel 1983).

Studies specifically mentioning ectomycorrhizas (total standing crop of active
plus inactive mycorrhizas) indicate that they comprise up to 6% of the total tree
biomass of Douglas fir (Fogel & Hunt 1983). Mycorrhiza standing crop estimates
range from 1 to 25 Mg ha^{-1} (Fogel 1983), but the reports are difficult to compare
due to differences in separation techniques, species, and whether total or active
biomass is reported.

Root research in agro-ecosystems has focused on modelling nutrient absorption
and competition among species. Root length rather than biomass has been the
preferred parameter in many of these studies (Nye & Tinker 1977). Consequently
very few standing crop estimates are available for comparison with forest ecosystems. Root biomass was estimated to be 16·2 Mg ha^{-1} in a native grassland studied
by Dahlman (1968). Total dry weight averages 1·2 Mg ha^{-1} in a number of winter
wheat cultivars, nearly equalling the shoot biomass of 1·25 Mg ha^{-1} (Gregory *et al.*
1978). Root biomass in soybean was considerably less than graminoid biomass;
only reaching 0·6 Mg ha^{-1} 85 days after sowing (Sivakumer, Taylor & Shaw 1977).

Increment in standing crops

Annual biomass increment (ΔB) of large structural roots of trees has been estimated from changes in measured standing crops by the monolith method (Harris,
Kinerson & Edwards 1977; Reichle *et al.* 1973), by large cores (Harris *et al.* 1977),
or indirectly by increments in annual standing crops predicted by allometric equations relating root mass to stem diameters (Fogel & Hunt 1979; Kira & Ogawa
1968). An assumption in these approaches is that mass of large roots increases with
time and that senescence and decomposition of large roots does not occur (Harris *et*

al. 1977). A report by McMinn (1963) that the total length of structural roots increased with increased age in 10–55 year-old Douglas fir stands supports this assumption. Fayle (1975) on the other hand, observed that the terminal metre or more was dead in many of the main horizontal roots of 17 and 30 year-old red pines. Kolesnikov (1968) reports that large structural roots of apple begin to die when development of the tree begins to slow. Harris *et al.* (1977) commonly encountered dead roots >25 mm in diameter apparently associated with living trees. Several studies employing sequential harvesting have failed to detect any significant seasonal differences in the biomass of large conifer roots (Persson 1978; Santantonio *et al.* 1977). Monthly core data showed no consistent pattern of large root (>5 mm) production in a *Liriodendron* stand (Harris *et al.* 1977). When the root core data were grouped in quarterly periods, however, standing crop of *Liriodendron* roots increased sharply during spring and summer then declined during autumn. The difference between autumn and winter standing crops of roots >5 mm diameter was 800 kg ha^{-1} or 10·4% of the root standing crop. Production of roots >5 mm diameter in a Douglas fir stand, as estimated by difference in annual allometric estimates, was 2·1 Mg ha^{-1} or 4% of the root biomass (Fogel & Hunt 1983).

An alternative technique for estimating annual production in large roots involves measuring annual volume increment represented by growth rings—an extremely laborious process, but probably more acurate than estimates determined by difference in annual allometric estimates or in harvested standing crops (Fayle 1975; Deans 1981). Fayle (1975) found that current annual increment (CAI) of horizontal lateral roots in red pine reaches an initial plateau of over 0·6 dm^3 year^{-1} per tree around 15 years and that vertical root CAI levels off at 0·55 dm^3 year^{-1} around year 20. This volume increment represents an annual production of 1·4 Mg ha^{-1} or 4·8% of the root biomass if one assumes a specific gravity of 0·474 for wood and that stocking density is 2600 trees ha^{-1} (2 × 2 m spacing). Annual production of roots >5 mm in diameter, in a 16 year-old sitka spruce plantation (3800 trees ha^{-1}), was slightly more than double that of red pine at 3·2 Mg ha^{-1} (15·7% of root biomass) (Deans 1981). CAI of the red pine plantation was 0·54 kg per tree compared to 0·83 kg per tree in the sitka spruce plantation, despite the greater density of trees in the latter.

Production of fine roots and mycorrhizas of forest trees has been estimated primarily by summing significant increments in standing crops as estimated by soil coring. Production of conifer fine roots (Table 1) has been estimated to range from 3·5 to 11·0 Mg ha^{-1}. The few reports available for deciduous trees indicate a similar production of 5·4–9·0 Mg ha^{-1} (Table 1). These estimates are difficult to compare directly given the differences in stand ages, species composition, stocking density, and sampling methods. On a proportional basis, production of non-mycorrhizal fine roots ranges from 40 to 84% of root biomass.

Similar values have been reported in some herbaceous systems. Production in a native tall grass prairie (Table 1) was 5·1 Mg ha^{-1} compared to lower values of 0·6–4·3 Mg ha^{-1} in cultivated annuals. The difference between perennial and

TABLE 1. Annual production (Mg ha^{-1}) of fine roots in different
ecosystems

Ecosystem	Age (years)	Production	Reference[*]
Coniferous forest			
Douglas fir	55	4·1–11·0	4
Loblolly pine	14	8·6	6
Red pine	53	4·1	7
Scots pine	5	3·5	8
Pacific silver fir	23	4·0	10
	180	6·5	10
Sitka spruce	16	5·3	3
Deciduous forest			
Yellow poplar	?	9·0	6
Oak–maple	80	5·4	7
Herbaceous			
Corn	<1	1·2–4·2	1
Soybean	<1	0·6	9
Tall grass prairie	?	5·1	2
Winter wheat	<1	1·1–1·2	5

[*] References: (1) Barber 1971; (2) Dahlman 1968; (3) Deans 1981; (4)
Fogel & Hunt 1983; (5) Gregory *et al.* 1978; (6) Harris, Kinerson &
Edwards 1977; (7) McClaugherty, Aber & Melillo 1982; (8) Persson
1978; (9) Sivakumar, Taylor & Shaw 1977; (10) Vogt *et al.* 1980.

annual herbaceous production might be due to lower plant densities in agro-
ecosystems or perhaps greater utilization of soil volume by perennials in poly-
cultures.

Production estimates for mycorrhizas are even more scarce than for other root
classes. Annual biomass increment of live ectomycorrhizas in 23 and 180-year old
stands of Pacific silver fir was 75 and 70% (1·5–1·9 Mg ha^{-1} of mycorrhiza biomass
respectively (Vogt *et al.* 1980). Slightly lower proportions for live plus dead
ectomycorrhizas of 61–68% (15·3–18·1 Mg ha^{-1}) were obtained in a 55 year-old
Douglas fir stand (Fogel & Hunt 1983). Deans (1981) reported production of roots
<5 mm diameter to be 5·3 Mg ha^{-1} or 108% of biomass. This estimate presumably
includes a large proportion of mycorrhiza biomass since nearly all (99%) of the
production was by roots 0–2 mm in diameter.

Biomass losses

Root losses

Several studies have failed to detect any significant decreases in the standing crops
of large conifer and *Liriodendron* roots (Harris *et al.* 1977; Persson 1978; Santantonio *et al.* 1977), despite the reported presence of large dead roots in the *Lidrioden-
dron* stand, an apple orchard, and a red pine plantation (Fayle 1975; Harris *et al.*
1977; Kolesnikov 1968).

Large annual decreases or 'die-back' of the order of 30–90% occur in the

TABLE 2. Annual losses (% total wt) of fine root
biomass in different forests

Ecosystem	Loss	Reference[*]
Deciduous forest		
European beech	80–92	4
Oak	52	5
Yellow poplar	42	2
Walnut	90	1
Coniferous forest		
Douglas fir	40–47	3,7
Scots pine	66	6

[*] References: (1) Bode 1959 in Hermann
1977; (2) Edwards & Harris 1977; (3) Fogel &
Hunt 1983; (4) Gottsche 1972 in Hermann 1977;
(5) Ovington, Heitkamp & Lawrence 1963;
(6) Perrson 1978; (7) Santantonio 1979.

length, volume, or number of fine root tips for a number of quite different forest
tree species, as has been known for some time (Fogel 1983). Graphs of 'white' root
length versus time show decreases similar to those in forest trees for raspberry
grown under irrigation (Atkinson 1973), apple trees (Head 1966), and pear trees
(Head 1968). Annual biomass loss of fine roots, estimated by summing the significant decreases between peaks and troughs in monthly or seasonal standing crops,
has revealed annual losses ranging from 40 to 92% in a number of forests (Table 2).
In a Douglas fir ecosystem, combined annual fine root/mycorrhiza loss ranges from
$14 \cdot 6$ to $18 \cdot 8$ Mg ha^{-1} (Fogel & Hunt 1983). These results indicate that fine roots
and mycorrhizas are two to five times more important in returning organic matter
to the soil than is leaf and branch litter (Fogel 1983; Edwards & Harris 1977).

Exudate losses

Root exudates include solubilized mucigel produced by the root cap, droplets sec-
reted by root hairs, soluble materials from living cells, and sloughed fragments of
cell walls. Carbohydrates, amino acids, organic acids, enzymes, auxins, vitamins,
and other compounds (some of which stimulate or inhibit fungi, bacteria, and
nematodes) have all been reported present in exudates (Russell 1977). Trees
appear to exude more amino acids/amides than crop plants and comparable
amounts of carbohydrates (Smith 1977). These compounds are important in the
nutrition of rhizosphere organisms, but the amount of net primary production lost
in exudates is difficult to estimate. The quantity of exudates is influenced by nitrogen levels, phosphorus supply, age of roots, rate of shoot extension, temperature,
water stress, and the presence of micro-organisms (Russell 1977). Collection of
exudates has been done in solution culture, with or without ballotini, under sterile
conditions since microbes would metabolize compounds as they are released. An
alternative approach in which the tops of plants are supplied with ^{14}C, and the

proportion of labelled carbon collected in exudates measured, produces a minimum estimate since the amount depends on the balance between exudation and re-absorption of the label. Solution culture experiments with barley indicate that exudates may equal 7–10% of the shoot weight (Russell 1977). Under laboratory conditions, wheat plants supplied with ^{14}C for 21 days exuded 8·3% of the label in non-sterile soil compared to 6·2% in sterile soil (Russell 1977). The solution cul-ture technique has been used under field conditions by Smith (1976) to collect exudates from unsuberized root tips of birch, beech, and maple. The roots were first severed and allowed to produce new tips. The tips were then inserted into tubes filled with sterile distilled water, and allowed to grow for 14 days. Only samples free of microbes were analysed for exudates. Exudates equalled 0·5–1·4% of the root weight. After adjusting for estimated number of root tips, length of growing season, and species composition of the stand under study, Smith calculated that annual production of exudates totalled 66·5 kg ha^{-1} in a northern hardwood forest. Exudates represented 43% of the new woody root tip mass in this study, or 1·1% of the live root biomass <3 mm diameter in a similar stand studied by McClaughtery, Aber & Melillo (1982). The results are interesting, but difficult to interpret since the technique employed permits collection only from axenic, unsub-erized woody roots and does not include exudates from non-woody or mycorrhizal roots. In addition, exudates have only been collected late in the growing season and the growth of roots in a non-aerated solution is quite different from that in soil. The annual production of exudates by herbaceous species apparently has not been calculated on an ecosystem basis (Smith 1977).

Herbivory losses

Very few estimates exist for photosynthate losses below-ground to herbivory (Magnusson & Sohlenius 1980). Herbivory by nematodes and cicada in a mesic hardwood stand dominated by *Liriodendron* has been calculated by Ausmus *et al.* (1977). In this stand, 40% of the nematode population is composed of root-feeding species. Nematodes annually consume 8·5% of the living roots <5 mm in diam.: cicada larvae consume 1·4% of the fine root mass.

   Phytophagous nematodes have also been studied in a 15–20 year-old Scots pine stand by Magnusson & Sohlenius (1980). In their study, root/fungal-feeding nematodes comprised 42% on average of the total number of nematodes and reached a maximum of 50–55%. Miscellaneous and bacteria-feeding nematodes dominated on a biomass basis, however. After assuming that half of the respiration of nematodes was by root/fungal-feeders and that half of the carbon utilized by this group was of fungal origin, consumption was estimated to be about 0·3% of the annual fine root production, an order of magnitude less than in the *Liriodendron* study. These losses are comparable to the 0.5–3.5% leaf weight losses to herbivory reported for deciduous forests (Bray & Dudkiewicz 1963; Reichle *et al.* 1973; Whittaker 1975).

NET PRIMARY PRODUCTION

Estimates from studies employing the harvest approach indicate that a large proportion of total net primary production (NPP) is channelled below-ground (Table 3). In forest ecosystems, below-ground production ranges from 40 to 73% of NPP. Below-ground processes account for 65–66% of NPP in shrub steppe communities and similar percentages of 50–85% have been reported for native grasslands.

The assumptions used in estimating the contribution of roots to net primary production should be examined to determine the reliability of the estimates and to identify possible areas for refinement. Estimating production by positive increments and loss by decreases in the standing crop of all roots irrespective of physiological status produces estimates generally regarded as conservative because steady-state conditions are assumed and mortality is not taken into account. This method of calculation assumes there is no acceptable method of distinguishing active (live) from inactive (dead) roots. In theory, production and loss estimates could be refined considerably by balancing changes in the standing crops of active and inactive roots (McClaugherty et al. 1982; Santantonio 1979). In practice, however, classification of the physiological status of roots has proved difficult. For instance, ectomycorrhizas pose a problem because of their small size (0·5–3·0 × 0·15–0·6 mm in diameter) and the possibility that the host tissue may continue to be physiologically active after the death of the fungal symbiont. Criteria for classifying active mycorrhizas are similar to those used for fine roots: turgidity, colour, integrity of root apex, association with mycelia, and the colour of sectioned host

TABLE 3. Contribution of roots (% total) to net primary production in different ecosystems

Ecosystem	Age (years)	Contribution	Reference[*]
Coniferous forest			
Douglas fir	55	73	3
Pacific silver fir	23	60	4
	180	71	4
Scots pine	14	60	1
Deciduous forest			
Yellow poplar	?	40	5
Shrub			
Atriplex steppe	?	65	2
Ceratoides steppe	?	66	2
Herbaceous			
Tall grass prairie	?	50	6
Short grass prairie	?	85	7

[*]References: (1) Ågren et al. 1980; (2) Caldwell & Camp 1974; (3) Fogel & Hunt 1983; (4) Grier et al. 1981; (5) Harris, Santantonio & McGinty 1980; (6) Kucera, Dahlman & Koelling 1967; (7) Sims & Singh 1971.
? = Not given or inappropriate.

tissues (Fogel 1983). Ectomycorrhizas from soil cores generally exhibit gradations in these characters, especially colour and turgidity, thus greatly increasing the possibility for errors in classification. Further errors may arise from incomplete recovery and improper handling, e.g. exposure to dry air (Lyford 1975). An alternative technique employing mitotic indexing of root meristems (Dunsworth & Kumi 1982) is unusable for field material due to the lack of intact root or mycorrhiza apices in many core samples.

Distinguishing active from inactive roots has also been a problem in herbaceous systems. Presence and absence of dividing cells in the root apex of grasses has been a criterion, but roots from soil cores often lack intact tips (Singh & Coleman 1973). Physiological status of some 'weeds' has been determined by measuring the electrical capacitance and resistance of roots, but this method is only suitable for roots of sufficient diameter to allow insertion of a probe (Singh & Coleman 1973). Vital stains such as congo red and tetrazolium have also been employed, but roots of various plant species react quite differently, apparently due to the age and physical properties of the roots as well as root cation exchange capacity (Böhm 1979). A modification of the vital staining technique involving the extraction of formazon (reduced tetrazolium) suffers the same problems as the tetrazolium vital staining technique (Sator & Bommer 1971; Singh & Coleman 1973). Carbon-14 labelling of roots under field conditions has also been tried (Sator & Bommer, 1971; Singh & Coleman 1973; Ueno, Yoshihara & Okada 1967). Singh and Coleman (1973) enclosed shoots of grasses in a plastic bag, exposed the tops to $^{14}CO_2$, extracted roots by coring, and then autoradiographed root segments for 4 weeks. Disadvantages of this approach include the time required to mount root segments and develop autoradiographs. The use of scintillation counting of pelleted root fragments reduces the time required to assay roots for activity but requires a highly significant correlation between disintegrations $min^{-1} mg^{-1}$ dry weight of roots and percentage of active roots (Singh & Coleman 1973). Another carbon-14 technique provides an index of the productivity of root systems by measuring the reduction in the $^{14}C/^{12}C$ ratio for structural carbon of the roots following pulse labelling of the shoots with $^{14}CO_2$ (Caldwell & Camp 1974). The advantage of this approach is that the movement of labile carbon into and out of the roots does not affect the estimate. Under field conditions underestimation of production may result from a change in the proportion of living and dead roots between the initial and final root harvests. Underestimation may also occur if ^{14}C-compounds are translocated between harvests or by the release and subsequent reincorporation of ^{14}C into new structural material during the decomposition of labelled roots.

The tops of herbs and shrubs are fairly easily contained in plastic bags, but forests pose a more formidable problem. Herbs, shrubs, and tree branches in a forest stand could be isolated in bags and $^{14}CO_2$ injected into these bags. Many of the roots under a single tree would not be labelled, however, due to the intermingling of non-labelled roots from adjacent trees (McMinn 1963). An attractive solution would involve labelling roots directly in soil cores and then measuring vitality by scintillation counting after separation of roots from the cores.

Below-ground NPP estimates could be further refined by making adjustments for other loss parameters. Exudation is presently ignored due to the problems in its estimation, but the research by Smith (1977) indicates a possible underestimation of root production by $0·5–1·4\%$; experiments under laboratory conditions indicate the potential for much larger errors (Russell 1977). Volatilization is another possible source of underestimation, but is poorly understood and extremely difficult to estimate under field conditions. Herbivory estimates have little potential impact on NPP estimates since herbivory is accounted for in the root loss term as calculated by decrements in standing crops. Estimates of herbivory are important, however, in clarifying our knowledge of the structure and function of below-ground ecosystems.

CONCLUSIONS

Roots have been shown to play a major role in NPP in a number of different ecosystems. Root NPP ranges from 40 to 85% of total NPP with most of the available estimates falling in the range of 60–70%. The estimated fluxes represent a sizeable drain on carbohydrate reserves of plants for root production, and via senescence provide most of the organic material entering decomposition. The importance of roots in NPP has implications at several levels of ecological organization. If mechanisms controlling carbon cycling are to be understood, below-ground processes cannot be ignored. Similarly, below-ground processes may be crucial in nutrient cycling if the cycles are closely coupled to organic matter, i.e. nitrogen, phosphorus and potassium. Nutrient absorption models for either trees or annual crops are flawed if they are based solely on root production and fail to account for potentially large nutrient losses in root senescence and decomposition given the rapid turnover of fine roots and mycorrhizas. The magnitude of below-ground NPP has further implications beyond the process and ecosystem level. Many earlier regional and global carbon budgets (e.g. Reiners 1973; Rodin & Bazilevich 1967; Whittaker 1962; Whittaker & Likens 1973) are based on the questionable assumption that the ratio of root production to mass is $0·2–1·0$ times that of shoot systems whereas the harvest approach suggests a more realistic factor would be 2·8 for *Liriodendron* (Harris *et al.* 1977) and 6·5–7·4 for Douglas fir (Fogel, pers. obs.).

The impact of the below-ground NPP research on our conception of ecosystem structure and function, while impressive, only represents a rough first approximation. Further refinement will depend on the development of techniques for measuring root vitality, exudation, and hyphal biomass of mycorrhizal fungi in the soil to supplement our understanding of production losses.

ACKNOWLEDGMENTS

Preparation of this paper was supported in part by National Science Foundation grant BSR 8215265.

REFERENCES

Ågren, G.I., Axelsson, B., Flower-Ellis, J.G., Linder, S., Persson, H., Staaf, H. & Troeng, E. (1980). Annual carbon budget for a young Scots pine. *Ecological Bulletin*, (*Stockholm*), **32**, 307–313.

Andersson, F. (1970). Ecological studies in a Scanian woodland and meadow area, southern Sweden. II. Plant biomass, primary production and turnover of organic matter. *Botaniska Notiser*, **123**, 8–51.

Atkinson, D. (1973). Seasonal changes in the length of white unsuberized root on raspberry plants grown under irrigated conditions. *Journal of Horticultural Science*, **48**, 413–419.

Ausmus, B.S., Ferris, J.M., Reichle, D.M. & Williams, E.C. (1977). The role of primary consumers in forest root dynamics. *The Belowground Ecosystem: A synthesis of plant-associated processes* (Ed. J. K. Marshall), pp 261–265, Range Science Department Science Series No. 26, Colorado State University, Fort Collins, Colorado.

Barber, S.A. (1971). Effect of tillage practice on corn (*Zea mays* L.) root distribution and morphology. *Agronomy Journal*, **63**, 724–726.

Böhm, W. (1979). *Methods of Studying Root Systems.* Springer-Verlag, New York.

Bray, J.R. (1963). Root production and the estimation of net productivity. *Canadian Journal of Botany*, **41**, 65–72.

Bray, J.R. & Dudkiewicz, L.A. (1963). The composition, biomass and productivity of two *Populus* forests. *Bulletin of the Torrey Botanical Club*, **90**, 298–308.

Caldwell, M.M. & Camp, L.B. (1974). Belowground productivity of two cool desert communities. *Oecologia*, **17**, 123–130.

Dahlman, R.C. (1968). Root production and turnover of carbon in the root–soil matrix of a grassland ecosystem. *Methods of Productivity Studies in Root Systems and Rhizosphere Organisms* (Eds M.S. Ghilarov, V. A. Kovda, L. N. Novichkova-Ivanova, L. E. Rodin & V. M. Sveshnikova), pp.11–21, USSR Academy of Sciences, Leningrad.

Deans, J.D. (1981). Dynamics of coarse root production in young plantation of *Picea sitchensis. Forestry*, **54**, 139–155.

Deans, J. & Ford, E.D. (1983). Modelling root structure and stability. *Plant and Soil*, **71**, 189–196.

Dice, S.F. (1970). *The biomass and nutrient flux in a second growth Douglas-fir ecosystem (a study in quantitative ecology).* Ph.D. thesis, University of Washington, Seattle, Washington.

Dunsworth, B.G. & Kumi, J. W. (1982). A new technique for estimating root system activity. *Canadian Journal of Forest Research*, **12**, 1030–1032.

Edwards, N.T. & Harris, W.F. (1977). Carbon cycling in a mixed deciduous forest floor. *Ecology*, **58**, 431–437.

Eis, S. (1974). Root system morphology of western hemlock, western red cedar, and Douglas–fir. *Canadian Journal of Forest Research*, **4**, 28–38.

Fayle, D.C.F. (1975). Distribution of radial growth during the development of red pine root systems. *Canadian Journal of Forest Research*, **5**, 608–625.

Fogel, R. (1983). Root turnover and productivity of coniferous forests. *Plant and Soil*, **71**, 75–85.

Fogel, R. & Hunt, G. (1979). Fungal and arboreal biomass in a western Oregon Douglas–fir ecosystem: distribution patterns and turnover. *Canadian Journal of Forest Research*, **9**, 245–256.

Fogel, R. & Hunt, G. (1983). Contribution of mycorrhizae and soil fungi to nutrient cycling in a Douglas–fir ecosystem. *Canadian Journal of Forest Research*, **13**, 219–232.

Ford, E.D. & Deans, J.D. (1977). Growth of Sitka spruce plantation: spatial distribution and seasonal fluctuations of lengths, weights and carbohydrate concentrations of fine roots. *Plant and Soil*, **47**, 463–485.

Gray, T.R. & Williams, S.T. (1971). *Soil Micro-organisms.* Oliver & Boyd, Edinburgh.

Gregory, P.J., McGowan, M., Biscoe, P.V. & Hunt, B. (1978). Water relations of winter wheat. 1. Growth of the root system. *Journal of Agricultural Science*, (Cambridge), **91**, 91–102.

Grier, C.G., Vogt, K.A., Keyes, M.R. & Edmonds, R.L. (1981). Biomass distribution and above- and belowground production in young and mature *Abies amabilis* zone ecosystems of the Washington Cascades. *Canadian Journal of Forest Research*, **11**, 155–167.

Harris, W.F., Kinerson, R.S., Jr & Edwards, N.T. (1977). Comparison of belowground biomass of natural deciduous forest and loblolly pine plantations. *Pedobiologia*, **7**, 369–381.

Harris, W.F., Santantonio, D. & McGinty, D. (1980). The dynamic below-ground ecosystem. *Forests: Fresh Perspectives from Ecosystem Analysis* (Ed. R. H. Waring), pp. 119–129, Oregon State University Press, Corvallis.

Head, G.C. (1966). Estimating seasonal changes in the quantity of white unsuberized root on fruit trees. *Journal of Horticultural Science*, **41**, 197–206.

Head, G.C. (1968). Seasonal changes in the amount of white unsuberized root on pear trees on quince rootstock. *Journal of Horticultural Science*, **43**, 49–58.

Heilman, P. & Gessel, S.P. (1963). Nitrogen requirements and the biological cycling of nitrogen in Douglas-fir stands in relationship to the effects of nitrogen fertilization *Plant and Soil*, **18**, 386–402.

Hermann, R.K. (1977). Growth and production of tree roots: a review. *The Belowground Ecosystem: A Synthesis of Plant-Associated Processes* (Ed J. K. Marshall), pp. 7–28. Range Science Department Science Series No. 26, Colorado State University, Fort Collins, Colorado.

Kira, T. & Ogawa, H. (1968). Indirect estimation of root biomass increment in trees. *Methods of Productivity Studies in Root Systems and Rhizosphere Organisms* (Eds M. S. Ghilarov, V. A. Kovda, L. N. Nivochkova-Ivanova, L. E. Rodin & V. M. Sveshnikova), pp 96–101. USSR Academy of Sciences, Leningrad.

Kolesnikov, V.A. (1968). The cycling renewal of roots in fruit plants. *Methods of Productivity Studies in Root Systems and Rhizosphere Organisms.* (Eds M. S. Ghilarov, V. A. Kovda, L. N. Novichkova-Ivanova, L. E. Rodin & V. M. Sveshnikova), pp. 102–106, USSR Academy of Sciences, Lenningrad.

Kucera, C.L., Dahlman, R.C. & Koelling, M.R. (1967). Total net productivity and turnover on an energy basis for tallgrass prarie. *Ecology*, **48**, 536–541.

Lyford, W.H. (1975). Rhizography of non-woody roots of trees in the forest floor. *The Development and Function of Roots.* (Eds J. G. Torrey & D. T. Clarkson). pp. 179–196, Academic Press, New York.

Lyr, H. & Hoffman, G. (1967). Growth rates and growth periodicity of tree roots. *International Review of Forestry Research*, Vol.2. (Eds J. A. Romberger & P. Mikola), pp. 181–236. Academic Press, New York.

McClaugherty, C.A., Aber, J.D. & Melillo, J.M. (1982). The role of fine roots in the organic matter and nitrogen budgets of two forested ecosystems. *Ecology*, **63**, 1481–1490.

McMinn, R.G. (1962). Characterization of Douglas-fir root systems. *Canadian Journal of Botany*, **41**, 155–122.

Magnusson, C. & Sohlenius, B. (1980). Root consumption in a 15–20 year old Scots pine stand with special regard to phytophagous nematodes. *Ecological Bulletin (Stockholm)*, **32**, 261–268.

Moir, W.H. & Bachelard, E.P. (1969). Distribution of fine roots in three *Pinus radiata* stands near Canberra. *Ecology*, **50**, 658–662.

Newbold, P.J. (1967). *Methods for Estimating the Primary Production of Forests.* IBP handbook No. 2, Blackwell Scientific Publications, Oxford.

Nihlgard, B. (1972). Plant biomass, primary production and distribution of chemical elements in a beech and planted spruce forest in South Sweden. *Oikos*, **23**, 69–81.

Nye, P.H. & Tinker, P.B. (1977). *Solute Movement in the Soil–Root System.* Studies in Ecology, Vol. 4. Blackwell Scientific Publications, Oxford.

Odum, E.P. (1971). *Fundamentals of Ecology* (3rd edn). W.B. Saunders Company, Philadelphia.

Ovington, J.D., Heitkamp, D. & Lawrence, D.B. (1963). Plant biomass of prairie, savanna, oak wood, and maize field ecosystems in central Minnesota. *Ecology*, **44**, 52–63.

Ovington, J.D. & Madwick, H.A.I. (1959). Distribution of organic matter and plant nutrients in a plantation of Scots pine. *Forest Science*, **5**, 344–355.

Pastor, J. & Bockheim, J.G. (1981). Biomass and production of an aspen-mixed hardwood-spodosol ecosystem in northern Wisconsin. *Canadian Journal of Forest Research*, **11**, 132–138.

Persson, H. (1978). Root dynamics in a young Scots pine stand in central Sweden, *Oikos*, **30**, 508–519.

Persson, H. (1980). Spatial distribution of fine-root growth, mortality and decomposition in a young Scots pine stand in central Sweden. *Oikos*, **4**, 77–87.

Reichle, D.E., Dinger, B.E., Edwards, N.T., Harris, W.F. & Sollins, P. (1973). Carbon flow and storage in a forest ecosystem. *Carbon and the Biosphere* (Eds E.G. Woodwell, & E.V. Pecan), pp. 345–365, National Technical Information Services, US Department of Commerce, Springfield, Virginia.

Reiners, W.A. (1973). Terrestrial detritus and the carbon cycle. *Carbon and the Biosphere* (Eds E.G.

Woodwell & E. V. Pecan), pp. 303–326, National Technical Information Service, US Department of Commerce, Springfield, Virginia.

Roberts, J. (1976). A study of the root distribution and growth in a *Pinus silvestris* L. (Scots pine) plantation in East Anglia. *Plant and Soil*, **44**, 607–621.

Rodin, L.E. & Bazilevich, N.I. (1967). *Production and Mineral Cycling in Terrestrial Vegetation*. Oliver and Boyd, London.

Russell, R.S. (1977). *Plant Root Systems: Their Function and Interaction with the Soil*. McGraw-Hill, London.

Santantonio, D. (1979). Seasonal dynamics of fine roots in mature stands of Douglas-fir of different water regimes: a preliminary report. *Root Physiology and Symbiosis* (Eds A. Reidacker & J. Gagnaire–Michard). Proceedings IUFRO Symposium on Root Physiology and Symbiosis, Nancy, France, Sept. 1978, Vol.6. CNRF, Champenoux, France.

Santantonio, D., Hermann, R.K. & Overton, W.S. (1977). Root biomass studies in forest ecosystems. *Pedobiologia*, **17**, 1–31.

Sator, C. & Bommer, D. (1971). Methodological studies to distinguish functional from nonfunctional roots of grassland plants. *Integrated Experimental Ecology* (Ed. H. Ellenberg), pp. 72–74, Ecological Studies No. 2. Springer-Verlag, New York.

Shaver, G.R. & Billings, W.D. (1975). Root production and root turnover in a wet tundra ecosystem, Barrow, Alaska. *Ecology*, **56**, 401–409.

Sims, P.L. & Singh, J.S. (1971). Herbage dynamics and net primary production in certain ungrazed and grazed grasslands in North America. *Preliminary Analysis of Structure and Function in Grasslands* (Ed. N.R. French), pp. 59–124, Range Science Department Science Series No. 10, Colorado State University, Fort Collins, Colorado.

Singh, J.S. & Coleman, D.C. (1973). A technique for evaluating functional root boimass in grassland ecosystems. *Canadian Journal of Botany*, **51**, 1867–1870.

Sivakumar, M.V.K., Taylor, H.M. & Shaw, R.H. (1977). Top and root relations of field-grown soybeans. *Agronomy Journal*, **69**, 470–473.

Smith, W.H. (1976). Character and significance of forest tree root exudates. *Ecology*, **57**, 324–331.

Smith, W.H. (1977). Tree root exudates and the forest soil ecosystem: Exudate chemistry, biological significance and alteration by stress. *The Belowground Ecosystem: A Study in Plant-Associated Processes* (Ed. J. K. Marshall), pp. 289–302, Range Science Department Science Series No. 26, Colorado State University, Fort Collins, Colorado.

Ueno, M., Yoshihara, K. & Okada, T. (1967). Living root systems distinguished by the use of carbon-14. *Nature, (London)*, **213**, 530–532.

Vogt, K.A., Edmonds, R.L., Grier, C.C. & Piper, S.R. (1980). Seasonal changes in mycorrhizal and fibrous-textured root biomass in 23- and 180-year-old Pacific silver fir stands in western Washington. *Canadian Journal of Forest Research*, **10**, 523–529.

Whittaker, R.H. (1962). Net production relations of shrubs in Great Smoky Mountains. *Ecology*, **43**, 357–377.

Whittaker, R.H. (1975). *Communities and Ecosystems*. MacMillan, New York.

Whittaker, R.H. & Likens, G.E. (1973). Carbon in the biota. *Carbon and the Biosphere* (Eds G.M. Woodwell & E.V. Pecan), pp. 281–300. National Technical Information Service, US Department of Commerce, Springfield, Virginia.

Whittaker, R.H., Likens, G.E., Bormann, F.H., Eaton, J.S. & Siccama, T.G. (1979). The Hubbard Brook ecosystem study: forest nutrient cycling and element behaviour. *Ecology*, **60**, 203–220.

Whittaker, R.H. & Marks, P.L. (1975). Methods of assessing terrestrial productivity. *Primary Productivity of the Biosphere* (Eds H. Leith and R. H. Whittaker). Ecological studies, No. 14, 55–118.

Methods of calculating fine root production in forests

R. I. FAIRLEY* AND I. J. ALEXANDER

Department of Botany, The University, Aberdeen, Scotland AB9 3UD

INTRODUCTION

Recent studies have indicated the importance of the fine root compartment in forest productivity and nutrient cycling (Fogel & Hunt 1979, 1983; Vogt *et al.* 1982; Cox *et al.* 1977). Fine root biomass up to 12 Mg ha^{-1} occurs in coniferous forests (Santantonio, Hermann & Overton 1977) and seasonal fluctuations in biomass can be large; inded some estimates suggest that they may account for up to 40% of total dry matter production (Cox *et al.* 1977; Ågren *et al.* 1980). However different methods of calculation can lead to different estimates of production and mortality, and although some are liable to underestimate others may lead to unrealistically high estimates (McClaugherty, Aber & Melillo 1982). This paper considers three methods and indicates the method likely to provide the best estimate of root production and mortality.

METHODS OF CALCULATION

Calculation of biomass flux generally depends upon the determination of differences between root biomass estimated on different sampling occasions. At the simplest level, production can be calculated as the difference between observed annual maximum and minimum biomasses. Recently more sophisticated methods have been proposed, prompted by the observation of frequent fluctuations in fine root biomass in some stands (e.g. Persson 1978) or in an attempt to provide estimates of throughput (cumulative annual mortality) and decomposition (cumulative annual disappearance of dead roots) (Santantonio 1979).

All these methods originate from a basic equation adapted from population dynamics:

$$B_{t+1} = B_t + \text{production} - \text{mortality} \qquad (1)$$

where B are measures of biomass at two times (t). This can be re-arranged.

$$\text{production} = B_{t+1} - B_t + \text{mortality} \qquad (2)$$

However, the equation cannot be used without a measure of mortality so in the

---

*Present address: Countryside Commission for Scotland, Battleby, Redgorton, Perth PH1 3EW.

past an approximation has been made by assuming mortality to be zero or by using the inequality

$$\text{production} \geqslant B_{t+1} - B_t \tag{3}$$

Over a sequence of sampling dates the estimate is found by summing these positive differences.

$$\text{production} \geqslant \sum_{j+1}^{k} (\Delta B)^+_j \tag{4}$$

where k is the number of samplings during the investigation $- 1$; $(\Delta B)^+_j$ are observed positive values of biomass change from the jth to the $j + 1$th sampling.

However, if both live and dead populations are measured this underestimate can be minimized. Thus,

$$B_{t+1}^{\text{living}} = B_t^{\text{living}} + \text{production} - \text{mortality} \tag{5}$$

and

$$B_{t+1}^{\text{dead}} = B_t^{\text{dead}} + \text{mortality} - \text{disappearance} \tag{6}$$

where disappearance is a term accounting for loss of roots from the dead root biomass, both through fragmentation to unrecognizable or irretrievable sizes and through weight loss during decomposition. Solving eqns (5) and (6) simultaneously we get

$$\text{production} = \Delta B^{\text{total}} + \text{disappearance} \tag{7}$$

where B^{total} is the total biomass of fine root material, alive plus dead; and

$$\text{mortality} = \Delta B^{\text{dead}} + \text{disappearance} \tag{8}$$

In this way estimates of production and mortality can be made independently of each other. Again, without a measure of, in this case 'disappearance' these equations cannot be used. In an attempt to provide optimal estimates of biomass flux Persson (1978 and other papers) used the inequalities:

$$\text{production} \geqslant \Sigma(\Delta B^{\text{total}})^+_j \tag{9}$$

$$\text{mortality} \geqslant \Sigma(\Delta B^{\text{total}})^+_j \tag{10}$$

and demonstrated that production estimated in this way was greater than when estimated by eqn (4). These equations however do not provide the best estimate derivable from such data and have an inherent tendency to underestimate.

If we envisage the flow of biomass such that all decreases in live biomass pass as increments into the dead biomass, and all decreases in dead biomass pass as 'inputs' into the measure of dead root disappearance (i.e. we ignore all possible losses by respiration, consumption or translocation), then decreases in dead biomass between sampling dates or decreases in live biomass which are greater than any simultaneous increase in dead biomass both provide some information about the rate of root disappearance, which can lead to better estimates of production and mortality.

When a decrease occurs in dead biomass between sampling dates at the same time as an increase in live biomass, an estimate of disappearance is given by the inequality

$$\text{disappearance} \geqslant -\Delta B^{\text{dead}} \qquad (11)$$

This is only a minimum estimate because the precise inputs into the dead biomass through mortality are unknown. Equation (11) can then be used in eqns (7) and (8) to provide more precise estimates of production and mortality under these conditions.

$$\text{production} \geqslant \Delta B^{\text{live}} \qquad (12)$$

$$\text{mortality} \geqslant 0 \qquad (13)$$

Similarly, when the decrease in the live biomass exceeds the observed increase in dead biomass or when simultaneous decreases occur in both live and dead biomasses between sampling dates, the minimum estimate of disappearance is found from the inequality

$$\text{disappearance} \geqslant -\Delta B^{\text{live}} - \Delta B^{\text{dead}} \qquad (14)$$

When this is used in eqns (7) and (8), then

$$\text{production} \geqslant 0 \qquad (15)$$

$$\text{mortality} \geqslant -\Delta B^{\text{live}} \qquad (16)$$

Incorporation of this information in calculating production and throughput estimates has been termed 'balancing transfers' (Santantonio 1979) which usefully describes the mathematical process of balancing observed changes in live and dead biomass with transfers of biomass from one compartment to another. The use of eqns (7)–(16) to balance these transfers is summarized in Fig. 1. This is similar to the decision matrix drawn up by McClaugherty et al. (1982) but differs in two important aspects. Firstly, these authors decide arbitrarily which of the two groups of equations to use in the event of a dead biomass increase simultaneous to a live

LIVE

			increase	decrease	
				$\Delta B^{\text{dead}} > \Delta B^{\text{live}}$	$\Delta B^{\text{live}} > \Delta B^{\text{dead}}$
D E A D	increase		$P = \Delta B^{\text{live}} + \Delta B^{\text{dead}}$ $M = \Delta B^{\text{dead}}$ $D = 0$	$P = \Delta B^{\text{live}} + \Delta B^{\text{dead}}$ $M = \Delta B^{\text{dead}}$ $D = 0$	$P = 0$ $M = -\Delta B^{\text{live}}$ $D = -\Delta B^{\text{live}} - \Delta B^{\text{dead}}$
	decrease		$P = \Delta B^{\text{live}}$ $M = 0$ $D = -\Delta B^{\text{dead}}$	$P = 0$ $M = -\Delta B^{\text{live}}$ $D = -\Delta B^{\text{live}} - \Delta B^{\text{dead}}$	

FIG. 1. Decision matrix illustrating the equations used for estimating fine root production, mortality and fine dead root disappearance. The appropriate quadrant is chosen according to the direction of change in the live and dead standing biomasses (B) during the interval between two sampling dates. Production (P), mortality (M) and disappearance (D) are calculated using the equations in the indicated quadrants. In the top right quadrant a choice of equations is given depending on whether the observed change is larger in the dead or in the live biomass. Derivation of the equations are discussed in the text. Annual estimates are calculated by summing the estimates from all sampling intervals within the year.

biomass decrease; secondly, in the event of a live biomass increase and dead biomass decrease they estimate mortality as equalling the change in live biomass, which appears to be illogical.

In all three of the above methods (eqn (4); eqns (9) and (10); balancing transfers), cumulative estimates depend upon the sum of differences observed between a series of observations. Clearly, it is possible for considerable error to accumulate in this process, depending upon the precision of biomass estimates from month to month. It is therefore important that only significant differences between individual biomass estimates should be included in the cumulative estimates of production, mortality and disappearance. This has generally been adhered to by workers using the simplest of these methods (eqn (4)). Persson (1978; see also errata in *Oikos*, **31**, 136) accepted all differences but applied a correction factor to remove overestimation introduced by random variation in the means from the different sampling occasions. When balancing transfers has been carried out no attempt has been made to differentiate between significant and insignificant changes (Santantonio 1979; McClaughtery *et al.* 1982) and the latter authors discovered that estimates so derived were unrealistically high.

EXAMPLE

In recent work on fine root dynamics in spruce humus (Fairley 1983; Alexander & Fairley 1983) data were collected which illustrates these points. Fluctuations, of which only some were significant ($P \leqslant 0.05$), were recorded in live and dead

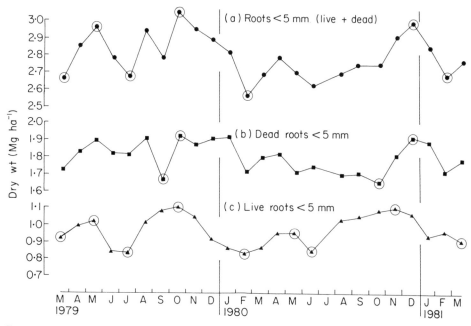

FIG. 2. Seasonal fluctuations in (a) all roots <5 mm diameter (live plus dead), (b) dead roots <5 mm diameter and (c) live roots <5 mm diameter in the humus layer of a 35-year old plantation of Sitka spruce (*Picea sitchensis* (Bong.) Carr.). Each point represents the mean oven dry weight (90°C) of fifteen samples from each of three replicate blocks. Maxima and minima significantly different from each other (Duncan's least significant range: $P \leqslant 0.05$) are indicated by circles.

biomass (Fig. 2). Estimates of production and where possible, mortality and disappearance, from these data using the three methods outlined above are given in Table 1. With balancing transfers, inclusion of all changes gave an estimate of production 1·4 times larger than that using statistically significant changes. Both these estimates are higher than those reached by eqns (4) or (9). An independently measured rate of fine root decay made using the litter bag method is in agreement

TABLE 1. Production and mortality of fine roots and their disappearance after death estimated by a number of different methods. For details of calculation see text

	Production $(\text{kg ha}^{-1}\text{yr}^{-1})$	Mortality $(\text{kg ha}^{-1}\text{yr}^{-1})$	Disappearance $(\text{kg ha}^{-1}\text{yr}^{-1})$
Balancing transfers (all fluctuations used)	879 ± 201	862 ± 193	856 ± 228
Balancing transfers (significant fluctuations only)	636 ± 175	644 ± 176	620 ± 152
Eqns (9) & (10)	567 ± 153	552 ± 141	No estimate possible
Eqn (4)	393 ± 44	Not quantified	No estimate possible
Litter bag measurement	–	–	689 ± 82

with the rate of disappearance of fine roots found from balancing transfers. Standard errors are large primarily because of considerable spatial variation found in the dead biomass.

PRACTICAL CONSIDERATIONS AND CONCLUSIONS

Balancing transfers requires some measurement of all the retrievable fraction of the fine root biomass. Methods of doing this have been discussed by Fairley (1983) and Fogel & Hunt (1979, 1983), where emphasis is made of its predominantly finely fragmented nature which restricts the use of sieves during extraction procedures. Balancing transfers also requires the differentiation of roots into live and dead categories. Where this proves to be impossible or impracticable eqn (9) provides the best estimate of production with loss of information only when live root increases are hidden by a simultaneous decrease in dead biomass. The advantages of separating live and dead, however, are considerable. It allows improvements of production estimates and estimation of throughput and disappearance.

ACKNOWLEDGMENTS

This work was carried out during the tenure of a Natural Environment Research Council studentship by R.I.F.

REFERENCES

Agren, G.I., Axelsson, B., Flowers-Ellis, J.G.K., Linder, S., Persson, H.,& Troeng, E. (1980). Annual carbon budget for a young Scots pine. *Structure and Function of Northern Coniferous Forests — An Ecosystem Study* (Ed. T. Persson). *Ecological Bulletin (Stockholm)*, **32**, 307–313.

Alexander, I.J. & Fairley, R.I. (1983). Effects of N fertilisation on populations of fine roots and mycorrhizas in spruce humus. *Plant and Soil*, **71**, 49–53.

Cox, T.L., Harris, W.F., Ausmus, B.S. & Edward, N.T. (1977). The role of roots in biogeochemical cycles in eastern deciduous forests. *The Belowground Ecosystem: A Synthesis of Plant-Associated Processes* (Ed. J. Marshall), Colorado State University, Department of Range Sciences, Fort Collins, Science Series No. 26, pp. 321–326.

Fairley, R.I. (1983). *Mycorrhiza and fine root dymanics in Sitka spruce.* Ph.D. Thesis. University of Aberdeen.

Fogel, R. & Hunt, G. (1979). Fungal and arboreal biomass in a Western Oregon Douglas-fir ecosystem: distribution patterns and turnover. *Canadian Journal of Forest Research*, **9**, 245–256.

Fogel, R. & Hunt, G. (1983). Contribution of mycorrhizae and soil fungi to nutrient cycling in a Douglas-fir ecosystem. *Canadian Journal of Forest Research*, **13**, 219–232.

McClaugherty, C.A., Aber, J.D. & Melillo, J.M. (1982). The role of fine roots in the organic matter and nitrogen budgets of two forested ecosystems. *Ecology*, **63**, 1481–1490.

Persson, H. (1978). Root dynamics in a young Scots pine stand in Central Sweden. *Oikos*, **30**, 508–519.

Santantonio, D. (1979). Seasonal dynamics of fine roots in mature stands of Douglas fir of different water regimes–a preliminary report. In *Root Physiology and Symbiosis.* (Ed. A. Reidaker), pp. 190–203. Proceeding of IUFRO Symposium, Nancy.

Santanonio, D., Hermann, K. & Overton, W.S. (1977). Root biomass studies in forest ecosystems. *Pedobiologia*, **17**, 1–31.

Vogt, K.A., Grier, C.C., Meier, & Edmonds, R.L. (1982). Mycorrhizal role in net primary production and nutrient cycling in *Abies amabilis* ecosystems in Western Washington. *Ecology*, **63**, 370–380.

Spatial and temporal aspects of root distribution as indicated by the use of a root observation laboratory

D. ATKINSON

Pomology Department, East Malling Research Station, East Malling, Maidstone, Kent ME19 6BJ

SUMMARY

1 Root production is important both to plants, for nutrient and water absorption, and to the soil fauna and microflora, as carbon and mineral sources. Accurate root productivity estimates are difficult to obtain because high spatial variability confuses estimates of changes with time. In addition, most methods are destructive so cannot be used frequently for long periods.

2 Root observation laboratories have been used to document temporal changes in root length but have rarely been used to give quantitative data. The use of root laboratories to provide quantitative data and the assumptions which need to be made to do this are discussed.

3 It is concluded that observation data for roots <2 mm can be used in this way, and the method is probably the only way to estimate production over short periods in long-term experiments. Spatial variability, systematic and random, mean that estimates of changes in the biomass of larger roots will be very much less reliable.

4 Values for production and potential decomposition of the roots of apple, strawberry and a grass sward are presented and discussed as examples of species types found in more natural plant communities. The amount and the timing of production and of the release of root material to the ecosystem varies between plant species, as does its contribution to total root density in the soil.

5 Although the density of apple roots in the soil is low ($L_A \simeq 7$ cm cm^{-2}) production can be comparable with other species and high during active periods, $\simeq 150$ kg ha^{-1} week^{-1}.

INTRODUCTION

The root system supplies the plant with water and mineral nutrients and anchors it in the soil. The supply of soil resources to the above-ground part of the plant is dependent upon the length of root available for absorption at particular times in relation to demand, modified by root distribution with depth, soil condition, etc. Some of these factors have been discussed for fruit tree roots by Atkinson (1983) who showed that new root production varied from year to year; periodicity of root growth can also vary greatly between species. With increasing age suberized and woody roots make up a higher proportion of the available root length, and roots of this type have been shown to function in the uptake of water and mineral nutrients.

43

Atkinson (1983) also found that a number of root system characters were altered when trees were grown in association with other species rather than under bare soil management. Other aspects of the dynamics of fruit tree root systems have been reviewed by Atkinson (1980a,b).

Root growth is important within the soil ecosystem not only to plants, but also to the flora and fauna, which depend upon roots as a food source. In this respect spatial and temporal aspects of root distribution and the transfer of root system production to herbivores, detritivores and decomposers will be vital.

For many years root laboratories have been used as a means of characterizing the seasonal periodicity of new root growth (Rogers & Head 1969) and the use of these laboratories has recently been reviewed by Huck & Taylor (1982). Estimating the seasonal periodicity of root production by other methods, e.g. the sequential removal of soil cores, is difficult because of the high spatial variability found with tree roots in the soil (Atkinson 1980b) and hence the high coefficient of variation associated with root sampling. This spatial element confounds temporal measurements. Despite this many estimates of root productivity have been published: Alexander & Fairley (1983) and Fogel (1983) are recent examples. The root laboratory should give a means of obtaining frequent estimates of temporal variation based on a constant fixed sample of the root system.

Changes in root length, usually assessed by intersection counting (Head 1966), can be used to characterize the seasonal periodicity of white root length and to compare treatments. To use root laboratory data to calculate root length for a tree or a mixed species community, the root length adjacent to the glass window (cm cm^{-2}) must be converted to root length in a volume (cm cm^{-3}). Window root lengths must therefore be multiplied by a factor representing the depth of soil in which roots can be seen. In addition, to use root laboratory values to calculate production for a whole tree or a larger soil volume a number of assumptions have to be made and factors allowed for. The main ones are that:

(i) the presence of the observation window does not make the root system atypical;

(ii) root density adjacent to the glass is typical of that in comparable parts of the soil volume away from the glass;

(iii) the position of the observation panel in relation to the horizontal spread of the root system is likely to give an acceptable representation, in relation to root density, periodicity, etc., of the whole soil volume being exploited.

(iv) the sample of the root system being observed is sufficiently large;

(v) the periodicity of root growth observed is similar to that elsewhere in the root system;

(vi) the distribution of growth on the window is representative of the remainder of the system.

These points are examined in detail in this paper, using both published and new data, to test the validity of using root laboratory measurements to give quantitative estimates of the performance of the whole root system.

Data is presented here showing the seasonal periodicity of root growth by apple trees, strawberry plants and a grass sward as models of some of the more common species types found in woodland communities. Making some of the assumptions detailed above, this data is used to calculate root productivity. The survival and decomposition of roots is discussed in relation to both the carbon and nutrient supply to the soil flora and fauna and the effectiveness of the root system.

MATERIALS AND METHODS

Materials

The results presented come mainly from experiments carried out using one of the East Malling Research Station Root Observation Laboratories (Rogers 1969), using the following species, cultivars and experimental designs.

Experiment 1

Trees of apple *Malus pumila* Mill. cv Worcester/MM. 104 were planted in 1961 at a spacing of 3.6×5 m but with their trunks planted 50 cm from an observation window (Head 1966). Trees were grown from 1961 to 1962 in cultivated soil and subsequently with a range of soil management treatments. In this paper detailed data are given for only two trees, 15S and 18S, which from 1963 to 1981 were grown under a sown sward of *Phleum pratense* L. ssp *bertolonii* (DC) Bornm, (= *P.nodosum*), irrigated to maintain a high soil water potential, $\simeq -30$ kPa, and with and without respectively an annual application of 100 kg ha^{-1} N.

Experiment 2

Trees of Cox and Golden Delicious/M.9 were planted in 1970 at densities of 0·09, 0·36, 1·44 and 5·76 m^2 tree^{-1}. Soil was maintained weed-free using herbicides. So measurement would always be made on the periphery of the root system, trees were planted 15, 30, 60 and 120 cm from the observation windows respectively.

Experiment 3

Strawberries (*Fragaria* $\times$ *ananassa*) cv Cambridge Favourite were planted in 1973 at a spacing of 0.6×0.6 m, either on their own or around apple trees, Cox/MM.106 at a spacing of 2.4×2.4 m. The strawberries were 0·3 m from the observation windows.

Experiment 4

Trees of Cox/M.26 were planted in 1971 at a spacing of 2.4×2.4 m with one of the following: (a) soil maintained weed-free using herbicides, (b) with an overall, S50 timothy, *Phleum pratense* spp. *bertolonii*, grass sward, (c) as in (b) but with irrigation applied to maintain the soil water potential greater than -30 kPa.

Experiment 5

Trees of Fortune/M.9 were planted in 1945 at a spacing of 4·2 × 4·2 m and grown under a number of soil management regimes but from 1965 approximately onwards in grass, except for a 2·2 m² herbicide-treated square around each tree base.

Methods

Root length was determined for both apple and strawberry by counting the number of intersections with a 1·2 × 1·2 cm grid on the observations windows (Head 1966). For grass, length was determined by counting the number of intersections with eleven horizontal lines at a range of representative depths and converting to total length using a factor derived from a plot of number of intersections against the length in the known areas around the lines (Atkinson 1977). Root survival (strip record) was assessed directly by recording with coloured marks on six paper strips per window, the dates new roots were produced and when they turned brown (Rogers 1969). The growth in diameter of woody roots was recorded at intervals using a microscope with a micrometer eye piece, for a sample of twenty roots from each of the treatments in Experiment 2.

In Experiment 5, root density was assessed by extracting the roots from 3·4 cm diameter soil cores taken both 50 and 150 cm from the trunk, either (a) in an intensive pattern of twelve cores around each of four trees or (b) from two positions around twenty-five trees. Cores were taken to a depth of 75 cm and divided into 15-cm increments. Nutrient determinations were made on root material from Experiments 2 and 5 using standard chemical methods.

At the conclusion of Experiment 1, the density of roots both adjacent to and away from the observation windows were compared. Four 220 ml soil cores were removed either from the 5 cm of soil adjacent to the glass or 10–15 cm behind the glass at four depths and at two positions in relation to the trunk, 50 and 200 cm. Soil dry bulk density was also assessed on these samples.

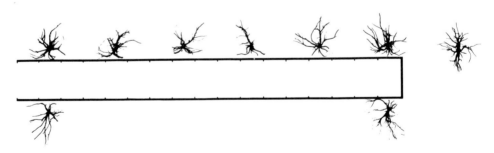

FIG. 1. The form of part of the root systems of eight trees grown adjacent to a root laboratory (Experiment 1) from 1961 to 1972 and of a similar tree grown near the laboratory. The recovery of only the central portion of the tree is a function of the tree-pulling method used.

RESULTS AND DISCUSSIONS

Interpretation of root laboratory results

Root system form

The forms of the root systems of some of the trees in Experiment 1, excavated in 1972 (Fig. 1), suggest that although they were compressed on the laboratory side they were not atypical, particularly when compared with a similar tree not adjacent to the glass. The root systems varied greatly from tree to tree. In Experiments 2–4 the trees were planted so the observation windows were at the periphery of the root system and not therefore asymetrically close to the glass.

Root density

The density of roots measured in 1982 adjacent to a window and 10–15 cm from the window (Table 1) indicated that although density varied between depths, overall density was similar beside and away from the glass. A comparison of density 50 and 200 cm from the trunk also showed no significant difference nor any interaction with distance from the glass. Bulk density was slightly lower adjacent to the glass. This was probably a residual effect of the repacking of soil adjacent to the glass, which was done when the laboratory was built in 1960 (Rogers 1969). Values obtained for a group of young trees (C.M.S. Thomas, pers. comm.) suggested higher root densities at the glass/soil interface. For established trees root densities at the window seem typical of those in comparable soil zones.

TABLE 1. (a) The density of apple roots (Worcester/MM.104) in core samples taken adjacent to or 10 cm from a root laboratory window 50 or 200 cm from the tree trunk and at a number of depths (mg ml^{-1}). (b) Soil density (g ml^{-1}) adjacent to and away from the glass

Distance from trunk (cm)	Position	Depth (cm) 10	25	50	95	Mean	Grand means
(a) Root density							
50	Beside	0·32	0·44	0·33	0·58	0·42	0·44
	Away	0·50	0·92	0·18	0·45	0·51	0·44
200	Beside	0·74	0·50	0·30	0·33	0·47	
	Away	0·61	0·42	0·20	0·24	0·37	
Standard error			0·31			0·13	0·12
(b) Soil density							
	Beside	1·67	1·60	1·56	1·54		1·59
	Away	1·68	1·66	1·62	1·66		1·65
Standard error			0·046				0·029

Root density: No significant effect of distance, position or depth.
Soil density: No significant effect of distance; depth significant at $P < 0.05$.

Sampling position

At the conclusion of Experiment 2, four replicate trees—but not those adjacent to the observation windows—were excavated from each treatment. The length of root recovered during this excavation (Atkinson, Naylor & Coldrick 1976) and an estimate of root length, calculated by assuming that root length at the window is representative of a layer of soil 0.3 cm thick, is shown in Table 2, for large roots.

TABLE 2. The length of larger diameter roots (>2 mm, root laboratory—fresh condition, >1 mm, excavation—dry condition), m, for the trees in Experiment 2 at a range of planting densities, m², either recovered by excavation (Atkinson, Naylor & Colrick 1976) in winter 1974/75 or estimated from root laboratory measurements in November 1974. Values in parenthesis show the range of measurements

Method	Tree density				
	0·09	0·36	1·44	5·76	Mean
Excavation	24	45	95	151	79
	(20–27)	(30–65)	(86–115)	(125–177)	
Root laboratory	40	124	86	32	71
	(21–62)	(80–192)	(0–212)	(0–80)	

The measurement of dry roots in the excavation study and fresh roots in the laboratory gives problems in comparing precisely the same diameter samples. Values for roots were broadly comparable, especially in the context of the size of coefficients of variation (CVs) usually associated with this type of measurement (Atkinson 1980b), e.g. for the trees in Experiment 5, the estimate of roots >2 mm diameter was 2.17 kg with a SE of 1.14 kg. Despite the large number of samples, the estimate of weight derived from the cores was below (about 80%) true weight. For roots >5 mm CVs could be as high as 2615%. The values in Table 2 suggest a systematic error with trees at the two higher densities, where the trunk and so many perennial roots are close to the glass, giving an overestimate and trees at the lower densities, where few major roots are near the glass, an underestimate. The basis of this is illustrated, for root weight, by the data from Experiment 5 (Table 3). The density of roots >2 mm was much higher, absolutely and relatively, close to the tree. For fine roots, lengths in Experiment 2 recovered by excavation, were about 10% of those estimated from the root laboratory. As 70–80% of root system weight is composed of large roots (approximately 2 mm diameter) (Atkinson *et al.* 1976)

TABLE 3. The weight of roots (g per core) of different diameter at a range of depths and distances from 26 year-old trees of Fortune/M.9

Distance from tree (m)	Root diameter (mm)	Depth (cm)				
		0–15	15–30	30–45	45–60	60–75
0·5	< 2	29·7	20·9	7·9	7·3	2·7
	> 2	47·3	162·4	20·5	21·1	3·8
1·5	< 2	2·8	12·2	9·1	3·6	4·0
	> 2	2·9	49·4	65·8	2·7	0

and usually only 63–78% of root is recovered by excavation (Dudney 1972), very high losses of fine root can be expected. Although the values of root length per unit soil surface area (L_A) calculated from root laboratory windows varied with tree density (0·5–14) they were similar to the range of published values for apple, (1·9–9·1 cm cm^{-2}) reviewed by Atkinson (1980b). Data for roots <2 mm diameter from Experiment 5, except for the values for 0–15 cm which may have been affected by treatment, were much more comparable at the two distances than those for the larger roots.

Estimates of root length obtained from a root laboratory seem acceptable and better for fine roots than those from excavations. Although the length of the larger roots can be estimated, the bias introduced by the position of the window in relation to the trunk, suggests that in some circumstances excavation may be the only way to obtain accurate total biomass estimates.

Sample size

Coefficients of variation for measurements made on roots are always high (Atkinson 1980b) so sample size is an important factor (Atkinson 1974a). The sample observed in the root laboratory can be large compared with samples studied in other methods. In Experiment 2 the laboratory sample represents 1% of soil volume for the 0·09 m^2 trees and 0·13% for the 5·76 m^2 trees. The removal of eight 3·4 cm diameter cores per tree to a depth of 1·2 m for these densities would represent 2·6 and 0·04% of soil volume respectively. In the high density planting a sample of this size could only be examined infrequently. The size of sample examined in the laboratory seems as large as that likely to be examined with other methods.

Periodicity of root growth

Only observation methods allow continuous monitoring of changes in root length, which can be rapid during periods of active growth. High CVs make it difficult to

TABLE 4. The linear regression equations and correlation coefficients for the relationship between values for root length obtained by the observation tube method (Y) and those for the same tree by a root laboratory window for 1 year-old apple trees of Cox/M.9

Variable	Regression equation	Correlation coefficient
Observation tube v. root laboratory window	$Y = 13·52 + 1·28X$	0·631*
Adjacent root laboratory window	$Y = 0·865 + 0·178X$	0·594*

* Significant at $P < 0·05$.
Adapted from Gurung (1979).

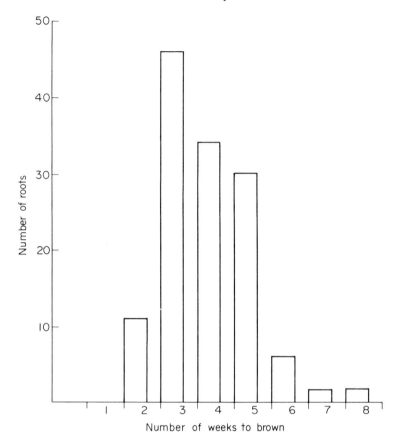

Fig. 4. The number of roots taking a given length of time (weeks) to brown: Experiment 2.

to be seen in the context of total root system weights of 3·5 Mg ha^{-1} for the fine roots of *Picea sitchensis* (Last *et al.* 1983). Here total annual production was estimated as 8·4 Mg ha^{-1} for roots of all types and 5·2 Mg ha^{-1} for roots <2 mm diameter. In contrast, but for a more limited soil depth, Alexander & Fairley (1983) estimated the productivity of their *P. sitchensis* at 879 kg ha^{-1} year^{-1}. Fogel (1983) reviewing the literature on a range of coniferous species put fine root

TABLE 5. Root production, 'potential for decomposition' (g m^{-3} week^{-1}) and contribution towards root length (cm cm^{-2} week^{-1}) for tree 15S in 1963

	Month							
	Apr	May	Jun	Jul	Aug	Sep	Oct	Nov
Production	3·2	7·7	3·3	4·1	17·1	13·6	5·7	3·6
Decomposition potential								
(a) loss of cortex	0·5	1·9	4·6	2·0	4·2	10·0	7·7	2·5
(b) root death	0·2	1·0	2·4	1·0	2·0	5·0	3·8	1·2
Contribution to L_A	0·2	0·5	0·2	0·2	1·0	0·8	0·3	0·2

TABLE 6. Total length of brown root at the end
of Experiment 2 and the estimated length of
white root produced during the experiment (cm
per window)

Density	Brown root	White root	% survival
0·36	473	1886	25
1·44	133	637	21

biomass at $1–12·6$ Mg ha^{-1} and production at $2·4–14·6$ Mg ha^{-1} year^{-1}. In this study annual production by apple would be approximately $2·5$ Mg ha^{-1} year^{-1}.

The pattern of the addition of new root to the root length available for absorption followed that of production. The maximum rate of addition (Table 5) was approximately 1 cm cm^{-2} week^{-1}, i.e. a contribution of 4 by white root to total L_A, given a 4-week period to 'browning'. Both white and brown roots function in absorption (Atkinson 1983), so on the basis of 25% of new root length surviving to become part of the permanent root system (Table 6) the total contribution of 1963 growth to perennial root L_A would be approximately 4 cm cm^{-2}. This falls within the range of the estimated increases in L_A between May and November 1974, which in Experiment 2 was $0·7–9·0$ cm cm^{-2} Fig. 5).

Woody root production

In Experiment 2 assessments were made at intervals of the lengths of woody root $<$ and > 2 mm diameter (Fig. 5). These are related to the lengths of white root produced over the same period. During 1970 a proportion of the new white roots produced survived into the next year as woody roots <2 mm diameter. This fraction increased slightly in 1971 when in addition, some of these woody roots increased in diameter and entered the >2 mm class. The onset of fruit production in 1972 reduced both new root production and the length of woody root. In 1974 both white and brown root lengths increased. The picture shown for trees at densities of $0·36$ and $1·44$ m^2 was similar.

The growth in biomass and volume of the woody root system depends, in part, on the increase in diameter of existing woody roots. The periodicity of increases in diameter, together with that of white root production in the same year, 1973, for trees at densities of $0·36$ and $1·44$ m^2 in Experiment 2, is shown in Fig. 6. New root growth was low in this year. Root thickening was most active between May and July when white root production was submaximal. Thickening and new root production occurred concurrently in the autumn. The difficulty of estimating woody root length from window measurements and the variation between roots in the extent of thickening (Head 1968) make it difficult to estimate productivity in this way. During 1973 trees at the $0·36$ m^2 density planting had 260 m m^{-3} woody root >2 mm diameter. An increment of 1 mm on 2 mm diameter roots would represent an addition of 690 g m^{-3} year^{-1} to the woody root system. This represents a value

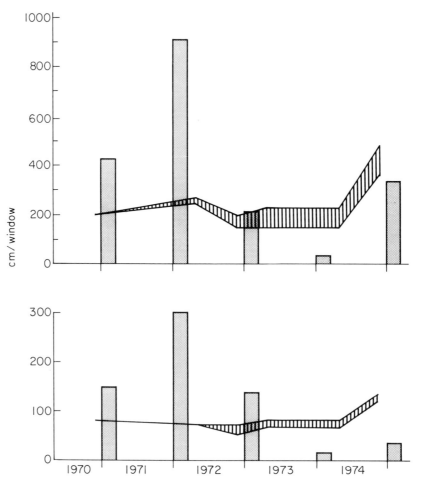

FIG. 5. The length of woody root for trees of Cox/M.9 at 0.36 m^2 spacing (Fig. 5a) and 1.44 m^2 spacing (Fig. 5b). Open area woody roots <2 mm, hatched area >2 mm. Bars are estimated white root production 1970–74.

TABLE 7. The concentration of mineral nutrients (% DW for N, P, K, Ca; ppm for Mn) in the roots of trees of Golden Delicious/M.9 (Experiment 2)

Root diameter (mm)	Depth (cm)	N	P	K	Ca	Mn
< 2	0–25	0·83	0·103	0·226	1·25	62
	50	0·71	0·087	0·279	1·21	56
	100	0·65	0·076	0·299	1·24	61
	> 100	0·66	0·080	0·475	1·14	63
> 2	0–25	0·78	0·110	0·204	1·58	38
	50	0·74	0·095	0·272	1·30	48
	100	0·67	0·078	0·235	1·40	46
	> 100	0·62	0·072	0·286	1·15	45

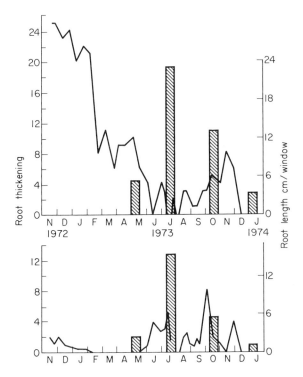

FIG. 6. The periodicity of white roots (cm per window) for trees of Cox/M.9 at spacings of 0·36 (Fig 6a) and 1·44 m² (Fig. 6b) in 1973. The rate of thickening of woody roots (μm week^{-1}) is shown as vertical bars.

higher than both that recorded by Atkinson *et al*. (1976) as the final weight of the root systems and the estimated productivity for roots >2 mm for *Picea sitchensis*, 3·3 Mg ha^{-1} year^{-1} (Last *et al*. 1983). This discrepancy indicates a need for an improved method of estimating the thickening of woody roots.

Mineral reserves in the root system

The root system is an important reservoir and sink for mineral nutrients. Concentrations vary in roots from different soil depths and with root diameter (Table 7). Generally N, P and Ca concentrations decreased with increasing depth and decreasing diameter. *K* increased with depth. The limited amount of data published on the mineral composition of tree roots make it difficult to assess the generality of these trends.

By combining data for the mineral content of roots with root weights at different positions within the soil volume (Table 3) a picture of nutrient storage within the perennial part of the root system and the soil volume can be obtained (Table 8). Most nutrients held in the root system are in large roots (>5 mm) and the upper 45 cm of soil depth. A tree occupying 17·6 m² (Experiment 5, 567 trees ha^{-1})

TABLE 8. The total quantity of nitrogen and phosphorus in the root system of trees of Fortune/M.9: N (g tree^{-1}); P (mg tree^{-1})

Depth (cm)	Root diameter (mm)				
Nitrogen	< 1	1–2	2–5	> 5	Total
0–15	1·25	0·63	0·27	1·63	3·8
30	1·90	1·95	1·99	17·5	23·2
45	0·94	1·66	0·87	14·4	18·2
60	0·85	0·51	0·44	0·35	2·3
75	0·43	0·46	–	0·04	0·9
Phosphorus					
0–15	118	37	18	220	393
30	183	171	197	2301	2852
45	89	200	79	2355	2723
60	57	51	34	93	235
75	36	42	–	1	79

contained 48·6 g N in its roots, 27·6 kg N ha^{-1}. Comparable values for other nutrients were 3·6 kg P, 8·8 kg K, 19·5 kg Ca and 1·8 kg ha^{-1} Mg. The root system thus stores substantial amounts of minerals but most are in parts unlikely to become available to soil organisms. In addition to this standing crop of perennial roots, the volume of production previously discussed, 2·5 Mg ha^{-1} year^{-1}, would contain about 20·8 kg N ha^{-1}. Much of this would be constantly recycled; the maximum 'potential decomposition' would release 1·5 kg N ha^{-1} week^{-1}. The amounts of nutrients held in the standing crop plus new root production are comparable to those found in leaves, e.g. 49 kg ha^{-1} for N (Greenham 1980) and more than those being moved annually into branches and fruit, e.g. 30 kg ha^{-1} for N approximately. Root growth represents a substantial component in nutrient cycles, which is often ignored.

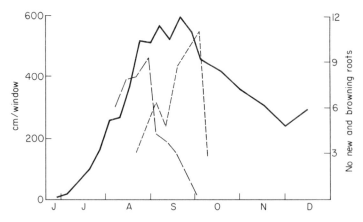

FIG. 7. The periodicity of white root (cm per window) (——) for strawberry cv. Cambridge Favourite in 1973 and an estimate of new root production (no. per 8 strips 0–60 cm depth) (— — —) and root browning (- - -).

Strawberry root growth

Strawberry roots are thinner than apple roots and similar to grass roots in many respects. Because of their smaller diameter, they were visible for a smaller distance from the glass, reducing the observation depth at the window. The periodicity of the presence of white strawberry roots, together with estimates of new root production and browning, obtained as 'strip' records (Rogers 1969) are shown in Fig. 7, for plants from Experiment 3, growing in the absence of apple. Root length development was rapid between June and August, as indicated both by the length present and estimated new production. Peaks of production and browning were about 4 weeks apart.

Effect of tree competition

In looking at data from experiments in monoculture it is important to remember that in natural communities plants occur in association with other plants. This effect is illustrated in Fig. 8. Root length in years 2 and 3, was relatively similar in periodicity for plants grown with and without apple trees, although length was greater for plants grown in monoculture. The length of white root fell during the experiment and in years 2 and 3, maximum growth occurred earlier than in year 1.

Root browning

The time for strawberry roots to turn brown (Fig. 9) was more variable than in apple (Fig. 4). A substantial number of roots went brown after just 1 week and over half had browned in 3 weeks. This time was therefore used in calculations. Some roots remained white as long as 11 weeks. Within the growing season the time of year did not affect survival.

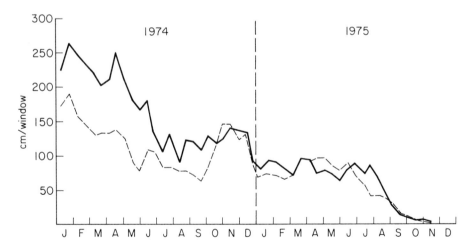

FIG. 8. The periodicity of white root length (cm per window) for strawberry plants in Experiment 3: plants grown in monoculture (——), plants grown with apple(– – –).

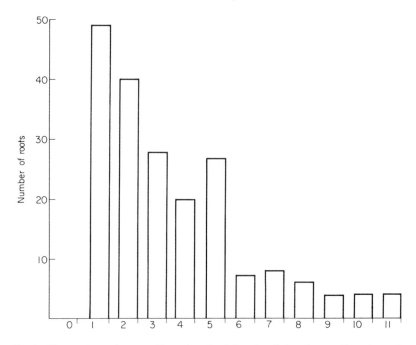

FIG 9. The numbers of roots taking a length of time (weeks) to brown. Experiment 3.

Root production and decomposition

Using the premises previously listed, estimates of root production and 'potential for decomposition' were calculated for 1973 (Table 9). Production was maximal in August and September and 'potential for decomposition' slightly later. Both rates were lower than the highest rates for apple (Table 5), as a result of the smaller root weight per unit length for this species. The maximum rate of production was 129 kg ha^{-1} week^{-1} and the annual rate 1·7 Mg ha^{-1} year^{-1}. The addition of new root to the total root density L_A, was much higher than for apple, a maximum addition of 9·7 cm cm^{-2} week^{-1}, 29·7 cm cm^{-2} over the period, allowing for the time taken to brown.

TABLE 9. Root production, potential for decomposition (g m^{-3} week^{-1}) and contribution to root length (cm cm^{-2} week^{-1})for newly planted Cambridge Favourite strawberry plants in 1973

	Month				
	July	Aug	Sept	Oct	Nov
Production	2·8	9·5	12·9	8·1	5·5
Decomposition potential	0·2	3·5	9·2	8·2	5·6
Contribution to L_A	2·1	7·1	9·7	6·1	4·1

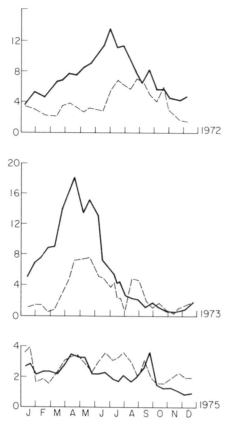

FIG. 10. The periodicity of white root length (cm per window) for grass grown with (– – –) and without (——) irrigation in Experiment 4.

Grass root growth

Observations were made on the growth of S50 timothy *Phleum pratense* ssp. *bertolonii* roots in Experiment 4. Growth was higher in 1973 than in 1972 and much lower in 1975 (Fig. 10). Data are presented for both irrigated and non-irrigated grass. Growth was higher without irrigation in 1972 and 1973 and similar in 1975. Data for 1974 has been presented by Atkinson (1977). The periodicity of root production was similar for the two treatments. In 1972 growth was highest from February to July, but in 1973 greatest during spring and early summer.

Root production and decomposition

Using data for 1972, the sward's second growing season, production and 'potential for decomposition' have been estimated using the assumptions stated. More information is needed on some of these, as little information is available on the length of time before browning occurs or the eventual fate of roots after browning has

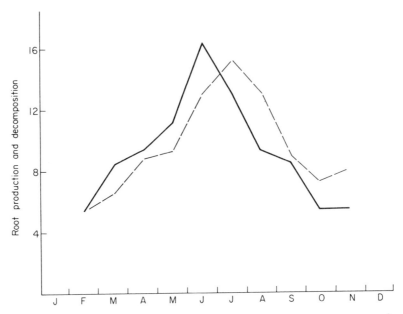

Fig. 11. Root production and potential for decomposition in Experiment 4 in 1972 (g m^{-3} week^{-1}): production (——) and decomposition (---) potential.

occurred. The assumptions made are therefore based on the information available for strawberry. In 1972 root production was maximal in June and decomposition in July (Fig. 11). The maximum rate of production was 160 kg ha^{-1} week^{-1} and the annual rate 4·1 Mg ha^{-1} year^{-1}. Total production was therefore higher than that for apple, mainly as a result of the longer period, February–November, over which production occurred. Potential decomposition was a maximum of 154 kg ha^{-1} week^{-1}.

CONCLUSIONS

Results from over two decades of root laboratory experimentation suggest that data from the observation windows can give acceptable estimates of root density and new root production. The confusion between spatial and temporal variation inherent in core sampling, and the high spatial variation inherent in tree root systems leading to high CVs, make sensitive tests of the relationship between growth at the window and elsewhere difficult to enact. A comparison of density beside and away from the glass, but at comparable positions (Table 1) where variance was not excessive (a SE of 0·13 on a grand mean of 0·44), showed no significant differences. Although reasonable estimates of fine root density are possible, the three-dimensional distribution of larger roots, their sparse distribution and high spatial variability mean that the estimates of biomass obtained for these are less good. This, however, will also apply to estimates obtained by any method other than total excavation.

TABLE 10. Relative maximum root density, (cm cm^{-2}) recorded adjacent to a root laboratory window for a range of species

Crop	L_A
Apple (Experiment 1)	6·7
Strawberry (Experiment 3)	30·1
Grass (Experiment 4)	90·2
Apple (Experiment 2)	30·1

To obtain quantitative data from window measurements a number of assumptions have to be made. This has shown areas where our knowledge is limited. For all of the species examined and probably most other species, the fate of roots following the loss of their primary cortex, the relationship between the time taken for a root to turn brown and its subsequent survival and growth, and the length of time for which roots of varying diameter survive need more complete documentation to assist in more accurate determinations of changes in size and productivity of the root system. Despite these difficulties observation methods are unique in their separation of spatial and temporal variation.

The estimates detailed here suggest that the annual productivity of fruit tree roots can be high, although lower than some published estimates for forest trees (Last *et al.* 1983). They also suggest that substantial amounts can be released and available for decomposition which may occur at high rates in a limited number of months during the year. Even in fruit trees much of this will occur in the surface 45 cm. The growth of new roots in fruit crops will 'fix' quantities of nutrients similar to those used in the production of crop and the growth of the branch system. Although in total the root system will hold similar amounts of nutrients to the leaves, a substantial fraction is associated with large woody roots and will be released only on the death of the tree.

In fruit species it has been shown (Atkinson 1983) that all roots can function, to some extent at least, in absorption. New growth thus increases available root length but is not the only immediate source of absorption potential. The three species examined varied in the contribution that new root made to total root density (cm cm^{-2}) (Table 10). The contribution of new growth was highest in grass although the estimate presented here was much smaller than the value published for grasses by Newman (1969). This may be influenced by the limited visibility obtained for soil close to the soil surface or to an error in the thickness of the volume of soil adjacent to the glass being studied. If a layer of only 1·5 mm is seen rather than 2 mm then estimated densities will be low by 25%.

Root growth can be affected by a range of factors, e.g. mineral nutrition, water supply and by competition from other species. Some of these effects, for apple, have been described by Atkinson (1983). For strawberry, competition with apple trees reduced productivity, although the total production by the apple and strawberry plants together would have been much higher than by either species on its own. The root laboratory is a useful means of quantifying some of these aspects.

ACKNOWLEDGMENTS

I am grateful to a number of present and former colleagues at East Malling for assistance in making some of the measurements reported, but especially to Caroline Thomas, Leslie Williams, Denise Naylor, Linda Watson and Graham Shaw.

REFERENCES

Alexander, I.J. & Fairley, R.I. (1983). Effects of N fertilization on populations of fine roots and mycorrhizas in spruce humus. *Plant and Soil*, **71**, 49–54.

Atkinson, D. (1974a). Field studies on root systems and root activity. *Report of East Malling Research Station for 1973*, p. 69.

Atkinson, D. (1974b). Some observations on the distribution of root activity in apple trees. *Plant and Soil*, **40**, 333–342.

Atkinson, D. Naylor, D. & Coldrick, G.A. (1976). The effect of tree spacing on the apple root system. *Horticultural Research*, **16**, 89–105.

Atkinson, D. (1977). Some observations on the root growth of young apple trees and their uptake of nutrients when grown in herbicide strips in grassed orchards. *Plant and Soil*, **49**, 459–71.

Atkinson, D. (1980a). The growth and activity of fruit tree root systems under stimulated orchard conditions. In *Environment and Root Behaviour* (Ed. D. N. Sen), pp. 171–85. Geobios International Jodhpur, India.

Atkinson, D. (1980b). The distribution and effectiveness of the roots of tree crops. *Horticultural Reviews*, **2**, 424–90.

Atkinson, D. (1983). The growth, activity and distribution of the fruit tree root system. *Plant and Soil*, **71**, 23–35.

Atkinson, D. & White, G.C. (1980). Some effects of orchard soil management on the mineral nutrition of apple trees. In *The Mineral Nutrition of Fruit Trees* (Eds D. Atkinson, J. E. Jackson, R. O. Sharples & W. M. Waller), pp. 241–254. Butterworth, Borough Green.

Barber, S.A. (1979). Growth requirements for nutrients in relation to demand at the root surface. In *The Soil—Root Interface* (Eds J. L. Harley & R. S. Russell), pp. 5–20. Academic Press, London.

Dudney, P.J. (1972). On the estimation of root biomass in a growth pattern experiment on apples. *Report of East Malling Research Station for 1971*, pp. 66–67.

Fogel, R. (1983). Root turnover and productivity of Coniferous forests. *Plant and Soil*, **71**, 75–86.

Gurung, H.P. (1979). *The influence of soil management on root growth and activity in apple trees.* M.Phil. Thesis, University of London.

Greenham, D.W.P. (1980). Nutrient cycling: The estimation of orchard nutrient uptake. In *Mineral Nutrition of Fruit Trees* (Eds D. Atkinson, J. E. Jackson, R. O. Sharples & W. M. Waller), pp. 345–52. Butterworth, Borough Green.

Head, G.C. (1966). Estimating seasonal changes in the quality of white unsuberized roots on fruit trees. *Journal of Horticultural Science*, **41**, 197–206.

Head, G.C. (1968). Seasonal changes in the diameter of secondarily thickened roots of fruit trees in relation to growth of other parts of the tree. *Journal of Horticultural Science*, **43**, 275–282.

Head, G.C. (1969). The effects of mineral fertilizer on seasonal changes in the amount of white root on apple trees in grass. *Journal of Horticultural Science*, **44**, 183–187.

Huck, M.G. & Taylor, H.M. (1982). The rhizotron as a tool for root research. *Advances in Agronomy*, **35**, 1–35.

Last, F.T., Mason, P.A., Wilson, J. & Deacon, J.W. (1983). Fine roots and sheathing mycorrhizas: their formation, function and dymanics. *Plant and Soil*, **71**, 9–21.

Newman, E.I. (1969). Resistance to water flow in soil and plant. I. Soil resistance in relation to amounts of roots: theoretical estimates. *Journal of Applied Ecology*, **6**, 1–12.

Rogers, W.S. (1968). Amount of cortical and epidermal tissue shed from roots of apple. *Journal of Horticultural Science*, **43**, 327–328.

Rogers, W.S. (1969). The East Malling root observation laboratories. In *Root Growth* (Ed. W. J. Whittington), pp. 361–376. Butterworth, London.

Rogers, W.S. & Head, G.C. (1969). Factors affecting the distribution and growth of roots of perennial woody species, *Root Growth* (Ed. W. J. Whittington), pp. 280–295. Butterworth, London.

An inexpensive rhizotron for research on soil and litter-living organisms

ALAN CARPENTER, J. M. CHERRETT, J. B. FORD, M. THOMAS
AND E. EVANS

School of Animal Biology, University College of North Wales, Bangor, Gwynedd LL57 2UW

INTRODUCTION

Rhizotrons, or root observation laboratories, have been used to study the root dynamics of a whole range of crops (Lyr & Hoffman 1967; Head 1973; Huxley & Turk 1975; Böhm 1979; Huck & Taylor 1982) since Rogers (1939a) described the first East Malling rhizotrons. He also reviewed the use of rhizotrons up to that time, concluding that little useful data had been collected from earlier models (Rogers 1939b). Few workers have used them to study pasture roots (Garwood 1967; Atkinson 1977). More recently, much effort has gone towards developing the minirhizotron, first described by Bates (1937), for use in annual crops (e.g. Bragg, Govi & Cannell 1983; Vos & Groenwold 1983). Most rhizotron research has been designed to study factors such as the periodicity of root growth, water relations, ageing and the effects of management practices.

Rhizotrons have been rarely used to study the role of invertebrates in root dynamics. Stansell *et al.* (1974) simulated the effects of root-feeding invertebrates by root pruning cotton plants. Some of the effects of invertebrate feeding on fruit tree roots have been reported by Pitcher & Flegg (1965) and Harding (1968). Huxley & Turk (1975) saw chafer grubs attack coffee roots but chose to apply insecticide rather than study their interactions with roots. The importance of the soil fauna in root decomposition has been reported by Rogers & Head (1969) and Atkinson & Wilson (1979). Atkinson & Wilson (1979) also showed that the soil fauna had a major effect on root–soil contact.

The usefulness of rhizotrons for studying plant/invertebrate interactions in the litter zone has not been exploited. This paper described a simple rhizotron designed to permit the study of plant/invertebrate interactions in the litter zone and in the soil.

THE RHIZOTRON

Our rhizotron is very simple and, as a result, inexpensive. The design is flexible and so the configuration described here could be readily modified. Many rhizotrons have been described in detail in the literature (see Böhm (1979) and Huck & Taylor (1982) for recent reviews). This paper stresses the simple design used. Further details and drawings are available from the authors.

The rhizotron is situated in amenity grassland in the Treborth Botanic Garden of the University College of North Wales. It is sited at the top of a slight slope to allow gravity drainage from the bottom of the rhizotron to the bottom of the slope. The rhizotron is 1.7×1.7 m square with the bottom 1.35 m below ground. It has a concrete floor draining to an outlet in the centre. The lower part of the side walls is made from four courses of breeze blocks ($430 \times 220 \times 75$ mm). Bolted to this are supports at the corners made of 60×60 mm right-angle steel and on three sides, in the middle, there are two pieces of 40×40 mm right-angle steel, with their right angles facing outwards and towards the corners. On the fourth side there is a door. On this side the two central pieces of right-angle steel are 600 mm apart, each 550 mm from the nearest corner. The door faces east.

The sides of the rhizotron rise above ground level to allow observation of the litter zone. The roof is sloping with the lowest side towards the prevailing westerly winds. The two low corners are 0.75 m and the front corners are 1.4 m above ground level.

The observation windows rise 1.3 m from the top of the breeze blocks and fit into the angles of the steel supports at the sides. The windows are made of 12.5 mm plate glass and rest on glaziers rubber strip, both on the bottom and between them and the steel supports. The top of the structure is made of 16 mm exterior grade plywood. The glass is held in place by the weight of soil at the bottom and by strips of wood 70×10 mm at the top and over the gap between the central pieces of angle steel on each side. The plywood is covered with builders felt to weatherproof it.

The temperature regime of the soil is buffered against the effects of the chamber by shutters made of 50 mm Venesta board, one completely covering each window. Internal transmission of light down the glass is prevented by external aluminium shutters that fit between the roof and the soil.

After completion of the structure, the soil around the outside was replaced and planted with various species of pasture plant. Total cost of excavating the hole, laying 30 m of drain and building the rhizotron was £480 sterling.

Design considerations

Normally rhizotrons protrude only slightly above soil level to minimize their effects on the microclimate. Originally we wished to observe events in the foliage as well as in the litter and soil zones so the rhizotron had to extend well above soil level. Unfortunately, the fitting of shutters to the outside of the observation windows to prevent the internal transmission of light down the glass means the rhizotron cannot be used to study above- and below-ground events in the way that we had hoped. Given that we can only usefully study the soil and litter zones because of the shutters disturbing the foliage, results to date show that too much of the glass is above soil level; 15–20 cm above soil level would have been enough to give adequate observation of the litter zone and would have allowed a greater depth of soil to have been studied as roots and earthworm burrows went deeper than anticipated. Once the value of studying the litter zone from a rhizotron has been established, it may be possible to redesign the Treborth rhizotron to minimize its effects on the microclimate.

Experience has shown that the internal shutters are very effective at maintaining normal soil temperature. Thinner glass would have increased the accuracy with which roots could be traced, although thinner glass might fracture if it was stressed by soil dynamics and changes in root volumes.

METHODS

Root growth and death and invertebrate activity are recorded weekly by tracing onto thin plastic sheeting stretched over wooden frames. From these records weekly increments in root growth can be estimated and root deaths recorded. At the same time any visible association between invertebrate activity and roots is noted and photographed if appropriate.

RESULTS AND DISCUSSION

Initially it was feared that light would disturb the invertebrate fauna while observations were made. Experience has shown that this does not occur as none of the animals seen have shown any reaction to light except for one specimen of *Arion ater* (L). which turned to align its dorsal surface to the light.

Earthworms and their burrows were the most common faunal components seen during 6 months observations. All earthworm burrows were permanent from September to February with few new ones being made. Earthworm burrows are utilized by all the invertebrates so far observed, three species of slug, *Deroceras reticulatus* (Muller), *Arion ater* and *Milax budapestensis* (Hazay) all utilize them (Table 1), particularly in hot and dry, or, cold and wet weather. Hunter (1966) similarly found slugs deeper in the soil in dry or frosty weather. All three slug species, and collembola, mites, diplura and enchytraeidae have been observed feeding on the material lining earthworm burrows. *M.budapestensis*, *A.ater*, collembola and diplura have also been seen eating living roots and root hairs where these are adjacent to earthworm burrows.

Table 1 shows that a third of the visible slug population may be below 100 mm and thus would escape detection in a programme that only samples the top 100 mm of soil. This may account in part for the variability of slug population data (Runham & Hunter 1970).

Slugs and a wide variety of adult beetles and spiders have been seen foraging in the litter zone, the beetles moving 1–2 cm down earthworm burrows. *Lumbricus* spp. was observed *in copula* in December and some earthworm burrows have been seen

TABLE 1. Depth of slugs in earthworm burrows over the period 9 Nov 83–31 Dec 83

| Depth (mm) | Species | | | |
	D.reticulatus	*A.ater*	*M.budapestensis*	Unidentified
0–100	29	9	8	0
100–200	4	2	5	11
200–450	1	1	0	0

with the green leaves of pasture plants pulled into them as well as the expected detritus. There appears to be scope for expansion of this aspect of rhizotron studies.

CONCLUSIONS

The Treborth Rhizotron is even simpler in construction than those described by Rogers (1939a) and much simpler than the current East Malling installation (Rogers 1969). The data presented here show that useful observations on the functioning of the soil ecosystem can be derived from an inexpensive installation and such data can indicate areas for experimental work that will help quantify our understanding of plant–herbivore interactions in the soil.

ACKNOWLEDGMENTS

We would like to thank Dr D. Atkinson of the East Malling Research Centre for useful discussions on rhizotron design. Nigel Brown, Curator of the Treborth Botanic Gardens, has provided considerable assistance with the construction and running of the rhizotron. The senior author is grateful to the New Zealand National Research Advisory Council for their financial support.

REFERENCES

Atkinson, D. (1977). Some observations on the root growth of young apple trees and their uptake of nutrients when grown in herbicided strips in grassed orchards. *Plant and Soil*, **46**, 459–471.

Atkinson, D. & Wilson, S.A. (1979). The root–soil interface and its significance for fruit tree roots of different ages. In *The Soil—Root Interface* (Ed. J. L. Harley R. S. Russell), pp.259–271. Academic Press, London.

Bates, G.H. (1937). A device for measuring root growth in soil. *Nature (London)*, **139**, 966–967.

Böhm, W. (1979). *Methods of Studying Root Systems.* Springer-Verlag, Berlin.

Bragg, P.L., Govi, G. & Cannell, R.Q. (1983). A comparison of methods, including angles and vertical minirhizotrons, for studying root growth and distribution in a spring oat crops. *Plant and Soil*, **73**, 435–440.

Garwood, E.A. (1967). Seasonal variation in appearance and growth of grass roots, *Journal of the British Grassland Society*, **22**, 121–130

Harding, D.J.L. (1968). A preliminary survey of the relationships between the soil fauna and roots of fruit trees. *Annual Report, East Malling Research Station for 1967*, pp. 169–172.

Head, G.C. (1973). Shedding of roots. In *Shedding of Plant Parts* (Ed. T. T. Kolowski), pp. 237–293. Academic Press, New York.

Huck, M.G. & Taylor, H.M. (1982). The rhizotron as a tool for root research. *Advances in Agronomy*, **35**, 1–35.

Hunter, P.J. (1966). The distribution and abundance of slugs on an arable plot in Northumberland. *Journal of Animal Ecology*, **35**, 543–557.

Huxley, P.A. & Turk, A. (1975). Preliminary investigations with *Arabica* coffee in a root observation laboratory in Kenya. *East African Agriculture and Forestry Journal*, **40**, 300–312.

Lyr, H. & Hoffman, G. (1967). Growth rates and growth periodicity of tree roots. *International Review of Forestry Research*, **2**, 181–236.

Pitcher, R.S. & Flegg, J.J.M. (1965). Observation of root feeding by the nematode *Trichodorus viruliferus* Hooper. *Nature (London)*, **207**, 317.

Rogers, W.S. (1939a). Root Studies, VIII. Apple root growth in relation to rootstock, soil, seasonal and climatic factors. *Journal of Pomology and Horticultural Science*, **17**, 99–129.

Rogers, W.S. (1939b). Root studies, VII. A survey of the literature on root growth, with special reference to hardy fruit plants. *Journal of Pomology and Horticultural Science*, **17**, 67–84.

Rogers, W.S. (1969). The East Malling root-observation laboratories. In *Root Growth* (Ed. W. J. Whittington), pp. 361–376. University of Nottingham, Nottingham.

Rogers, W.S. & Head, G.S. (1969). Factors affecting the distribution and growth of roots of perennial woody species. Ibid. pp. 280–292.

Runham, N.W. & Hunter, P.J. (1970). *Terrestrial Slugs*. Hutchinson University Library, London.

Stansell, J.R., Klepper, B., Browning, V.O. & Taylor, H.M. (1974). Effect of root pruning on water relations and growth of cotton. *Agronomy Journal*, **66**, 591–592.

Vos, J. & Groenwold, J. (1983). Estimation of root densities by observation tubes and endoscope. *Plant and Soil*, **74**, 295–300.

A new method for measuring nutrient supply rates in soils using ion-exchange resins

D. J. GIBSON*, I. A. COLQUHOUN AND P. GREIG-SMITH

School of Plant Biology, University College of North Wales, Bangor, Gwynedd

INTRODUCTION

There are many currently available methods of soil analysis. However, whether they measure total, plant-available or water-soluble nutrients they all suffer from a number of limitations. All of these procedures require removal from the study site of soil samples. Although repeated measures through time can be obtained using randomized block designs (e.g. Frankland, Ovington & Macrea 1963; Williams 1969; Usher 1970; Gupta & Rorison 1975; Taylor, De-Felice & Havill 1982), these methods require local destruction of the soil and cannot be adequately used to measure temporal changes within a site and take into account spatial variability (Ball & Williams 1968). Furthermore, these methods do not provide a measure of soil nutrient supply. These limitations are alleviated to some extent by the use of soil embedded lysimeters or ion-selective electrodes (Nair & Talibudeen 1973; Page & Talibudeen 1977; Page, Smalley & Talibudeen 1978). The method being proposed here is a repeatable, non-destructive measure of soil nutrient supply based on the use of ion-exchange resins. This will go some way towards assessing the differential resource supply rates experienced by species within communities (Tilman 1982).

Synthetic ion-exchange resins have been used successfully as a means of assessing soil nutrient status (e.g. Arnold 1958; Hislop & Cooke 1968; Barrow & Shaw 1977; El-Nennah 1978; Sibbesen 1977, 1978, 1979; Waterhouse & Bille 1978; Smith 1979). Sibbesen (1977) describes a simple procedure for assessing the phosphate status in soils by shaking nylon-net bags of resin with soil suspensions. Binkley & Matson (1983) have recently used *in-situ* ion-exchange resin bags for assessing forest soil nitrogen availability.

In this report the use of terylene-net bags of anion and cation exchange resins placed in the soil for a short period of time is suggested. The *in-situ* resins retain ions which become available through local soil supply. The resin bags are removed from the soil and nutrient levels subsequently measured on solutions eluted from the resins. Replacing the resin bags in the soil for a second period allows another, and subsequent, measure of the soil nutrient status.

The exchange potential of an ion-exchange resin is influenced by a number of factors, the most important being molecular size, valency and concentration (Kunin

*Present address: Division of Pinelands Research, Center for Coastal and Environmental Studies, Rutgers: the State University of New Jersey, Freid Station. P.O. Box 206, New Lisbon, New Jersey 08064, USA.

& Myers 1947; BDH Chemicals Ltd 1971). Thus, when in exchange with a soil the ions taken up by the resin will depend on these factors. If the exchange capacity of the resin is not exceeded then the concentration of ions on the resin corresponds to the concentration of ions in the exchange solution. On this basis, ion-exchange resins are used in water purification (e.g. Ayres 1947).

MATERIALS AND METHODS

Construction of resin bags

Rectangular resin bags ($6 \cdot 5 \times 7 \cdot 0$ cm) were constructed from terylene-net in a similar manner to that of Sibbesen (1977). The terylene had a mesh size ($0 \cdot 3$ mm) sufficient to retain the resin without loss and allow through-movement of moisture. The bags were sewn with non-rotting polyester thread and heat-sealed. Cation resin bags contained $7 \cdot 0$ g of moist cation-exchange resin ('Amberlite IR-120'), anion bags $5 \cdot 0$ g of anion-exchange resin ('Amberlite IRA-402').

Regeneration of the resin bags

To regenerate the resin bags prior to use the bags were initially washed with distilled water. They were then shaken for 30 minutes with 50 ml 5% (v/v) HCl per bag. This was repeated three times. After the final wash and a further rinse with distilled water the bags were ready for use.

Placement of resin bags in the soil

Resin bags were placed below the soil surface at intervals of 10 cm along two transects, one in the grassland and one in a dune slack, at Newborough Warren, Anglesey (SH 422643), and removed after 1 week. The grassland transect was sampled twice, once in the summer and again in early autumn of 1982. The resin bags were placed 10 cm below the soil surface. The dune slack transect was sampled once in the summer of 1982, with the resin bags placed at two depths. Parallel to the dune slack transect (10 cm away) soil samples had been collected in the spring for conventional analysis (ammonium acetate extraction at pH 9 of air-dried and sieved soil).

Elution of resin bags

Each resin bag was initially rinsed in distilled water to clean off attached soil and root material. Subsequently each resin bag was shaken for 30 minutes with 50 ml 5% (v/v) HCl. The eluate was filtered (Whatmans No. 1) into plastic vials and stored at $5 °C$ prior to analysis.

Analysis of resin bag eluates

The above procedure produced for every soil transect position two solutions: the cation-exchange resin bag eluate and the anion-exchange resin bag eluate. The metallic cations in the cation eluate were analysed using atomic absorption spectro-photometry. Nitrate and phosphate in the anion resin eluates, and ammonium in the cation eluates, were measured using colorimetric procedures (West & Lyles 1960; Sims & Jackson 1971; Allen *et al.* 1974).

RESULTS

The data from this type of procedure relate to the amount of ion-exchange resin used per bag and the period of time the resin bags are left in the ground. However, providing the same procedure is followed for subsequent trials then levels of ions

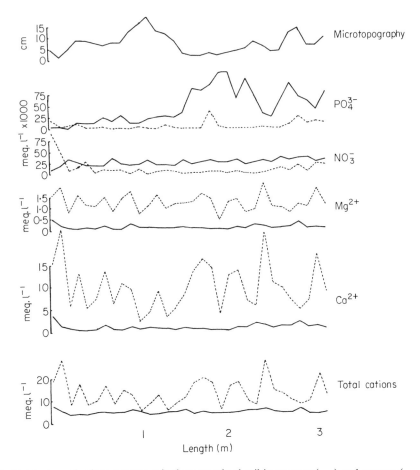

Fig. 1. Nutrient supply along a transect in dune grassland soil in summer (——) and autumn (– – –) as measured by ion-exchange resin method.

eluted from the resin bags can be compared both spatially and temporally. Expression of the ion levels in millequivalents per litre allows a direct comparison of ion levels on an exchange basis. This technique allows nutrient supply to be measured for any soil cation or anion. However, for the purpose of this report data are only presented for magnesium, calcium, potassium, nitrate and phosphate.

Data from the grazed dune grassland (Fig. 1) show the spatial variation in nutrient supply along a 3·2 m transect. Temporal changes in nutrient supply are also evident by comparing summer and autumn levels. Although the total cation supply levels, particularly levels of magnesium and calcium, were considerably lower in the summer, the anion supply of phosphate and nitrate within the soil was higher during this period than during the autumn. Nitrate supply although higher during the summer showed little spatial variation whereas phosphate appears to be

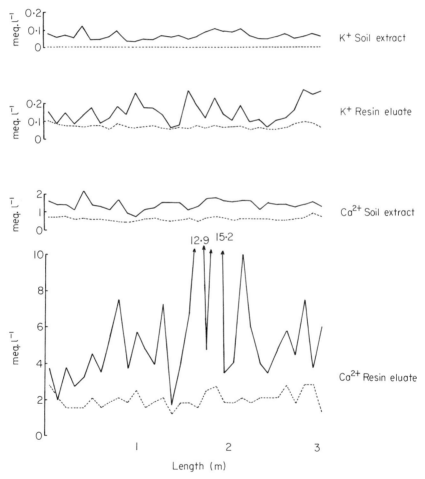

FIG. 2. Nutrient levels measured by conventional soil extraction and nutrient supply measured by ion-exchange resin method in dune slack soil: 3 cm (——) and 10 cm (– – –) depth.

partially related to microtopography. There was little spatial variation in cation supply patterns in the summer compared with in the autumn. Calcium was the ion in greatest supply, and its pattern closely matches that of the total cations. Magnesium supply, although not as high as calcium, was also similar in its spatial variation suggesting similar patterns of nutrient release from the soil.

Spatial variation in nutrient supply is also illustrated in the data from the dune slack soil (Fig. 2). Calcium and potassium both exhibit a wide spatial variation in the supply of the ions near the soil surface. However, at a depth of 10 cm not only is the supply of these ions lower, but the spatial variation is less. Similar conclusions may also be reached with conventional soil analysis of samples from a parallel transect. However, it should be emphasized that the conventional analysis provides data on potential plant-available nutrient levels, and not the actual supply rate as illustrated by the ion-exchange resin eluate. The supply rate of calcium, for example, near the soil surface showed a larger variation than did the exchangeable amount extracted from the soil using conventional methods.

DISCUSSION

Ion-exchange resins behave in a similar manner to a plant root in their ion uptake capacity (Sibbesen 1978; Smith 1979). Anion-exchange resin extracted phosphate levels and cation-exchange resin extracted potassium levels have been shown to have as good correlations with plant growth as have conventional soil extracted levels (Pratt 1951; Cooke & Hislop 1963; Smith 1979). Because of this, and the *in-situ* nature of the resin bag method proposed here the levels of ions eluted are considered to be a measure of soil nutrient supply. This is in contrast to the potential nutrient supply of a soil expressed by conventional exchange methods.

Tilman (1982) argues that the resource supply to individuals allows an explanation of competition and hence community structure. However, although there are data on spatial variation in soil nutrients, and there are data on temporal variation, there are few data on both spatial and temporal variation in nutrient supply. Nair & Talibudeen (1973) were able to measure potassium and nitrate supply rates under winter wheat. However, their technique using ion-selective electrodes, although useful, does not allow the large suite of ions to be measured repeatably that is possible using ion-exchange resins.

It is suggested that ion-exchange resin bags could be used routinely in many ecological and agricultural applications. The investigation described here was concerned primarily with the spatial variation in nutrient supply. However, other applications could include measuring the temporal changes induced in soil nutrient supply following specific treatments, such as fertilizer application. At present the procedure is being investigated for use in monitoring changes in soil sulphate supply induced by acid rain pollution. The importance of such applications is that nutrient supply can be repeatably monitored with minimal disturbance to the soil using unsophisticated equipment.

ACKNOWLEDGMENTS

We would like to thank the Nature Conservancy Council for permission to carry out field work at Newborough Warren and D.J.G. and I.A.C. would like to acknowledge NERC for financial support.

REFERENCES

Arnold, P.W. (1958). Potassium uptake by cation exchange resins from soils and minerals. *Nature*, **182**, 1594–1595.

Allen, S.E., Grimshaw, H.M., Parkinson, J.A. & Quarmby, C. (1974). *Chemical Analysis of Ecological Materials.* Blackwell Scientific Publications, Oxford.

Ayres, A.J. (1947). Purification of zirconium by ion-exchange columns. *Journal of the American Chemical Society*, **69**, 2879–2881.

Ball, D.F. & Williams, W.M. (1968). Variability of soil chemical properties in two uncultivated brown earths. *Journal of Soil Science*, **19**, 379–391.

Barrow, N.J. & Shaw, T.C. (1977). Factors affecting the amount of phosphate extracted from soil by anion-exchange resin. *Geoderma*, **18**, 309–323.

B. D. H. Chemicals Ltd (1971). *Ion Exchange Resins* (5th edn, 3rd revised impression). B. D. H. Chemicals Ltd., Poole, England.

Binkley, D. & Matson, P (1983). Ion-exchange resin bag method for assessing forest soil nitrogen availability. *Soil Science Society of America Journal*, **47**, 1050–1052.

Cooke, I.J. & Hislop, J. (1963). Use of anion-exchange resin for the assessment of available soil phosphate. *Soil Science*, **96**, 308–312.

El-Nennah, M. (1978). Phosphorus in soil extracted with anion-exchange resin. 1. Time-dissolution relationship. *Plant and Soil*, **49**, 647–651.

Frankland, J.C., Ovington, J.D. & Macrea, C. (1963). Spatial and seasonal variation in soil, litter and ground vegetation in some Lake District Woodlands. *Journal of Ecology*, **51**, 97–112.

Gupta, P. L. & Rorison, , I.H. (1975). Seasonal differences in the availability of nutrients down a podzolic profile. *Journal of Ecology*, **63**, 521–534.

Hislop, J. & Cooke, I.J. (1968). Anion exchange resin as a means of assessing soil phosphate status: a laboratory technique. *Soil Science*, **105**, 8–11.

Kunin, R. & Myers, R.J. (1947). The anion exchange equilibrium in an anion exchange resin. *Journal of the American Chemical Society*, **69**, 2874.

Nair, P.K.R. & Talibudeen, O. (1973). Dynamics of K^+ and NO_3^- concentrations in the root zone of winter wheat at Broadbalk using specific-ion electrodes. *Journal of Agricultural Science, Cambridge*, **81**, 327–337.

Page, M.B. & Talibudeen, O. (1977). Nitrate concentration under winter wheat and in fallow soil during summer at Rothamsted. *Plant and Soil*, **47**, 527–540.

Page, M.B., Smalley, J.L. & Talibudeen, O. (1978). The growth and nutrient uptake of winter wheat. *Plant and Soil*, **49**, 149–160.

Pratt, P.F. (1951). Potassium removal from Iowa soils by greenhouse and laboratory procedures. *Soil Science*, **72**, 101–117.

Sibbesen, E. (1977). A simple ion-exchange resin procedure for extracting plant-available elements from the soil. *Plant and Soil*, **46**, 665–669.

Sibbesen, E. (1978). An investigation of the anion-exchange resin method for soil phosphate extraction. *Plant and Soil*, **50**, 305–321.

Sibbesen, E. (1979). Anionbytter-resin metoden til ekstraktion af jordfosfat (Anion-exchange resin methods for soil phosphate extraction). *Tidsskrift for Planteavl*, **83**, 478–484.

Sims, J.R. & Jackson, G.D. (1971). Rapid analysis of soil nitrate with chromotropic acid. *Proceedings of the Soil Science Society of America*, **35**, 603–606.

Smith, V.R. (1979). Evaluation of a resin-bag procedure for determining plant available P in organic, volcanic soils. *Plant and Soil*, **53**, 245–249.

Taylor, A.A., De-Felice, J. & Havill, D.C. (1982). Seasonal variation in nitrogen availability and utilization in an acidic and calcareous soil. *New Phytologist*, **92**, 141–151.

Tilman, D. (1982). *Resource Competition and Community Structure*. Princeton, New Jersey.

Usher, M.B. (1970). Pattern and seasonal variability in the environment of a Scots pine forest soil. *Journal of Ecology*, **58**, 669–679.

Waterhouse, P.L. & Bille, S.W. (1978). Comparison of the short-term phosphate uptake by plants to the decrease in resin extractable phosphate in a cropped soil. *Plant and Soil*, **50**, 67–79.

West, P.W. & Lyles, G.L. (1960). A new method for the determination of nitrates. *Analytica Chemica Acta*, **23**, 227–232.

Williams, J.T. (1969). Mineral nitrogen in British grassland soils, 1. Seasonal patterns in simple models. *Oecologia Plantarum*, **4**, 307–320.

Grass root production in restored soil following opencast mining

R. I. FAIRLEY*

Department of Agriculture, The University, Newcastle upon Tyne, NE1 7RU

INTRODUCTION

In opencast mining, soil is returned to the site as separate subsoil and topsoil layers. Soil depth is, therefore, little altered from the original, though the physical and moisture characteristics may be altered substantially. Following restoration, land is generally grassed for a period of years prior to the installation of a drainage system. Even after drainage, low intensity farming is continued for some years. If the land were to be drained immediately following restoration, and the grass cropped more intensively with both increased fertilizer use and stocking rates, the organic cycles in the soil might be speeded up and the restoration process accelerated.

Beneficial effects of root production in restored soil may include the improvement of the physical structure of the soil, partly by the direct physical action of the roots creating root channels which aid water movement and partly by the effect of additional organic matter. At the same time current restoration procedures greatly alter the soil profile creating problems for root growth through compaction and the disruption of continuous pores.

METHODS

Experimental sites and design

Experiments have been established at three sites. Two of these are on the Butterwell opencast site, in Northumberland; one was drained immediately following restoration and the other was left undrained. The third site is on adjacent, unmined land of a similar soil type, to provide estimates of yield potentials of similarly managed but undisturbed land. The Butterwell site is restored progressively, and consequently the topsoil in the research area is not stored in heaps prior to restoration.

Two series of plots were established at all three sites: (a) grass ley (*Lolium perenne* L. S23) plots (30 × 30 m) managed for three silage cuts and receiving 300 kg N ha^{-1} year^{-1}, 100 kg P$_2$O$_5$ ha^{-1} year^{-1} and 100 kg K$_2$O ha^{-1} year^{-1} split between three applications; (b) grazed plots (30 × 33 m) under five nitrogen fertilizer rates—0, 100, 200, 300 and 400 kg N ha^{-1} year^{-1}—applied in five equal dressings during the growing season. Each plot received 60 kg P$_2$O$_5$ ha^{-1} year^{-1} and 60 kg K$_2$O ha^{-1} year^{-1} also in five equal dressings. Plots were managed with different numbers of sheep in an attempt to graze all plots to a uniform level.

*Present address: Countryside Commission for Scotland, Battleby, Redgorton, Perth PH1 3EW.

81

At each site the grazing plots were replicated twice making a total of ten plots per site. Silage plots are part of a larger cereal and grass rotation experiment. Sampling reported here took place in all grazing plots but in only one silage plot at each site.

Root measurements

Monthly assessments of root quantity and distribution were made by the core break method (Drew & Saker 1980). Six cores per treatment were extracted in 10 cm segments and broken at approximately half their length. The sum of root numbers on both exposed surfaces gave an estimate of root number in the core.

In mid-June 1982 six cores were removed from grass plots which had just received their first silage cut of the season. These were divided into 10 cm lengths, soaked in a solution of sodium pyrophosphate for 36 hours, agitated in water and the roots washed clean of soil over a 0·5 mm sieve. Live roots were removed from organic debris with forceps. Differentiation of live and dead roots was aided by vital staining with Trypan blue. Their length was estimated over a 1 cm grid (Tennant 1975) before being oven-dried to provide biomass estimates. Cores from grazing plots in September 1983 were similarly treated to provide estimates to compare with core break counts.

RESULTS

Length and biomass of roots in ungrazed plots differed between the three sites, most being found at the undisturbed site and least on the undrained restored site (Fig. 1). Penetration of roots barely exceeded 60 cm on the restored site whereas they penetrated the undisturbed profile to below 1 m. Roots on the restored site were particulary less abundant in the subsoil (below 20 cm). Drainage allowed increased production and allowed some increase in rooting in the subsoil. The quantities of root produced were related to above-ground production. Root–shoot

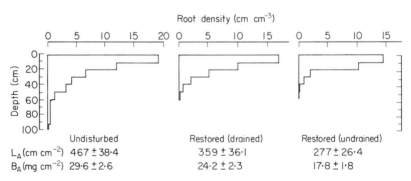

Fig. 1. Distribution of roots in ungrazed *Lolium perenne* S23 plots in June 1982 on drained and undrained land restored from opencast mining and on neighbouring undisturbed land. Total root lengths (L_A) and biomass (B_A) cm^{-2} are also given. Values are given ± 1 standard error.

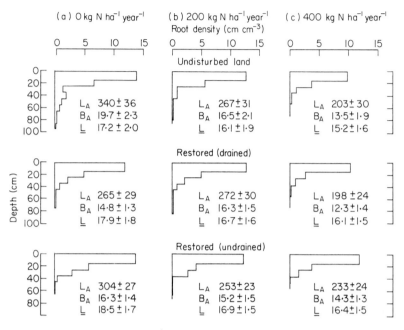

FIG. 2. Distribution of live roots (cm cm^{-3}) beneath grazed swards of *Lolium perenne* S23 on restored and undisturbed soil under three rates of nitrogen fertilization (0, 200 and 400 kg N ha^{-1} year^{-1}). Total root length (cm cm^{-2}) (L$_A$), biomass (mg cm^{-2}) (B$_A$), and the mean length per unit weight of root (cm mg^{-1}) (*L*) are also shown. Sampling took place on 13 September, 1983. Values are shown ±1 standard error.

ratios at the undisturbed site, the restored drained and undrained sites respectively were 0·34, 0·42 and 0·45.

Quantities of roots in grazed plots in September 1983 were considerably less (Fig. 2) and more concentrated in the surface horizons, particularly at high levels of nitrogen fetilization and at the undrained restored site. Site differences in total root quantities were small except in unfertilized plots.

Monthly counts suggested that root production was greatest in the spring (Fig. 3). A decline in number during the early summer may reflect root death or decay of dead roots which may have been counted whilst indistinguishable from live roots. Differences between sites were small, but in highly fertilized plots there was more root at the undrained site particularly in the spring. At this time too, differences between fertilizer treatments were most marked, showing a distinct inverse relationship between level of nitrogen applied and number of roots counted.

DISCUSSION

Root penetration was impeded in restored soil, particularly in the subsoil. This was most marked in ungrazed plots where roots did not grow below 60 cm. Similar restriction was discerned in grazing plots but to a less marked degree, probably due

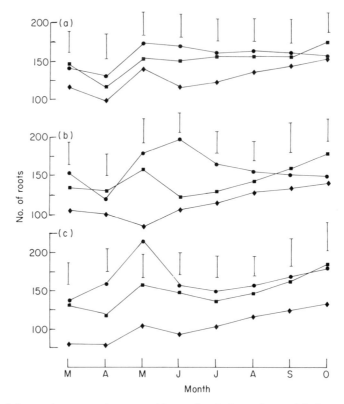

FIG. 3. Seasonal changes in root number counted in core breaks beneath grazed *Lolium perenne* S23 swards on (a) drained and (b) undrained restored land and on (c) neighbouring undisturbed land. Plots fertilized with 0 kg N ha^{-1} (●), 200 kg N ha^{-1} (■) and 400 kg N ha^{-1} (◆). I = LSD ($P = 0.05$) at each sampling date.

to the effect of grazing swards to a similar level at all sites. Root–shoot ratios were higher at restored sites, and the reduced above-ground yields there may in part be due to the increased allocation of assimilate below ground. Increased allocation below-ground probably arises from (a) unfavourable conditions for nutrient and water absorption decreasing the efficiency of individual roots, and (b) the decreased volume of soil utilized by roots leading to a more rapid depletion of nutrient within their vicinity. Under both these conditions, the maintenance of a functional balance between root and shoot would require decreased shoot growth and/or increase in the absorbing capacity of the root.

Grazing leads to a marked reduction in the quantity of root. This is an effect of the constant defoliation reducing above-ground photosynthetic area. Roots are also more concentrated in the surface under grazing which is probably a result of the return of nutrients to the surface in animal excreta.

Increased nitrogen applications led to reduced root densities despite considerably increased above-ground production. They also led to reduced penetration of the root system and increased proportions of the total root length and biomass in the surface horizons. These effects have been observed before (see Whitehead (1970) for references) and again arise through the maintenance of a functional balance between root and shoot activity.

Roots in the most favourable treatments were heavier per unit length (also seen in barley; Welbank *et al.* 1974), possibly because fertilized roots may live longer, reducing annual turnover and further decreasing the below-ground demand for assimilate.

Differences between treatments and sites in grazing plots were greatest in late spring when root growth seemed to be at its maximum. The decline in differences thereafter may in part be due to the weather. Spring and early summer of 1983 were particularly wet whereas mid-summer was particularly dry. Nutritional differences may have declined as water rather than nitrogen became limiting to plant growth.

If the biomass input and penetration of roots into restored soil profiles is important in returning the land to its full fertility potential then a high input system of farming early on in the post-restoration phase may not be wholly advantageous. However, it has yet to be seen whether the build-up of organic matter and its related effects on soil structure are not greater in the high input grazing system because of the recycling of organic matter and nutrients in animal excreta and the absence of silage or hay removal.

ACKNOWLEDGMENTS

This work forms part of the Northern Research Project of the NCB Opencast Executive and is financed by the NCB. I am grateful to P. Shotton for assistance in root coring and counting, and to A. Younger and A. M. Baker who designed and managed the experimental plots.

REFERENCES

Drew, M.C. & Saker, L.R. (1980). Assessment of a rapid method, using soil cores, for estimating the amount and distribution of crop roots in the field. *Plant and Soil*, **55**, 297–305.

Tennant, D. (1975). A test of a modified line intersect method of estimating root length. *Journal of Ecology*, **63**, 995–1001.

Welbank, P.J., Gibb, M.J., Taylor, P.J. & Williams, E.D. (1974). Root growth of cereal crops. *Annual Report of Rothamsted Experimental Station 1973*, pp. 26–66.

Whitehead, D.C. (1970). The role of nitrogen in grassland productivity. *Bulletin 48 of The Commonwealth Agricultural Bureaux*.

Functional significance of root morphology and root system architecture

A. H. FITTER

Department of Biology, University of York, York

SUMMARY

1 Individual roots in soil vary greatly in size, longevity, mineral content and activity, and their interaction with other organisms will depend on these variables, many of which are determined by the relationship of the root to the whole root system. Measurements of root morphology are therefore an important part of our understanding of the soil system.

2 Estimates of quotients of morphological attributes of root systems are easy to make and provide some information, primarily about root diameter.

3 An alternative is to study the topology of root systems, the theoretical background to which is explained here. This enables tests of deviation from randomness (or other growth models), and experiments suggest that in soil of low nutrient status or high moisture content, root growth is topologically random, but that high nutrient status and mycorrhizal infection on the one hand and low soil moisture on the other produce deviations from randomness, though in opposite directions.

4 These results are tentatively interpreted in relation to search theory.

ECOLOGICAL SIGNIFICANCE OF ROOT MORPHOLOGY

Root system architecture and morphology varies widely between species, between individuals of a species and even within individual root systems. Ecological explanations of the analogous variation in shoot systems have been advanced, in relation to light interception and growth rate (Leopold 1971; Horn 1971) or mechanical strength (McMahon & Kronauer 1976), and morphometric approaches to the analysis of branching structures have been widely applied to shoots (Whitney 1976; Fisher & Honda 1979a, b; Steingraeber, Kascht & Franck 1979; Pickett & Kempf 1980; Fisher & Kibbs 1982) to the point at which a system of classification of architectural models of trees has been advanced (Hallé, Oldeman & Tomlinson 1978), which is the first step towards an understanding of their functional significance. In contrast, studies of roots are still limited by the traditional botanical ordering system, whereby roots are classifed as axes and laterals of various orders.

I have pointed out elsewhere the deficiencies of such a system and the availability of alternatives (Fitter 1982). There are two ways of approaching the problem: one is to determine attributes of the whole system and to express the morphology in terms of these attributes, directly or as quotients. Thus weight, length, surface area and volume of a root system are all measurable with varying degrees of ease and accuracy. Such measures have the advantage of being relatively simple to obtain, but carry limited information. Diameters of root system members of various orders can be determined and will be closely related to, for example, the quotient of length to weight, but the problems of accurately determining large numbers of roots and of expressing the data in a useable form are considerable. Alternatively, it is possible to attempt to quantify the branching structure of the system in terms of both geometry and topology. I propose to assess these two approaches here.

Quantitative information on root morphology is important to an understanding of their importance to the soil ecosystem. Many studies have shown that knowledge of fine root dynamics is necessary to quantify energy and nutrient flows (Ford & Deans 1977; Persson 1983; Fogel, p. 23); the production of such fine roots is determined partly by root system architecture. The term 'fine roots' is deliberately vague, yet in truth root thickness is a continuous variable, and one which determines many aspects of root function. It directly affects the uptake abilities of the root, both in relation to the transport of solutes in soil (Nye 1973) and to the maintenance of contact with soil (Huck, Klepper & Taylor 1970). In addition, root size determines their suitability as food, since nutrient concentrations are probably normally higher in fine roots (approximately twice as high for N, P and K in 'small roots' as opposed to 'large roots' of *Quercus petraea*: Allen 1974; see also Atkinson, p. 43), and their ability to interact with other soil inhabitants, whether symbionts or pathogens.

Of particular significance in relation to symbiont function is the longevity of root system members, since the cost of establishing a symbiotic association must be recouped within the life-span of the root. The subject has been thoroughly investigated for leaves (Chabot & Hicks 1982), but little is known of the longevity in soil of different parts of root systems, although estimates of half-lives of 10–30 days for individual meristems have been made (Garwood 1967; Coleman, Reid & Cole 1983; Atkinson, p. 43).

For these reasons it is therefore important to investigate the control of fine root production and more generally of root system architecture, in order to understand the interaction of root systems with the rest of the soil ecosystem.

GROSS MORPHOLOGY

The two attributes most commonly determined in studies of root growth are length and weight. They have the advantage of being readily quantified and several workers have used the ratio of length to weight as an indicator of gross morphology (e.g. Christie & Moorby 1975; Fitter 1976; Lefroy 1982; Robinson & Rorison 1983a; Narayanan & Reddy 1983): assuming constant density, this ratio reflects the mean diameter of roots within the system. The ratio can be termed specific root length (SRL), by analogy with specific leaf area and other terms from classical growth analysis (Evans 1972). It is highly plastic and its behaviour is one possible index of root morphology.

Variation in specific root length

Genetic variation

Root morphology varies considerably between species, particularly in relation to the thickness of the finest roots (Baylis 1975), and the studies quoted above indicate a wide range of values of SRL. There is also variation between genotypes within a species: SRL for five maize genotypes ranged from 11·8 to 21·8 and hybrids had SRL values in three of four cases intermediate between the parents (Barber 1978).

Age

Very young root systems normally have high values of SRL, which then declines rapidly as thickening of roots occurs, or as the proportion of fine roots declines. Robinson & Rorison (1983a) found an initial decline followed by more or less constant values from 7 to 35 days in water culture for three grasses (*Holcus lanatus*, *Lolium perenne* and *Deschampsia flexuosa*), whereas Lefroy (1982) recorded more or less continuous declines in sand culture at four levels of P for rape and lettuce over 56 days, but a steady increase for barley. The same discrepancy is shown in Fig. 1, which illustrates changes in SRL for eight grasses grown for up to 50 days in sand culture at high and low fertility. Five of the eight species have very similar patterns, but *Briza media* has much higher values than the rest, while *Holcus lanatus* differs in having a constant value and *Anthoxanthum odoratum* in showing an increasing SRL with time.

   None of the SRL figures quoted in these three studies are consistent for a given species, further emphasizing the inconstancy of the ratio and the need to define precisely the conditions of growth, age and possibly genotype.

Soil conditions

SRL is highest in soil of low fertility for most species. This is clearly shown by Christie & Moorby (1975), Lefroy (1982) and Narayanan & Reddy (1983) and is true also in the study of eight grasses illustrated in Fig. 1, in which SRL is greatest at lower fertility levels at all times in five species (*Briza*, *Phleum*, *Poa*, *Holcus* and *Hordeum*) and at all but one harvest in the other three. Christie & Moorby (1975) grew three grasses differing greatly in their responsivenes to P in solution culture and found that SRL was increased by 19–104% in low P solutions (0·003 ppm) as compared to high P (3 ppm). The differences were proportional to the increase in plant weight brought about by a high P supply (Table 1). This correlation between plasticity of the root system and of growth rate has often been reported; D. Robinson (pers. comm.) found that SRL was 51% higher in infertile as opposed to fertile soil (3·7 *vs* 57·6 μg N g^{-1} soil) for *Holcus lanatus*, but no different for *Deschampsia flexuosa*. *D. flexuosa* is of course one of the least nutrient-responsive grasses. These differences presumably reflect differences in root diameter. Christie & Moorby

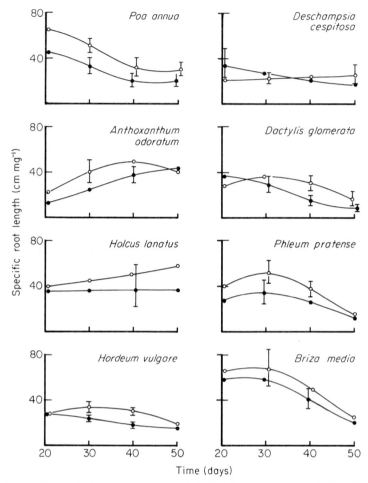

FIG. 1. Specific root length of eight grasses over a period of 30 days at two levels of soil fertility. Plants were grown individually in a glasshouse in 10 cm pots of sand enriched with 167·4 (●) or 18·6 (○) mg of a commercial fertilizer (14:14:14—N: P_2O_5:K_2O). Values are derived from polynomial regressions of ln W_R and ln L_R against time, and represent fitted means and 95% confidence limits.

(1975) showed that nodal (but not seminal) axes had smaller diameters in low P conditions—74% of the high P level for the least growth-responsive species and 48% for the most responsive—a result similar to that obtained by Anghinoni & Barber (1980) with *Zea*.

The effect of other soil factors has not been widely reported. Veresoglou & Fitter (1984) studied three co-existing grasses (*Agrostis capillaris*, *Holcus lanatus* and *Poa pratensis*) in three soil moisture and temperature regimes: low and high temperatures (12/8°C, day/night and 20/15°C respectively) combined with high soil moisture (field capacity: 26% w/w), and high temperature with low soil moisture (half field capacity: 13% w/w). In all three species low temperature raised the root weight ratio (W_R/W_{RS}), but low moisture was ineffective. In contrast, SRL

TABLE 1. Responsiveness to phosphate supply, specific root length and root diameter of three arid-zone grasses (Christie & Moorby 1975)

		Thyridolepis mitchelliana	*Astrebla elymoides*	*Cenchrus ciliaris*
Response to P $\left\{\dfrac{\text{Total dry weight at 3 ppm}}{\text{Total dry weight at 0·003 ppm}}\right.$		3·0	13·7	29·6
Specific root length (cm mg^{-1})				
3 ppm		16·2	16·0	13·6
0·003 ppm		19·2	20·5	27·8
ratio		1·10	1·28	2·04
Root diameter (μm)				
Nodal axes				
3 ppm		704	–	968
0·003 ppm		522	–	460

responses varied between treatments: *Agrostis*, in the field the most drought-resistant species, had its highest SRL in the low moisture treatment, where *Poa* had the lowest value. Both *Poa* and *Holcus* had highest SRL in the high temperature, high moisture conditions.

Other organisms

There is little published information on the impact of mycorrhizal infection on root morphology. In a series of trials I have been unable to find any consistent effect of VA mycorrhizal infection on the SRL of a range of species, although total root length may be greatly reduced (Fitter 1977). Structural changes do, however, occur and these will be described later.

Pathogens such as club-root can be seen to have visible and dramatic effects on root systems. Rovira (1978) gave data on root weight and length of wheat plants infected with the nematode *Heterodera avenae*. When expressed as SRL (Table 2) these show a halving of the ratio in the presence of nematodes, implying a great reduction in the proportion of fine root or a large change in morphology; phosphate fertilizer had almost no effect.

TABLE 2. Specific root length (cm mg^{-1}) of wheat plants as affected by nematodes and fertilizer (Rovira 1979)

		Phosphorus added	
		–	+
Nematicide	–	6·9	8·7
	+	14·3	15·1

Spatial and temporal heterogeneity

In soil, factors such as nutrient supply and moisture level are variable in time and space. A single expression of SRL for a whole root system may therefore conceal

These relationships have a number of uses in the study of root systems.

(i) It is possible to calculate expected values of p_e and d for networks developed following particular rules. Initially one may test to see whether systems deviate from random (i.e. probability of branching is equal at all links). Alternatively predictions for other rules of growth may be generated.

(ii) Root systems grown under different conditions (water or nutrient supply, soil physical conditions, other roots or other organisms) may be compared, with each other or with expected standards.

(iii) The functional efficiency of root systems of various topologies can be tested, e.g. in terms of water transport or of the probability of location of other organisms in soil.

(iv) The vulnerability of the system to disturbance can be assessed. Since high magnitude links carry materials to and from large numbers of sources, they are the most valuable to the system.

Random expectations

For a system of a given magnitude it is easy to calculate the maximum pathlength and diameter. These occur when the *tree* consists of a single main axis with simple side branches; then $d = n$ and

$$p_e = \tfrac{1}{2}(n^2 + 3n - 2) \text{ (Werner \& Smart 1973)} \tag{1}$$

Minimum values are also readily calculated, and are achieved when the system is perfectly dichotomous; then

$$d = \mid \log_2(n - 1) \mid + 2 \tag{2}$$

where $\mid x \mid$ represents the integer part of x, and

$$p_e = n(d^* + 1) - 2^{d-1} \tag{3}$$

where d^* is the value of d from eqn (2) (Werner & Smart 1973).

Expected values are more difficult to calculate, however, and are as yet only available for the situation where systems are grouped into TDN classes rather than ambilateral classes. Assuming all TDN to be equally probable, the expected p_e for magnitude n is (Knuth 1968; Werner & Smart 1973):

$$E(p_e; n) = \frac{2^{2(n-1)}}{N(n)} \tag{4}$$

Since there are more TDN in ambilateral classes where p_e is high than where it is low, these expected values will be overestimates for root systems.

Influence of soil conditions

These methods have not been widely applied to morphological systems in biology previously, although they underlie much recent evolutionary and phylogenetic mycorrhizal infection (which may be equivalent in some circumstances) both resulted in simpler branching patterns, approaching maximum p_e and d more closely, while low soil moisture had the opposite effect, tending to produce more nearly dichotomous branching. These responses can be interpreted as searching responses. The theory of search is mathematically well-developed (Stone 1975). In its simplest form, the probability of encounter between a randomly moving body and a fixed object depends upon the width of its search path (w), the length of the path L and the area A in which the object is placed (Koopman 1956):

$$p(x) = 1 - e^{-wL/A} \tag{5}$$

This simple formula implies that for a topologically random root system, the probability of encounter with a fixed object, such as a mycorrhizal spore, is a function of root length and diameter, and spore density. At normal values of root density of around 1 cm cm^{-3} and spore density of around 10 cm^{-3}, this requires a root diameter of 0·7 mm to achieve a 50% probability of encounter and 2·3 mm to achieve 90%. It is perhaps not surprising that mycorrhizal infection is most common from hyphal growth rather than from spores. Most importantly, it can be seen that root width and length both contribute equally to the probability of encounter, but that width of course contributes as its square in terms of the carbon costs of root growth. Increase in root density is therefore a more cost-effective way of ensuring encounter than increased root diameter.

In practice, many of the targets 'sought' by roots are themselves mobile, including hyphae, water molecules and ions. Search theory formulae are available for such a condition (Hellman 1970; Saretsalo 1973), but it is not yet practical to apply them to roots.

CONCLUSIONS

Although the data required to calculate single quotients, such as specific root length, are easily acquired, they give very limited information. Where a simple index of root 'thickness' is required, SRL is a useful parameter, particularly since its relationship to soil conditions, such as moisture and fertility, and to variations in those conditions, are reasonably well understood. Where a more detailed analysis of root system structure is required, however, topological analysis may give greater insights.

Previous methods of ordering root systems have been wholly descriptive, and have made difficult the formulation of testable hypotheses about the significance of root system structure. Recently Fowkes & Landsberg (1981) put forward a model based on the distribution of a given amount of assimilate between roots of varying dimensions, and on the relationship between root axial resistance and root radius. They demonstrated that systems comprising many short roots are initially more effective in depleting uniformly moist soil than those of few long roots, but they produce local depletion and are subsequently less effective. Their model can also be used to predict optimum inter-root spacing, which is dependent upon the geometry and topology of the system. The results described in this paper show that topology influences both the functioning of the system and its relationship to other soil organisms. In particular the compromise between exploitative, transport-efficient structures of low p_e and d, and the more explanatory structures characterized by high p_e and d, is likely to be of great significance.

The rate of production, behaviour, longevity and structural characteristics of the members of root systems are fundamental to an understanding of the soil system. I believe that a topological approach to root system structure may be of great value in determining many of these characteristics and in elucidating the interactions of roots with other soil inhabitants.

ACKNOWLEDGMENTS

I am grateful to Maureen Chapman for much assistance in the collection of topological data on root systems, to Dr M. Woldenberg and Professor M. H. Williamson for helpful discussions on trees in biology, and to Dr D. Atkinson for his comments on this paper.

REFERENCES

Allen, S.E. (1974), *Chemical Analysis of Ecological Materials*. Blackwell Scientific Publications, Oxford.

Anghinoni, I. & Barber, S.A. (1980). Phosphorus influx and growth characteristics of corn roots as influenced by phosphorus supply. *Agronomy Journal*, **72**, 685–688.

Barber, S.A. (1978). Growth requirements for nutrients in relation to demand at the root surface. *The Soil Root Interface* (Ed. by J. L. Harley & R. Scot Russell), pp. 5–20. Academic Press, London.

Baylis, G.T.S. (1975). The magnolioid mycorrhiza and mycotrophy in root systems derived from it. *Endomycorrhizas* (Ed. by F. E. Sanders, B. Mosse & B. P. Tinker), pp. 373–90. Academic Press, London.

Cayley, A. (1857). On the analytical forms called trees. *Philosophical Magazine*, **18**, 374–378.

Chabot, B.F. & Hicks, D.J. (1982). The ecology of leaf life spans. *Annual Review of Ecology and Systematics*, **13**, 229–259.

Christie, E.K. & Moorby, J. (1975). Physiological responses of arid grasses, I. The influence of phosphorus supply on growth and phosphorus absorption. *Australian Journal of Agricultural Research*, **26**, 423–436.

Coleman, D.C., Reid, C.P.P. & Cole, C.V. (1983). Biological strategies of nutrient cycling in soil systems. *Advances in Ecological Research*, **13**, 1–55.

Drew, M.C., Saker, L.R. & Ashley, T.W. (1973). Nutrient supply and the growth of the seminal root system in barley, I. The effect of nitrate concentration on the growth of axes and laterals. *Journal of Experimental Botany*, **24**, 1189–1202.

Drew, M.C. & Saker, L.R. (1980). Assessment of a rapid method, using soil cores, for estimating the amount and distribution of crop roots in the field. *Plant & Soil*, **55**, 297–305.

Duncan, W.G. & Ohlrogge, A.J. (1958). Principles of nutrient uptake from fertiliser bands, II. *Agronomy Journal*, **50**, 605–608.

Evans, G.C. (1972). *The Quantitative Analysis of Plant Growth*. Blackwell Scientific Publications, Oxford.

Fisher, J.B. & Honda, H. (1979a). Branch geometry and effective leaf area: a study of *Terminalia*-branching pattern, I. Theoretical trees. *American Journal of Botany*, **66**, 633–644.

Fisher, J.B. & Honda, H. (1979b). Branch geometry and effective leaf area: a study of *Terminalia*-branch pattern, II. Real trees. *American Journal of Botany*, **66**, 645–655.

Fisher, J.B. & Hibbs, D.E. (1982). Plasticity of tree architecture: specific and ecological variations found in Aubreville's model. *American Journal of Botany*, **69**, 690–702.

Fitter, A.H. (1976). Effect of nutrient supply and competition from other species on root growth of *Lolium perenne* in soil. *Plant and Soil*, **45**, 177–189.

Fitter, A.H. (1977). Influence of mycorrhizal infection on competition for phosphorus and potassium by two grasses. *New Phytologist*, **79**, 119–125.

Fitter, A.H. (1982). Morphometric analysis of root systems: application of the technique and influence of soil fertility on root system development in two herbaceous species. *Plant, Cell and Environment*, **5**, 313–322.

Ford, E.D. & Deans, J.D. (1977), Growth of Sitka spruce plantations: spatial distribution and seasonal fluctuations of lengths weights and carbohydrate concentrations of fine roots. *Plant and Soil*, **47**, 463–485.

Fowkes, N.D. & Landsberg, J.J. (1981). Optimal root systems in terms of water uptake and movement. *Mathematics and Plant Physiology* (Ed. by D. A. Rose & D. A. Charles–Edwards), pp. 109–128. Academic Press, London.

Garwood, E.A. (1967). Seasonal appearance and growth of grass roots. *Journal of the British Grassland Society*, **22**, 121–130.

Halle, F., Oldeman, R.A.A. & Tomlinson, T.B. (1978). *Tropical Trees and Forests: An Architectural Analysis*. Springer–Verlag, Berlin.

Hellman, O. (1970). On the effect of a search upon the probability distribution of a target whose motion is a diffusion process. *Annals of Mathematics and Statistics*, **41**, 1717–1724.

Horn, H. (1971). *The Adaptive Geometry of Trees*. University Press, Princeton.

Huck, M.G., Klepper, B. & Taylor, H.M. (1970). Diurnal variations in root diameter. *Plant Physiology*, **45**, 529–530.

Knuth, D.E. (1968). *The Art of Computer Programming, Vol. I: Fundamental Algorithms*. Addison–Wesley, New York.

Koopman, B.O. (1956). The theory of search, II. Target detection. *Operations Research*, **4**, 503–531.

Kormanik, P.P. & McGraw, A.C. (1982). Quantification of vesicular–arbuscular mycorrhizae in plant roots. *Methods and Principles of Mycorrhizal Research* (ed. by N.C. Schenck), pp. 37–46. American Phytopathological Society, St. Paul.

Lefroy, R.D.B. (1982). *The Supply and Utilisation of Phosphate in the Control of Plant Growth*. D.Phil. thesis, University of York.

Leopold, L.B. (1971). Trees and streams : the efficiency of branching. *Journal of Theoretical Biology*, **31**, 339–354.

MacDonald, N. (1983). *Trees and Networks in Biological Systems*. Wiley, London.

McMahon, T.A. & Kronauer, R.E. (1976). Tree structures, deducing the principle of mechanical design. *Journal of Theoretical Biology*, **59**, 443–466.

Murphy, J. & Riley, J.P. (1962). A modified single solution method for the determination of phosphate in natural waters. *Analytica Chimica Acta*, **27**, 31–36.

Narayanan, A. & Reddy, K.B. (1983). Effects on phosphorus deficiency on the form of the plant root system. *Proceedings of the Ninth International Plant Nutrition Colloquium* (Ed. by A. Scaife), pp. 412–417. Commonwealth Agricultural Bureaux, London.

Nye, P.H. (1973). The relation between the radius of a root and its nutrient absorbing power (alpha). *Journal of Experimental Botany*, **24**, 783–786.

Persson, H.A. (1983). The distribution and productivity of fine roots in boreal forests. *Plant and Soil*, **71**, 87–101.

Pickett, S.T.A. & Kempf, J.S. (1980). Branching patterns in forest shrubs and understorey trees in relation to habitat. *New Phytologist*, **86**, 219–228.

Robinson, D. & Rorison, I.H. (1983a). A comparison of the responses of *Lolium perenne* L., *Holcus lanatus* L. and *Deschampsia flexuosa* (L.) Trin. to a localised supply of nitrogen. *New Phytologist*, **94**, 263–273.

Robinson, D. & Rorison, I.H. (1983b). Relationships between root morphology and nitrogen availability in a recent theoretical model describing nitrogen uptake from soil. *Plant andCell and Environment*, **6**, 641–647.

Rovira, A.D. (1978). Biology of the soil–root interface. *The Soil Root Interface* (ed. by J. L. Harley & R. Scott-Russell), pp. 145–160. Academic Press, London.

Saretsalo, L. (1973). On the optimal search for a target whose motion is a Markov process. *Journal of Applied Probability*, **10**, 847–856.

Savage, H.M. (1983). The shape of evolution: systematic tree topology. *Biological Journal of the Linnean Society*, **20**, 224–244.

Smart, J.S. (1978). The analysis of drainage network composition. *Earth Surface Processes*, **3**, 129–170.

Steingraeber, D.E., Kascht, L.J. & Franck, (1979). Variation of shoot morphology and bifurcation ratio in sugar maple (*Acer saccharum*) saplings. *American Journal of Botany*, **66**, 441–445.

Stone, L.B. (1975). *The Theory of Optimal Search.* Academic Press, London.

Strahler, A.N. (1957). Quantitative analysis of watershed geomorphology. *Transactions of the American Geophysical Union*, **38**, 913–920.

Veresoglou, D.S. & Fitter, A.H. (1984). Spatial and temporal patterns of growth and nutrient uptake of five coexisting grasses. *Journal of Ecology*, **72**, 259–272.

Werner, C. & Smart, J.S. (1973). Some new methods of topologic classification of channel networks. *Geographical Analysis*, **5**, 271–295.

Whitney, G.G. (1976). The bifurcation ratio as an indicator of adaptive strategy in woody plant species. *Bulletin of the Torrey Botanical Club*, **103**, 67–72.

The rhizosphere: carbon sources and microbial populations

E. I. NEWMAN

Department of Botany, University of Bristol, Bristol BS8 1UG

SUMMARY

1 Available evidence suggests that, in bulk soil and in the rhizosphere, the biomass of fungi can be either greater or less than that of bacteria.

2 Soluble exudate production by living roots is usually within the range 1–10 g per 100 g root dry weight increase; root cap plus mucigel may provide a further 2–5 g per 100 g. Some published figures for total 'rhizodeposition' are much larger than this, but they include dead roots and perhaps CO_2 from root respiration.

3 A large proportion of fungi isolated from the rhizosphere can degrade cellulose and pectin, but few of the bacteria can do so.

4 There is evidence that, in grassland, roots are close enough together for motile bacteria to migrate and fungi to ramify from one to another. Since a quarter to a half of roots present at any one time in the field are dead or dying, fungi in the rhizosphere may well be using dead roots nearby as their main source of carbon substrate.

5 Thus, the rhizosphere is a clearly defined microhabitat for bacteria, most of them dependent on soluble exudate from the root; whereas many of the fungi are obtaining their substrate from outside the rhizosphere.

INTRODUCTION

It has been known for many years that in the soil immediately around a root—the rhizosphere—the abundance of bacteria is much greater than in soil further from the root (e.g. Table 1). Usually the soil which adheres to roots after their removal from the bulk soil is considered as the rhizosphere; numbers of bacteria in it have been determined by culture of serial dilutions on agar media. If similar techniques are applied to fungi, often they, too, show increased numbers in the rhizosphere, but usually their proportional increase, as indicated by the R:S ratio (Table 1) is less than for bacteria. Sometimes fungi show no clear difference in abundance between rhizosphere and bulk soil (e.g. Papavizas & Davey 1961).

It was also discovered some years ago that living and apparently healthy roots exude a wide range of soluble organic substances including sugars, organic acids and amino acids (Rovira 1969). This led naturally to the assumption that these exudates were the carbon substrates for the increased microbial population of the rhizosphere. A further step was to think of the rhizosphere as a region rich in organic materials which were easily assimilated by the microbes, and thus a region of rapid growth and metabolism, in contrast to the rest of the soil, where most of

107

the organic matter is not readily metabolized and microbial activity is mostly slow. Thus, the key features of the 'classical' rhizosphere may be summarized as follows:

(i) Each root has its own rhizosphere, effectively separate from that of neighbouring roots.

(ii) The rhizosphere is a region of high microbial abundance and activity, especially of bacteria.

(iii) Soluble exudates from the root are the main substrates for the micro-organisms.

The aim of this paper is to review evidence on the rhizosphere micro-organisms and their carbon substrates, especially from publications since the early 1970s, and to suggest ways in which the classical view of the rhizosphere needs to be modified in the light of this evidence.

MICROBIAL GROWTH RATE IN THE RHIZOSPHERE

The few sets of published data on bacterial growth rates in the rhizosphere or on the root surface show that growth is rapid only for the first few days (Bowen & Rovira 1973; Van Vuurde & Schippers 1980; Bennett & Lynch 1981; Turner & Newman 1984). For example, Van Vuurde & Schippers (1980) showed that during the first 5 days of the life of a particular segment of wheat root in soil, bacteria on the root surface increased about a hundredfold, but during the next 5 days only about twofold. The decline in growth rate could be due either to shortage of a substance required by the bacteria or to death caused by microfauna. However, since a plateau of numbers has been found when plants were grown in sand with no other organisms except the bacteria present, and in conditions where mineral nutrients were not limiting bacterial growth (Bennett & Lynch 1981; Turner & Newman 1984), it is likely that shortage of carbon substrate is the limiting factor. The bacteria multiply rapidly until they reach a population density where competition for carbon substrate is intense; this is illustrated by the model of Newman & Watson (1977). Thus, it is a mistake to think of the rhizosphere as always a region of ample substrate supply and rapid microbial growth: that is so only for short periods, for any particular root.

RELATIVE ABUNDANCE OF FUNGI AND BACTERIA

If the abundance of bacteria and fungi is determined by standard dilution plate methods, there is usually a markedly higher abundance of bacteria in the rhizosphere than in root-free soil, but a proportionately smaller increase of fungi (e.g. Table 1). However, such data cannot give a meaningful answer to the question, are bacteria more abundant than fungi? Such a comparison must be on the basis of biomass. Unfortunately very different values for microbial biomass in soil are obtained using different techniques, and the most reliable technique is still uncertain. Determination of bacterial numbers by plating usually gives much lower results than does direct observation (Lundgren 1981); and the results by direct

TABLE 1. Numbers of bacteria and fungi (million per g dry weight of soil) in rhizosphere of wheat and in root-free soil, determined by dilution plating. Data of Rouatt, Katznelson & Payne (1960)

	Rhizosphere	Root-free soil	R:S ratio*
Bacteria	1200	53	23
Fungi	1.2	0.1	12

* Ratio of number in rhizosphere: number in root-free soil.

observation, for either bacteria or fungi, can vary substantially depending on which stain is used (Söderström 1979; Lundgren 1981). The ratio biomass fungi:biomass bacteria was found to be 3–6 in soil at the Matador prairie site (Saskatchewan) and 5–10 in arable soil close by (Coupland 1979; Shields, Paul & Lowe 1973); although both bacteria and fungi were determined by direct observation, they were stained with different stains, which may have influenced the results. In my own research group we have made many measurements of the abundance of bacteria and fungi on the root surfaces of grassland plants, staining both groups with phenol acetic aniline blue and using direct microscopical observation. This stain will sometimes stain up dead cells, but Frankland (1975) found that it slightly underestimated the length of mycelium (from woodland soil) which contained protoplasm. If there is preferential stimulation of bacteria in the rhizosphere it should be maximal at the root surface. The overall conclusion from all our measurements is that fungi may cover either more or less of the root surface than bacteria. For example, on the roots of *Plantago lanceolata* collected from forty sites in England and South Wales, the ratio of cover per cent fungi:cover per cent bacteria on the root surfaces ranged from 0·28 to 14·0 (Newman, Heap & Lawley 1981). Table 2 shows results from experiments in which *Lolium perenne* was grown in pots of initially nutrient-deficient soil and supplied with different soluble nutrients. Fungi were measured by staining with phenol acetic aniline blue followed by direct observation, bacteria by that method in Experiment 1 but by dilution plating in Experiments 2 and 3. When all essential nutrients were in ample supply bacteria exceeded fungi, but when

TABLE 2. Ratio of percentage of root surface covered by fungi: percentage covered by bacteria. Data from experiments of Turner (1983) in which *Lolium perenne* was grown in soil with various mineral nutrients supplied weekly

	Cover % fungi / Cover % bacteria		
Experiment	1	2	3
Nutrients supplied to plants			
Complete*	0·62	0·61	0·06
Complete–N†	3·0	3·6	1·9
Complete–P‡	5·4	nd	2·3
None	1·25	3·6	1·8

* Modified Hoagland's solution.
† As Complete, but with N omitted.
‡ As Complete, but with P omitted.
nd, not determined.

either nitrogen or phosphorus was omitted, or no nutrients were supplied, fungi exceeded bacteria.

In addition to the technical problems with stains, there is uncertainty about what proportion of the mycelium is mycorrhizal or parasitic rather than saprophytic. In our own work we have found little correlation between internal mycorrhizal infection and root-surface mycelium; for example, in an experiment using the same soil and species as in Table 2 there was scarcely any difference between the Complete, Complete–N and Complete–P treatments in the percentage of root length that was mycorrhizal (H. Humfress, unpubl.). This suggests that the results in Table 2 were little influenced by changes in mycorrhizal mycelium. An alternative method, which estimates the metabolically active biomass of saprophytes indirectly, involves adding glucose to a soil sample and measuring CO_2 production during the next few hours (Anderson & Domsch 1975). Bacteria and fungi can be estimated separately by applying selective inhibitors. By this method the ratio of fungi:bacteria has been found to range from 1·5 to 9 in agricultural, grassland and forest soils, and from 0·13 to 1·5 in the rhizospheres of crop plants and prairie grasses (Anderson & Domsch 1975, 1980; Vancura & Kunc 1977; Nakas & Klein 1980).

Therefore, the available evidence indicates that in bulk soil fungi usually exceed bacteria in biomass, whereas in the rhizosphere either group may have the larger biomass. Clearly neither group should be ignored.

ORGANIC MATERIALS PRODUCED BY ROOTS

Although classical studies emphasized soluble exudates, in recent years there has been increased interest in insoluble organic materials produced by roots, which could potentially be substrate for micro-organisms. Although these, too, are sometimes called exudates, the term 'rhizodeposition' seems a better word to cover all organic materials produced by roots.

Measurement of rhizodeposition is fraught with technical problems, and all published data are open to criticism on some ground or other. Here I concentrate on those data which seem to me technically most satisfactory. Table 3 summarizes some data on amounts of rhizodeposition by axenic plants grown in solution culture (with in some cases glass beads as a solid medium). Here and later rhizodeposition is expressed per g of final root dry weight, and is approximately mg rhizodeposition per g root growth. Most values for soluble exudation determined by chemical methods lie in the range 10–100 mg g^{-1}. The value reported by Prikryl & Vancura (1980) is much higher, but since most of it was exuded between days 4 and 8 it may have come largely from the seed, whose weight would have greatly exceeded the final dry weight of the root. Insoluble organic materials, when measured, were much less than the soluble exudate (Table 3).

The only satisfactory measurements of rhizodeposition into soil have involved growing plants throughout their lives in an atmosphere containing $^{14}CO_2$, so that they become uniformly labelled. The soil is separated from the air-space so that ^{14}C can reach it only via the plant's roots. Such experiments have been carried out by

TABLE 3. Amounts of organic material ('rhizodeposition'), determined chemically, produced by roots of young axenic plants

Species	Age* (days)	Rhizodeposition (mg g^{-1} root dry wt)		Reference
		Soluble	Insoluble	
Wheat	14	515	nd	Prikryl & Vancura (1980)
Barley	16	62	nd	Barber & Lynch (1977)
	27	42–82	8–13	Barber & Gunn (1974)
Others	Various	6–247	3–8	See Newman (1978)

* At end of experiment.
nd, not determined.

Barber & Martin (1976), Martin (1977a), Johnen & Sauerbeck (1977), Whipps & Lynch (1983) and Whipps (1984). Some of their results are summarized in Table 4. These values are in units directly comparable with those of Table 3, provided that the specific activity of ^{14}C and the ratio of C:dry weight were the same in the root as in its products; deviations from these assumptions are likely to be small compared with the variations between experiments. The amount of rhizodeposition was little affected by the presence of micro-organisms. This is surprising: one might have expected micro-organisms to reduce substantially the concentration of soluble materials, though perhaps not of insoluble materials which would include microbes themselves. Martin (1977b) and Whipps & Lynch (1983) suggested that microbial activity in the non-sterile treatments was severely restricted by dryness of the soil. It was not possible to water the containers during the experiments, so uptake by the plants reduced the soil water content, but to what extent was not measured. In Experiment C the sand was reported by the authors to be very dry by the end, except near the base.

In the experiments of Barber & Martin (1976) and Martin (1977a) usually more $^{14}CO_2$ was recovered from the soil when microbes were present than from sterile soil. They attributed this to respiration by microbes using rhizodeposition as substrate. They therefore added the amount of extra $^{14}CO_2$ to the other recovered ^{14}C to calculate total rhizodeposition in non-sterile conditions, and concluded that rhizodeposition was increased by the presence of micro-organisms. This may well be true. However, most of their $^{14}CO_2$ recovery rates in both sterile and non-sterile conditions were extremely low: measurements by several other methods show that growth-respiration (i.e. that part of respiration associated with root growth processes) is usually 500–1500 mg g^{-1} root and maintenance respiration must be added to that (Hansen & Jensen 1977; Lambers & Steingröver 1978; Farrar 1981). Unless some explanation can be found for the much lower recorded respiration rates in Table 4 and other experiments by these authors, no conclusions can be drawn from them.

Soluble exudate assessed by ^{14}C was in the soil experiments (Table 4) in a similar range to the more reliable of the values obtained by chemical methods (Table 3).

TABLE 4. Amount of ^{14}C recovered from soil in which uniformly ^{14}C-labelled wheat plants were grown for 3 weeks at 18°C, 16 h photoperiod. Experiment A, data of Barber & Martin (1976); Experiments B and C, data of Whipps & Lynch (1983)

Growth medium	$\dfrac{^{14}\text{C recovered}}{^{14}\text{C in root}}$ (μCi mCi^{-1})		
	Soluble	Insoluble	CO$_2$
Experiment A			
Soil, not sterilized	19	240	241
Soil, sterilized then reinoculated	38	241	194
Soil, sterile	36	223	42
Experiment B			
Soil, not sterilized	156	462	930
Experiment C			
Sand, not sterilized	306	329	39
Sand, sterile	393	245	5

Soluble material was more abundant in the sand experiment (Table 4, Experiment C), but this may have been due to severe droughting. The large amount of insoluble rhizodeposition indicated by the ^{14}C method (Table 4) has aroused great interest but there has been little discussion of what it may comprise. The following possibilities need to be considered: (i) carbon dioxide respired by the root which was fixed by soil bacteria; (ii) mycorrhizal hyphae and spores; (iii) mucigel; (iv) sloughed root cap cells; (v) tissues of the roots themselves. Judging from data for broad bean and soya bean (Kucey & Paul 1982; Bethlenfalvay, Brown & Pacovsky 1982) external VA mycorrhizal tissue could be up to 50 mg g^{-1} root. However, in the experiments featured in Table 4 the soil was well fertilized so mycorrhizal development was probably poor. In any case neither mycorrhizal tissue nor CO$_2$ fixation by micro-organisms could occur in sterile soil. Of the six experiments (Barber & Martin 1976; Martin 1977a; Whipps & Lynch 1983) where axenic and non-axenic plants were compared, only two showed markedly more insoluble ^{14}C under non-axenic conditions, so these two possible contributors are probably not major components. Mucigel and root cap cells are the sources commonly cited for insoluble rhizodeposition; some approximate estimates of the amounts of organic matter involved are given in Table 5. The amounts of root cap material are in agreement with insoluble material recovered from axenic plants in solution culture (Table 3). Mucigel appears to contribute more dry matter production than the root cap, but the two together are, on these figures, unlikely to contribute more than about 60 mg g^{-1}, far less than the insoluble rhizodeposition measured by the ^{14}C method (Table 4). Therefore, we are driven to the conclusion that most of the 'rhizodeposition' is in fact root tissue. A key question then is whether this tissue was already dead at the time of harvest, and if so whether it was killed by the drying of the soil. Holden (1975) and Henry & Deacon (1981) have shown that by the time roots of wheat and barley are 3 weeks old a substantial proportion of their epidermal and cortical cells no longer have stainable nuclei. These plants had not been

TABLE 5. Amount of sloughed root cap tissue and of mucigel produced, expressed as mg dry wt g^{-1} dry wt root

	Amount (mg g^{-1})	Notes and source of data
Root cap		
Zea mays	7	(1)
Convolvulus arvensis	4	(2)
Mucigel		
Zea mays	48	(3)
	11–17	(4)
Triticum aestivum	29–47	(4)

(1) Volume of root cap tissue produced per day from number of new cells per day × final volume per cell (Juniper & Clowes 1965; Clowes 1971, 1976; Clowes & Woolston 1978). Root assumed 1 mm diameter, extending 25 mm day^{-1}; dry wt:volume ratio assumed same for root cap and root.

(2) By ratio of elongation rate root cap:elongation rate root. Data of Phillips & Torrey (1971).

(3) Data of Floyd & Ohlrogge (1971). Roots not axenic. Assumes roots have diameter 1 mm and dry wt:volume ratio 0·1 g cm^{-3}.

(4) Data of Vancura *et al.* (1977). Roots axenic.

subjected to drought, so evidently death of cortical cells within a few weeks is a normal occurrence. Early disintegration of the cortex has been reported for the grass *Bouteloua gracilis* (Beckel 1956) and for apple trees (Head 1973). The other key question concerning the insoluble ^{14}C in the soil (Table 4) is whether it includes whole roots, which either died before harvest or were broken off by the harvesting procedure. Whipps & Lynch (1983) and Whipps (1984) determined separately the ^{14}C in the rhizosphere and bulk soil. Whereas the distribution of soluble ^{14}C was variable, consistently a high proportion of the insoluble ^{14}C was in the bulk soil (Table 6). This cannot be insoluble material from the harvested roots, since that could not move far; unless it was fixed CO_2 (which it cannot have been in sterile pots) it must have been whole roots. It is not known whether these roots were alive until harvest but broken off by the harvesting procedure, nor whether they were influenced by the lack of watering. Thus, the total ^{14}C recovered from the soil overestimates rhizodeposition, and the most satisfactory values come from the data for ^{14}C *in the rhizosphere* of soil-grown plants, obtained by Whipps & Lynch (1983)

TABLE 6. Amounts of ^{14}C recovered from rhizosphere and root-free soil in which ^{14}C-labelled wheat plants were grown for 3 weeks. Data of Whipps & Lynch (1983)

	$\dfrac{^{14}\text{C recovered}}{^{14}\text{C in root}}$ (μCi mCi^{-1})			
	Soluble		Insoluble	
	Rhizosphere	Soil	Rhizosphere	Soil
Experiment B				
Soil, not sterilized	48	108	113	349
Experiment C				
Sand, not sterilized	283	23	33	296
Sand, sterile	376	16	27	217

and Whipps (1984). Taking all their experiments, involving both barley and wheat, soluble ^{14}C in the rhizosphere was 39–84 μCi per mCi in the root, and insoluble ^{14}C 113–253 μCi per mCi (except for one value of 340).

On the basis of all the available information, it appears that soluble exudation is most often in the range 10–100 mg g^{-1} root. Insoluble root-derived material in the rhizosphere of still functional roots is often 100–250 mg g^{-1}; if root cap plus mucigel do not exceed 60 mg g^{-1} of this (Table 5) the remainder must be from dying root cells. Much of the insoluble 'rhizodeposition' measured by ^{14}C labelling is in fact whole roots which are not in the rhizosphere. Except in the 'litter' layer of any soil, root tissue is likely to be the largest source of organic matter. If soluble exudates total 10–100 mg g^{-1} root and root cap plus mucigel 20–50 mg g^{-1}, they together contribute only 3–15% as much organic matter as the root itself, since it must die sooner or later.

USE OF INSOLUBLE ORGANIC MATERIALS BY MICROBES

Since insoluble organic materials from rhizodeposition and dead roots provide a substantial source of potential substrate, it is important to know what proportion of the rhizosphere population can degrade them. Turner (1983) grew *Lolium perenne* in pots of grassland soil, isolated bacteria from the root surfaces, and tested randomly selected isolates for ability to degrade three major insoluble plant products, cellulose, pectin and starch. While many of the bacteria could degrade starch, few could degrade pectin and none cellulose (Table 7). This agrees with previous work on bacteria from wheat rhizosphere (Rouatt, Katznelson & Payne 1960; Neal, Larson & Atkinson 1973) which showed less than 0·1% of the bacteria able to degrade cellulose, and less than 2% able to degrade pectin. Christie (1976) grew *L. perenne* in the same soil as was used by Turner (1983) and isolated fungi from the root surfaces. Of 102 isolates which he identified to species level, most belong to species recorded by Domsch, Gams & Anderson (1980) as being able to degrade cellulose, pectin and starch (Table 7). Of the fungi isolated by other workers from *L. perenne* roots at field sites in England, N. Ireland and New Zealand, fourteen were identified to species level, and about half of these have been reported to degrade the

TABLE 7. Percentage of bacteria and fungi, isolated from root surfaces of *Lolium perenne*, which can degrade cellulose, pectin and starch

	Cellulose	Pectin	Starch	Notes
Bacteria: per cent of isolates	0	6	57	(1)
Fungi: per cent of isolates	94	91	87	(2)
Fungi: per cent of species	50	64	43	(3)

(1) Data from 100 randomly selected isolates. Plants grown in pots. Data of Turner (1983).

(2) Data from 102 isolates identified by Christie (1976). Plants grown in pots of same soil as for bacterial isolates.

(3) Data from 14 species isolated from field-grown *L.perenne* by Waid (1957), Thornton (1965) and McIlwaine & Malone (1976).

three substrates (Table 7); for many of the remainder Domsch *et al.* (1980) give no information about their ability to degrade these substrates. Thus there appears to be a contrast between bacteria and fungi in the rhizosphere of ryegrass; many of the fungi can degrade cellulose and pectin but few of the bacteria can. There is no reason to suppose that ryegrass has a fungal flora which is unusual in this respect. For example, out of more than 1000 named fungal isolates from root surfaces of *Phaseolus vulgaris* (Taylor & Parkinson 1965) more than 90% have the ability to degrade cellulose, pectin and starch. Waid (1957) isolated fungi from within the cortex of ryegrass roots, even when they were white and apparently healthy. The fungi could grow on a standard agar medium, Czapek-Dox, and included species with no known pathogenic ability, e.g. *Trichoderma viride* and *Penicillium* spp. Ability to degrade cellulose and pectin, two major constituents of plant cell walls, may well be crucial to the success of these fungi. Pectin is also a major constituent of mucigel. Root-surface bacteria are often embedded in the mucigel (Foster & Rovira 1978; Greaves & Darbyshire 1972) but it seems that few of them may be able to degrade it. Electron micrographs sometimes show bacteria embedded within root cell walls (Foster & Rovira 1976) and it is assumed they are in the process of decomposing the wall around them. Possibly such bacteria are too firmly attached to be isolated by normal procedures, so they do not comprise part of the rhizosphere population, as normally defined. It appears that within the rhizosphere itself many of the fungi can degrade cellulose and pectin, but few of the bacteria can do so.

INTERACTIONS BETWEEN ROOTS

Most studies of the rhizosphere have been confined to fairly young plants, so that the roots and their adhering soil can easily be isolated. Much of the earth's land surface is, however, dominated by perennial plants; even-aged stands of young, widely-spaced plants are the exception. We should therefore consider whether neighbouring roots are usually far enough apart to be effectively separate, as far as their associated micro-organisms are concerned, or whether, on the contrary, there can be significant interactions between the rhizospheres of one root and the next.

Densities of roots in the upper soil layers under permanent grassland have been found to range from 30 to 340 cm cm^{-3} (Dittmer 1938; Pavlychenko 1942; Barley 1970; Ares 1976). Fine roots in the upper soil under conifer forest can range from 0·1 to 5 cm cm^{-3} (Reynolds 1970; Roberts 1976; Ford & Deans 1977; Persson 1978); 180 cm cm^{-3} has been reported for mixed hardwood forest (Lyford 1975). If the root density is 100 cm cm^{-3} we may, for the sake of illustration, imagine the roots all lying parallel, with distance between nearest neighbours about 1 mm. Similarly, a root density of 1 cm cm^{-3} would give an inter-root distance of 10 mm. We may take these two values as illustrative of grassland and forest, respectively, though in practice the roots will be irregularly arranged and will sometimes pass closer to each other.

Standard methods of studying the rhizosphere by sampling 'closely adhering soil' give no indication of how far from the root the rhizosphere extends. Foster & Rovira (1978), by counts of bacteria in transmission electron micrographs of *Trifolium subterraneum*, found a tenfold drop in numbers between the root surface and the soil 20 μm away, suggesting a very narrow rhizosphere. However, motile bacteria can move at 20–80 μm s^{-1} (Vaituzis & Doetsch 1969), which if maintained would be about 70–300 mm h^{-1}. In artificial chemical gradients a motile bacterium can move several mm in less than an hour (Smith & Doetsch 1969; Adler 1973). Filamentous fungi can translocate ^{14}C-labelled compounds several cm (Thrower & Thrower 1961) and some can grow about 10 mm across bare plastic, away from a nutrient source (Howard 1978). (These examples do not include strand-forming fungi, which can translocate over much larger distances.) Thus, there is potential for bacteria and fungi to move from one root to another through soil, and for fungi to translocate substrate obtained from one root into the vicinity of another. Vesicular-arbuscular mycorrhizal fungi can interconnect two host plants of different species and can obtain ^{14}C from both (Francis & Read 1984).

In experiments conducted in my laboratory, various combinations of grassland plants were grown in soil, two species either in the same pot or in separate pots. Table 8 shows an example of the results. When the two species were grown separately *Trifolium repens* had markedly more fungal mycelium on the root surface than *Lolium perenne*, but when they were together they had the same fungal abundance, about mid-way between their monoculture values. Internal mycorrhizal infection, although differing between the species, showed in this case no response to the neighbouring plant in the pot. Bacteria showed some difference between mixed-species pots and monoculture, though proportionately less than fungi. In a separate experiment the composition of the bacterial microflora of the rhizosphere was investigated (Lawley, Campbell & Newman 1983). Several species or groups differed significantly between mixture and monoculture pots; one motile species gave results (Table 8) strongly suggesting that it migrated from *L. perenne* to *T. repens* roots when they were intermingled. In all, seven pairs of plant species were investigated (Christie, Newman & Campbell 1978; Lawley, Newman &

TABLE 8. Abundance of micro-organisms associated with roots of *Lolium perenne* and *Trifolium repens* which were grown in pots of soil either separately or together. Data of Christie, Newman & Campbell (1978) and Lawley, Campbell & Newman (1983)

	Lolium perenne		*Trifolium repens*	
	alone	mixed	mixed	alone
Fungal mycelium on root surface (mm mm^{-2})	1.4a	4.1b	4.2b	9.2c
Internal mycorrhizal infection	0.60a	0.62a	1.09b	1.07b
Bacteria on root surface (cover %)	0.31a	0.67b	0.24a	0.48ab
Pseudomonas sp. (% of total bacteria)	1.6b	0.3a	3.7b	1.7b

Measures of abundance: root-surface fungi, mycelium length per unit root surface area; mycorrhiza, estimated intensity, scale 0–4; bacteria, percentage of root surface covered; *Pseudomonas*, numbers as percentage of total bacterial numbers in rhizosphere, separate experiment from other data.

Within any row, any two figures not followed by the same letter are significantly different ($P < 0.05$).

TABLE 9. Percentage (by weight) of roots which were dead, in upper layer of soil under several vegetation types. Live and dead roots separated by hand sorting, except (2) where ^{14}C-labelling used

Site	Dominant species	Percentage of roots dead	Notes and source of data
Prairie, Colorado	*Bouteloua gracilis*	11–42	(1)
		24–31	(2)
Forest, Sweden	*Pinus sylvestris*	34,72	(3)
Chaparral, California	*Adenostoma fasciculatum*	43–63	(4)
	Ceanothus greggii	46–64	

(1) Ares (1976).
(2) Singh & Coleman (1974). $^{14}CO_2$ fed to leaves, labelled roots taken as alive.
(3) Persson (1979). Fine roots only.
(4) Kummerow, Krause & Jow (1978). Fine roots only.

Campbell 1982; Lawley *et al.* 1983). In five out of these seven the root-surface fungal abundance in mixture tended towards the mean of the two monoculture values. No changes in the composition of the fungal flora between mixture and monoculture were detected. In contrast, total bacteria changed less than fungi between mixture and monoculture, but there were changes in composition of the bacterial population. These results support the hypothesis that fungi can ramify from one root to another and can translocate far enough to use materials from both hosts; bacteria are more directly dependent on substrate supply from the nearest root but some of them can migrate from one root to another. As far as I know, no equivalent data are available for woody plants.

These experiments were with plants of the same age, usually about 2–3 months old. In the field, where plants will mostly be older and of varied age, some roots are likely to be dying or dead. Table 9 shows results from three vegetation types which, if representative, indicate that often dead roots comprise a quarter to more than half of all the roots present in the upper layer of soil. Therefore there will be a dead root within a few mm of most living roots in grassland, and within a few cm in forest. If fungi can ramify and bacteria migrate from root to root, then the fungi and bacteria in the rhizosphere of a living root may well be close enough to a dead root to be influenced by it. Fungi isolated from the surface of a living root may be largely dependent for carbon substrate not on that root but on a dead one a few mm away.

SYNTHESIS

The data upon which I have drawn in this paper are limited to a few soils and plant species, and many come from young crop plants. Also there are technical problems which make interpretation of some of the data difficult. The synthesis now presented about the relations between saprophytic micro-organisms and roots is therefore more a hypothesis than a set of conclusions.

In mineral soils a major source of organic materials is dead roots. For every 100 g dry weight of roots dying, soluble exudates from living roots commonly comprise

1–10 g, and root cap plus mucigel from living roots perhaps 2–5 g. At any time dead roots are sufficiently abundant in the upper layers of soil under perennial vegetation to ensure that any living root is only a few mm, or at most a few cm, from a dead one.

In the rhizosphere the biomass of fungi may be either greater or less than that of bacteria. Of the fungi growing close to roots, many can degrade cellulose, pectin and starch. These fungi are probably obtaining most of their carbon substrate not from rhizodeposition by the living root but from dead roots nearby, and from individual dying cells in roots which are still partly alive and functional. For such fungi the habitat spans from one root to another and probably beyond. In contrast, the habitat of a non-motile bacterium extends only micrometres around it; a motile bacterium can alter its environment by swimming along the root or from one root to another, but must ultimately use as substrate the substances which reach its cell surface. Most bacteria living on the root surface or in the rhizosphere are unable to degrade cellulose or pectin and are presumably much dependent on soluble organic materials from the root, or released by exoenzymes of fungi. The growth rate supported by soluble exudates is rapid for only short periods: bacteria soon become so abundant that competition is intense. Bacteria within root tissue, although only a fraction of a mm away, are in a different microenvironment from the rhizosphere and may well be different species able to attack different substrates.

Thus, the classical picture of the rhizosphere applies well to bacteria, occupying a microhabitat a fraction of a mm wide, and dependent on soluble exudates from the root for their main substrate. In contrast fungi, although they form a substantial part of the biomass within the rhizosphere and may well influence other organisms there, e.g. by producing antibiotics or acting as food for animals, are probably obtaining most of their substrate from elsewhere.

ACKNOWLEDGMENTS

I am grateful to Dr J. M. Whipps for sending me data in advance of publication, and to the Natural Environment Research Council for financial support.

REFERENCES

Adler, J. (1973). A method for measuring chemotaxis and use of the method to determine optimum conditions for chemotaxis by *Escherischia coli*. *Journal of General Microbiology*, **74**, 77–91.

Anderson, J.P.E. & Domsch, K.H. (1975). Measurement of bacterial and fungal contributions to respiration of selected agricultural and forest soils. *Canadian Journal of Microbiology*, **21**, 314–322.

Anderson, J.P.E. & Domsch, K.H. (1980). Quantities of plant nutrients in the microbial biomass of selected soils. *Soil Science*, **130**, 211–216.

Ares, J. (1976). Dynamics of the root system of blue grama. *Journal of Range Management*, **29**, 208–213.

Barber, D.A. & Gunn, K.B. (1974). The effect of mechanical forces on the exudation of organic substances by the roots of cereal plants grown under sterile conditions. *New Phytologist*, **73**, 39–45.

Barber, D.A. & Lynch, J.M. (1977). Microbial growth in the rhizosphere. *Soil Biology and Biochemistry*, **9**, 305–308.

Barber, D.A. & Martin, J.K. (1976). The release of organic substances by cereal roots into soil. *New Phytologist*, **76**, 69–80.

Barley, K.P. (1970). The configuration of the root system in relation to nutrient uptake. *Advances in Agronomy*, **22**, 159–201.

Beckel, D.K.B. (1956). Cortical disintegration in the roots of *Bouteloua gracilis* H.B.K. (Lag). *New Phytologist*, **55**, 183–190.

Bennett, R.A. & Lynch, J.M. (1981). Bacterial growth and development in the rhizosphere of gnotobiotic cereal plants. *Journal of General Microbiology*, **125**, 95–102.

Bethlenfalvay, G.J., Brown, M.S. & Pacovsky, R.S. (1982). Relationships between host and endophyte development in mycorrhizal soybeans. *New Phytologist*, **90**, 537–543.

Bowen, G.D. & Rovira, A.D. (1973). Are modelling approaches useful in rhizosphere biology? *Modern Methods in the Study of Microbial Ecology* (Ed. by T. Rosswall). pp. 443–450. Swedish Natural Science Research Council, Stockholm.

Christie, P. (1976). *Interactions between root micro-organisms and grassland plant species in mixtures and monocultures*. Ph.D. Thesis, University of Bristol.

Christie, P., Newman, E.I. & Campbell, R. (1978). The influence of neighbouring grassland plants on each others' endomycorrhizas and root-surface micro-organisms. *Soil Biology and Biochemistry*, **10**, 521–527.

Clowes, F.A.L. (1971). The proportions of cells that divide in root meristems of *Zea mays* L. *Annals of Botany*,**35**, 249–261.

Clowes, F.A.L. (1976). Cell production by root caps. *New Phytologist*, **77**, 399–407.

Clowes, F.A.L. & Woolston, R.E. (1978). Sloughing of root cap cells. *Annals of Botany*, **42**, 83–89.

Coupland, R.T. (Ed.) (1979). *Grassland Ecosystems of the World : Analysis of Grasslands and their Uses*. Cambridge University Press, Cambridge.

Dittmer, H.J. (1938). A quantitative study of the subterranean members of three field grasses. *American Journal of Botany*, **25**, 654–657.

Domsch, K.H., Gams, W. & Anderson, T. (1980). *Compendium of Soil Fungi*. Academic Press, London.

Farrar, J.F. (1981). Respiration rate of barley roots: its relation to growth, substrate supply and the illumination of the shoot. *Annals of Botany*, **48**, 53–63.

Floyd, R.A. & Ohlrogge, A.J. (1971). Gel formation on nodal root surfaces of *Zea mays*. Some observations relevant to understanding its action at the root–soil interface. *Plant and Soil*, **34**, 595–606.

Ford, E.D. & Deans, J.D. (1977). Growth of a Sitka spruce plantation: spatial distribution and seasonal fluctuations of lengths, weights and carbohydrate concentrations of fine roots. *Plant and Soil*, **47**, 463–485.

Foster, R.C. & Rovira, A.D. (1976). Ultrastructure of wheat rhizosphere. *New Phytologist*, **76**, 343–352.

Foster, R.C. & Rovira, A.D. (1978). The ultrastructure of the rhizosphere of *Trifolium subterraneum*. *Microbial Ecology* (ed. by M.W. Loutit & J.A.R. Miles), pp. 278–290. Springer-Verlag, Berlin.

Francis, R. & Read, D.J. (1984). Direct transfer of carbon between plants connected by vesicular-arbuscular mycorrhizal mycelium. *Nature*, **307**, 53–65.

Frankland, J.C. (1975). Estimation of live fungal biomass. *Soil Biology and Biochemistry*, **7**, 339–340.

Greaves, M.P. & Darbyshire, J.F. (1972). The ultrastructure of the mucilaginous layer on plant roots. *Soil Biology and Biochemistry*, **4**, 443–449.

Hansen, G.K. & Jensen, C.R. (1977). Growth and maintenance respiration in whole plants, tops and roots of *Lolium multiflorum*. *Physiologia Plantarum*, **39**, 155–164.

Head, G.C. (1973). Shedding of roots. *Shedding of Plant Parts* (Ed. by T.T. Kozlowski), pp. 237–293. Academic Press, New York.

Henry, C.M. & Deacon, J.W. (1981). Natural (non-pathogenic) death of the cortex of wheat and barley seminal roots, as evidenced by nuclear staining with acridine orange. *Plant and Soil*, **60**, 255–274.

Holden, J. (1975). Use of nuclear staining to assess rates of cell death in cortices of cereal roots. *Soil Biology and Biochemistry*, **7**, 333–334.

Howard, A.J. (1978). Translocation in fungi. *Transactions of the British Mycological Society*, **70**, 265–269.

Johnen, B.G. & Sauerbeck, D.R. (1977). A tracer technique for measuring growth, mass and microbial breakdown of plant roots during vegetation. *Soil Organisms as Components of Ecosystems* (Ed. by U. Lohm & T. Persson), pp. 366–373. Swedish Natural Science Research Council, Stockholm.

Juniper, B.E. & Clowes, F.A.L. (1965). Cytoplasmic organelles and cell growth in root caps. *Nature*, **208**, 864–865.

Kucey, R.M.N. & Paul, E.A. (1982). Biomass of mycorrhizal fungi associated with bean roots. *Soil Biology and Biochemistry*, **14**, 413–414.

Kummerow, J., Krause, D. & Jow, W. (1978). Seasonal changes of fine root density in the southern Californian chaparral. *Oecologia*, **37**, 201–212.

Lambers, H. & Steingröver, E. (1978). Growth respiration of a flood-tolerant and a flood-intolerant *Senecio* species: correlation between calculated and experimental values. *Physiologia Plantarum*, **43**, 219–224.

Lawley, R.A., Campbell, R. & Newman, E.I. (1983). Composition of the bacterial flora of the rhizosphere of three grassland plants grown separately and in mixtures. *Soil Biology and Biochemistry*, **15**, 605–607.

Lawley, R.A., Newman, E.I. & Campbell, R. (1982). Abundance of endomycorrhizas and root-surface micro-organisms on three grasses grown separately and in mixtures. *Soil Biology and Biochemistry*, **14**, 237–240.

Lundgren, B. (1981). Fluorescein diacetate as a stain of metabolically active bacteria in soil. *Oikos*, **36**, 17–22.

Lyford, W.H. (1975). Rhizography of non-woody roots of trees in the forest floor. *The Development and Function of Roots* (Ed. by J.G. Torrey & D.T. Clarkson), pp. 179–196. Academic Press, London.

McIlwaine, R.S. & Malone, J.P. (1976). Effects of chloropicrin soil treatments on the microflora of soil and ryegrass roots and on ryegrass yield. *Transactions of the British Mycological Society*, **67**, 113–120.

Martin, J.K. (1977a). Factors influencing the loss of organic carbon from wheat roots. *Soil Biology and Biochemistry*, **9**, 1–7.

Martin, J.K. (1977b). The chemical nature of the carbon-14-labelled organic matter released into soil from growing wheat roots. *Soil Organic Matter Studies*, Vol. 1, pp. 197–203. International Atomic Energy Agency, Vienna.

Nakas, J.P. & Klein, D.A. (1980). Mineralization capability of bacteria and fungi from the rhizosphere–rhizoplane of a semiarid grassland. *Applied and Environmental Microbiology*, **39**, 113–117.

Neal, J.L., Larsen, R.I. & Atkinson, T.G. (1973). Changes in rhizosphere populations of selected physiological groups of bacteria related to substitution of specific pairs of chromosomes in spring wheat. *Plant and Soil*, **39**, 209–212.

Newman, E.I. (1978). Root microorganisms: their significance in the ecosystem. *Biological Reviews*, **53**, 511–554.

Newman, E.I., Heap, A.J. & Lawley, R.A. (1981). Abundance of mycorrhizas and root-surface micro-organisms of *Plantago lanceolata* in relation to soil and vegetation: a multi-variate approach. *New Phytologist*, **89**, 95–108.

Newman, E.I. & Watson, A. (1977). Microbial abundance in the rhizosphere: a computer model. *Plant and Soil*, **48**, 17–56.

Papavizas, G.C. & Davey, C.B. (1961). Extent and nature of the rhizosphere of *Lupinus*. *Plant and Soil*, **14**, 215–236.

Pavlychenko, T.K. (1942). *Root systems of certain forage crops in relation to management of agricultural soils.* Publication of the National Research Council of Canada, no. 1088.

Persson, H. (1978). Root dynamics in a young Scots pine stand in Central Sweden. *Oikos*, **30**, 508–519.

Persson, H. (1979). Fine-root production, mortality and decomposition in forest ecocystems. *Vegetatio*, **41**, 101–109.

Phillips, H.L. & Torrey, J.G. (1971). Deoxyribonucleic acid synthesis in root cap cells of cultured roots of *Convolvulus*. *Plant Physiology*, **48**, 213–218.

Prikryl, Z. & Vancura, V. (1980). Root exudates of plants, VI. Wheat root exudation as dependent on growth, concentration gradient of exudates and the presence of bacteria. *Plant and Soil*, **57**, 69–83.

Reynolds, E.R.C. (1970). Root distribution and the cause of its spatial variability in *Pseudotsuga taxifolia* (Poir.) Britt. *Plant and Soil*, **32**, 501–517.

Roberts, J. (1976). A study of root distribution and growth in a *Pinus sylvestris* L. (Scots pine) plantation in East Anglia. *Plant and Soil*, **44**, 607–621.

Rouatt, J.W., Katznelson, H. & Payne, T.M.B. (1960). Statistical evaluation of the rhizosphere effect. *Soil Science Society of America Proceedings*, **24**, 271–273.

Rovira, A.D. (1969). Plant root exudates. *Botanical Review*, **35**, 35–57.

Shields, J.A., Paul, E.A. & Lowe, W.E. (1973). Turnover of microbial tissue in soil under field conditions. *Soil Biology and Biochemistry*, **5**, 753–764.

Singh, J.S. & Coleman, D.C. (1974). Distribution of photo-assimilated ^{14}carbon in the root system of a shortgrass prairie. *Journal of Ecology*, **62**, 359–365.

Smith, J.L. & Doetsch, R.N. (1969). Studies on negative chemotaxis and the survival value of motility in *Pseudomonas fluorescens*. *Journal of General Microbiology*, **55**, 379–391.

Söderström, B.E. (1979). Seasonal fluctuations of active fungal biomass in horizons of a podzolised pine-forest in Central Sweden. *Soil Biology and Biochemistry*, **11**, 149–154.

Taylor, G.S. & Parkinson, D. (1965). Studies on fungi in the root region, IV. Fungi associated with the roots of *Phaseolus vulgaris* L. *Plant and Soil*, **22**, 1–20.

Thornton, R.H. (1965). Studies of fungi in pasture soils, I. Fungi associated with live roots. *New Zealand Journal of Agricultural Research*, **8**, 417–449.

Thrower, S.L. & Thrower, L.B. (1961). Transport of carbon in fungal mycelium. *Nature*, **190**, 823–824.

Turner, S.M. (1983). *Effects of Nitrogen and Phosphorus on Micro-organisms of the Rhizoplane*. Ph.D. Thesis, University of Bristol.

Turner, S.M. & Newman, E.I. (1984). Growth of bacteria on roots of grasses: influence of mineral nutrient supply and interations between species. *Journal of General Microbiology*, **130**, 505–512.

Vaituzis, Z. & Doetsch, R.N. (1969). Motility tracks: technique for quantitative study of bacterial movement. *Applied Microbiology*, **17**, 584–588.

Van Vuurde, J.W.L. & Schippers, B. (1980). Bacterial colonization of seminal wheat roots. *Soil Biology and Biochemistry*, **12**, 559–565.

Vancura, V. & Kunec, F. (1977). The effect of streptomycin and actidione on respiration in the rhizosphere and non-rhizosphere soil. *Zentralblatt für Bakteriologie*, II, **132**, 472–478.

Vancura, V., Prikryl, Z., Kalanchova, L. & Wurst, M. (1977). Some quantitative aspects of root exudation. *Soil Organisms as Components of Ecosystems* (Ed. by U. Lohm & T. Persson), pp. 381–386. Swedish Natural Science Reseach Council, Stockholm.

Waid, J.S. (1957). Distribution of fungi within the decomposing tissues of ryegrass roots. *Transactions of the British Mycological Society*, **40**, 391–406.

Whipps, J.M. (1984). Environmental factors affecting the loss of carbon from the roots of wheat and barley seedlings. *Journal of Experimental Botany*, **35**, 767–773.

Whipps, J.M. & Lynch, J.M. (1983). Substrate flow and utilisation in the rhizosphere of cereals. *New Phytologist*, **95**, 605–623.

The microbial biomass in soil

P. C. BROOKES, D. S. POWLSON AND D. S. JENKINSON
Soils and Plant Nutrition Department, Rothamsted Experimental Station, Harpenden, Herts, AL5 2JQ

The soil microbial biomass consists of many species of bacteria and fungi, together with larger soil organisms such as yeasts, algae and protozoa. Although each species has a particular role in soil, for many purposes the biomass can be considered as a single compartment.

Jenkinson (1966) and Jenkinson & Powlson (1976) showed that increased CO_2-C evolution and O_2 consumption occurred when soil was fumigated with $CHCl_3$, the fumigant removed and the soil then incubated aerobically. The increase was due to mineralization of the killed microbial cells by the microbial population subsequently recolonizing the soil. The amount of biomass carbon in the original soil could then be estimated from the relationship: Biomass $C = F/k_c$ where F is the CO_2-C evolved from the fumigated soil, less that from a similarly incubated, but non-fumigated, control soil, and k_c is the fraction of carbon in the killed cells evolved as CO_2-C in the 10-day incubation period (determined experimentally and taken as 0·45 for soils incubated at 25°C). Close agreement was found between the amount of biomass measured by the chloroform-fumigation method and by direct observation of stained microbial tissue in soil, except in strongly acid soils (<pH 4·5), where the chloroform-fumigation method gave low values, for reasons which are still uncertain.

The chloroform-fumigation method showed that the soil microbial biomass is much larger than previously realized. For example, the plough layer of the unmanured plot of the Broadbalk Continuous Wheat Experiment, which has grown wheat for about 140 years, contains more than 0·5 Mg ha^{-1} of microbial biomass carbon. Converted into fresh weight, this depth of soil contains about 5 Mg ha^{-1} of living microbial biomass, approximately the weight of 100 sheep.

Both because of the large amount of biomass in soil and also because it is a much more labile component of soil organic matter than most other fractions (Jenkinson & Ladd 1981) the biomass is an important reservoir of plant nutrients, especially N and P. In order to use N fertilizer most efficiently it is necessary to understand the biological processes controlling the release of native nitrogen from soil. The biomass plays a key role in this and accurate measurements of the size and turnover of the biomass are important in understanding the supply of soil N to crops.

Methods for measuring soil microbial biomass N (Shen, Pruden & Jenkinson 1984), P (Brookes, Powlson & Jenkinson 1982; Hedley & Stewart 1982) and S (Saggar, Bettany & Stewart 1981) have been developed and used to calculate both the amounts in, and the 'flux' of these nutrients through, the biomass. In grassland soils, the annual flux of N and P through the soil biomass is equal to, or greater than, the offtakes of these nutrients by the grass (Brookes, Powlson & Jenkinson

123

1984). In most arable soils the flux tends to be less. Not all of this flux is available to plants. Some will be utilized by the next generation of microbes, some will be in forms unavailable to plants and some (especially P) will be fixed by the soil. We do not yet know the relative importance of these processes. However, models for the turnover of plant nutrients in soil only match experimental data successfully when the flux of nutrients through the biomass is taken into account (e.g. Halm, Stewart & Halstead 1972).

The biomass is a sensitive indicator of changes in soil due to changing management practices. For example, when grassland is ploughed in, much of the organic P subsequently mineralized appears to be derived from the biomass (Brookes, Powlson & Jenkinson 1984). Long-term application of metal-contaminated sewage-sludge to agricultural land has been shown to reduce the size of the biomass considerably (Brookes & McGrath 1984). The significance of this effect for long-term soil fertility is not yet known.

The large size of the biomass is surprising in view of the relatively low carbon inputs into soil. For example, the annual input of organic carbon into the plough-layer of the unmanured plot on Broadbalk is about $1\cdot2$ Mg ha^{-1}, little more than twice the carbon contained in the cells of the microbial biomass (Jenkinson & Ladd 1981). As a response to the low substrate availability in soil, the biomass exhibits many characteristics typical of a mainly dormant population, e.g. low maintenance energy, low respiration rate and slow mean cell-division rate (about once every $2\cdot5$ years: Jenkinson & Ladd 1981). Yet the biomass has an ATP content (Jenkinson, Davidson & Powlson 1979) and adenylate energy charge (AEC) (Brookes, Tate & Jenkinson 1983) characteristic of organisms in active growth *in vitro*. The AEC takes into account the relative proportions of ATP, ADP and AMP in the living cell. The AEC of actively growing cells *in vitro* is high ($0\cdot8$–$0\cdot95$), compared to that of spores (about $0\cdot1$). The AEC measured in a grassland soil was high ($0\cdot85$), again, as with ATP, comparable to that found in actively growing micro-organisms. There is a paradox here, as the largely dormant soil population appears to behave, at least in terms of ATP content and AEC, similarly to organisms living in conditions where food is plentiful.

Possibly the maintenance of a high AEC and a high ATP content is an evolutionary response to the way substrate enters soil. Soils contain much organic matter that is highly resistant to microbial attack, though there is some evidence of a slow trickle of nutrients from this material. At intervals, this meagre supply is supplemented by more substantial pulses of food from dead roots and root exudates. Perhaps under these conditions, a survival strategy based on high AEC and ATP may be more efficient than (say) a spore strategy for the soil biomass.

REFERENCES

Brookes, P.C. & McGrath, S.P. (1984). Effects of metal-toxicity on the soil microbial biomass. *Journal of Soil Science*, **35**, 341–346.

Brookes, P.C., Powlson, D.S. & Jenkinson, D.S. (1982). Measurement of microbial biomass phosphorus in soil. *Soil Biology and Biochemistry*, **14**, 319–329.

Brookes, P.C., Powlson, D.S. & Jenkinson, D.S. (1984). Phosphorus in the soil microbial biomass. *Soil Biology and Biochemistry*, **16**, 169–173.

Brookes, P.C., Tate, K.R. & Jenkinson, D.S. (1983). The adenylate energy charge of the soil microbial biomass. *Soil Biology and Biochemistry*, **15**, 9–16.

Halm, B.J., Stewart, J.W.B. & Halstead, R.L. (1972). The phosphorus cycle in a native grassland ecosytem. *Isotopes and Radiation in Soil–Plant Relationships Including Forestry*, pp. 571–586. SM151/7, IAEA, Vienna.

Hedley, M.J. & Stewart, J.W.B. (1982). Method to measure microbial phosphate in soils. *Soil Biology and Biochemistry*, **14**, 377–385.

Jenkinson, D.S. (1966). Studies on the decomposition of plant material in soil, II. Partial sterilization of soil and the soil biomass. *Journal of Soil Science*, **17**, 280–302.

Jenkinson, D.S., Davidson, S.A. & Powlson, D.S. (1979). Adenosine triphosphate and microbial biomass in soil. *Soil Biology and Biochemistry*, **11**, 521–527.

Jenkinson, D.S. & Ladd, J.N. (1981). Microbial biomass in soil: measurement and turnover. *Soil Biochemistry*, Vol.5 (Ed. by E.A. Paul and J.N. Ladd), pp. 415–471. Marcel Dekker, New York.

Jenkinson, D.S. & Powlson, D.S. (1976). The effects of biocidal treatments on metabolism in soil, V. A method for measuring soil biomass. *Soil Biology and Biochemistry*, **8**, 209–213.

Saggar, S., Bettany, J.R. & Stewart, J.W.B. (1981). Measurement of microbial sulphur in soil, *Soil Biology and Biochemistry*, **13**, 493–498.

Shen, S.M., Pruden, G. & Jenkinson, D.S. (1984). Mineralization and immobilization of nitrogen in fumigated soil and the measurement of mocrobial biomass nitrogen. *Soil Biology and Biochemistry*, (in press).

Nitrogen fixation in the rhizosphere: significance in natural and agricultural systems

K. E. GILLER AND J. M. DAY

Rothamsted Experimental Station, Harpenden, Herts

SUMMARY

1 Over long periods many ecosystems accumulate nitrogen at a rate which cannot be attributed to abiotic and symbiotic sources alone. Nitrogen-fixing bacteria are ubiquitous in soils, but the significance of their contribution to the soil nitrogen balance in natural and agricultural systems is a subject of controversy.

2 In this paper we assess evidence concerning the contribution of bacterial rhizosphere nitrogen fixation to the nitrogen balance of soil and vegetation in both temperate and tropical regions. The specificity of plant–bacterial associations, potential energy sources for nitrogen fixation and estimates of the amount of nitrogen fixed in different environments are considered.

3 Accumulating evidence suggests that the contribution of rhizosphere nitrogen fixation to the nitrogen budget of plants is small. In systems where levels are low enough to limit plant production a small input may be valuable.

INTRODUCTION

The atmosphere is the primary source of nitrogen in soils. Despite increasing application of nitrogen fertilizers the principal route for the entry of atmospheric nitrogen into soil is biological nitrogen fixation (Stevenson 1982). Nitrogen is a major nutrient required for plant growth and low nitrogen availability is frequently the predominant factor limiting production in plant communities (Lee, Harmer & Ignaciuk 1982) or governing the rate of ecosystem development (Marrs *et al.* 1982). Where nitrogen is scarce, its availability may depend on the rate of biological nitrogen fixation.

Gorham, Vitousek & Reiners (1979) suggested that the importance of nitrogen fixation within a particular ecosystem may be linked to the stage of ecosystem development. Phototrophic nitrogen fixation is likely to be important in early seres, especially in the colonization of skeletal habitats where the efficiency of nutrient capture is low (Reiners 1981). It may also play an important role in secondary succession (e.g. after fire) where nitrogen has been lost but other nutrients are abundant (Reiners 1981). Nitrogen fixation by asymbiotic heterotrophic microorganisms may be important when there is an accumulation of organic matter which is utilizable as a carbon source, for instance in mature forests with an abundance of decaying wood.

TABLE 1. Gains of nitrogen to soil under different vegetation; after Moore (1966) with some additional values

Vegetation	Location	Duration of study (years)	Net positive nitrogen balance (kg ha^{-1} year^{-1})	Reference
Grasslands				
Bluegrass	Missouri, USA	8	114	Whitt (1941)
Kentucky bluegrass	Pennsylvania, USA	18	17	White, Holben & Richer (1945)
Lolium rigidum	Morredin, Australia	3	25	Parker (1957)
Dactylis glomerata	Hurley, UK	9	78	Clement & Williams (1967)
Cynodon plechostachys	Ibadan, Nigeria	3	90	Jaiyebo & Moore (1963)
Savanna	Ghana	Calculated	2–16	Greenland & Nye (1959)
Mixed herbaceous veg.	Broadbalk Wilderness, UK	81	49	Jenkinson (1977)
Agriculture				
Wheat	Broadbalk, Rothamsted, UK	115	41	Jenkinson (1977)
Rice	Thailand	2	40	Firth *et al.* (1973)
Woodlands				
Pinus sylvestris	Tentsmuir sands, UK	11–19	50	Ovington (1951)
Monocultural tree stands	Bedgebury, UK	20	58	Ovington (1956)
Mixed woodland	Broadbalk Rothamsted, UK	81	53	Jenkinson (1971)
Mixed woodland	Geescroft Rothamsted, UK	82	15	Jenkinson (1971)
Regenerating rainforest	Ibadan, Nigeria	3	594	Jaiyebo & Moore (1963)
Lowland rainforest	Ghana	Calculated	6–168	Greenland & Nye (1959)
Highland rainforest	Ghana	Calculated	17–45	Greenland & Nye (1959)

In many mid-successional stages where legumes are absent, appreciable amounts of nitrogen nevertheless accumulate (Table 1). Under grasslands, for instance, there is often a concomitant gradual accumulation of organic carbon but this is normally not sufficient to maintain the very high C:N ratios required to favour the growth of free-living nitrogen-fixing bacteria in the soil. The situation is similar in agricultural soils where an accumulation of carbon and nitrogen has been measured in the absence of legumes (Jenkinson 1977). Growth of agricultural crops may be likened to a rotation of short-term secondary successions, where there are often flush releases of organic matter which favour nitrogen fixation by free-living bacteria. The plant rhizosphere provides a potentially suitable niche for nitrogen fixation due to the enrichment of the soil with carbon from roots.

Nitrogen balance studies indicate that annual net gains of nitrogen in the absence of legumes may be as high as $114 \, kg \, N \, ha^{-1}$ in grasslands and $594 \, kg \, N \, ha^{-1}$ in forest ecosystems (Table 1). This latter figure is exceptionally high but several instances are described where annual gains of nitrogen amount to $50 \, N \, kg \, ha^{-1}$ in woodlands. Nitrogen is accumulated in these ecosystems in excess of the amounts lost by plant removal, leaching and denitrification. Abiotic sources of nitrogen include precipitation, dust and dry ammonia absorption and the amount gained may vary with successional development. The impaction and absorption of dust or ammonia for example will be greater in the presence of a more extensive plant canopy (Gorham et al. 1979). Contributions from biological nitrogen fixation may come from free-living heterotrophic bacteria or from phototrophic micro-organisms such as lichens and blue-green and other phototrophic bacteria, whose activity is limited to environments where they are able to intercept light. A possible large contribution of nitrogen from rhizosphere-associated nitrogen fixation was suggested by the discovery of high rates of acetylene reduction activity by soil cores containing roots of Paspalum notatum (Dobereiner, Day & Dart 1972). Heterotrophic nitrogen-fixing bacteria were subsequently isolated from the rhizospheres of many tropical grasses. Further indications of the possible role of rhizosphere stimulated or 'associative' nitrogen fixation came from significant although variable increases in dry matter and nitrogen content in response to the inoculation of crops with nitrogen-fixing bacteria (e.g. Smith et al. 1976). This evidence led to considerable interest in the exploitation of heterotrophic nitrogen-fixing bacteria in agriculture because of the potential economic benefits.

In order to assess or enhance contributions from nitrogen fixation, accurate measurements of the nitrogen inputs are required. The lack of methods which can accurately discriminate between contributions from various nitrogen sources in the field has limited our understanding and led to the present controversy (McBride 1983) over the importance and potential of nitrogen fixation in the rhizosphere. The magnitude of the nitrogen contribution by free-living nitrogen-fixing bacteria in various environments is considered here. Particular attention is paid to the interactions between free-living nitrogen-fixing bacteria and the roots of plants.

salt marsh grass *Spartina* (McClung & Patriquin 1980). Evidence of specific associations between bacteria and plants from bacterial strain and plant species interactions observed in inoculation responses (Reynders & Vlassak 1982) is also only
indicative of a casual assocation. The pH of the rhizosphere has been shown to vary
considerably between genotypes of barley (Brown & Bell 1969) which may cause
the composition of rhizosphere flora to vary within species.

ROOT CARBON AS AN ENERGY SOURCE FOR NITROGEN-FIXING BACTERIA

Amounts of nitrogen fixed in the soil are limited by the availability of carbohydrate.
Carbon lost from both living and dead plant roots is a major potential substrate for
soil micro-organisms. The amount of nitrogen fixed in soil is governed by both the
amount of carbon available and by the ability of the nitrogen-fixing bacteria to
capture and utilize the energy source. Addition of carbon substrates is frequently
found to increase nitrogen fixation by heterotrophic bacteria in soil (e.g. Rao 1978)
and increase the amount of fixed nitrogen incorporated into tropical grasses (De-
Polli *et al.* 1977) as measured with $^{15}N_2$.

Carbon losses from roots

Carbon is lost from roots as simple sugars, alcohols and polysaccharides as well as
more complex structural polysaccharides, hemicelluloses and celluloses. Losses
occur by root exudation, by degradation and lysis of cortical cells by material
sloughed off from growing roots and by root death. Mechanisms of loss of root
carbon and the amounts and composition of carbon available from roots are
reviewed by Newman (p. 107).

It is estimated that up to 30% of total photosynthate is translocated below
ground in wheat and barley (Barber & Martin 1976), and in maize (Helal &
Sauerbeck 1983). In grasslands 50% of total photosynthate may be transferred to
the roots (Warembourg & Paul 1977). It is difficult to estimate the quantity of
carbon available for microbial growth in the rhizosphere of plants. Recent evidence
suggests that little of this carbon is present as soluble root exudates (Whipps &
Lynch 1983) and the main carbon addition to the soil appears to be from root
death (Newman, p. 107).

Efficiency of nitrogen fixation

The efficiency of carbon utilization for nitrogen fixation by free-living bacteria
varies considerably and is dependent on the carbon source and incubation conditions. Measured efficiencies vary from 4 g C g N fixed^{-1} in *Azospirillum brasilense* to
174 g C g N fixed^{-1} in *Aerobacter aerogens* (Table 2). Of the free-living nitrogen-
fixing bacteria studied none can utilize polysaccharides or other more complex
carbon compounds for growth directly (Brown 1982). The activity of nitrogen-

TABLE 2. Efficiency of free-living nitrogen-fixing bacteria on various carbon sources. (Incubations conducted under aerobic conditions, usually with reduced pO_2)

	Efficiency g C consumed/ g N fixed	Carbon source	Reference
Aerobacter aerogenes	89–174	glucose	Jensen (1956)
Derxia gummosa	16–20	glucose	Jensen *et al.* (1960)
Azotobacter chroococcum	11–32	mannitol	Dalton & Postgate (1969)
Azospirillum brasilense	4–14	lactate	Dobereiner & Day (1976)
Azospirillum brasilense	8–34	malate	Dobereiner & Day (1976)
Azospirillum brasilense	36–45	malate	Okon *et al.* (1977)

fixing bacteria in the rhizosphere will depend on their ability to capture and utilize simple carbon compounds in exudates and root carbohydrates which have been broken down by other micro-organisms.

There is an added complication in understanding carbon flow in the rhizosphere in that little is known of the interactions between mycorrhizas and the amounts of carbon available for utilization by soil bacteria.

Maximal rates of nitrogen fixation

Whilst estimates of the maximum possible nitrogen fixation for a given amount of root carbon involve assumptions and inaccuracies, they can provide an indication of the potential for nitrogen fixation. If it is assumed that: (i) the efficiency of conversion of root carbon is 10 g C g N fixed^{-1}; (ii) nitrogen-fixing bacteria comprise 10% of the total rhizosphere population (a high estimate) and acquire an amount of available carbon in proportion to their numbers; (iii) *all* of the carbon translocated below ground is *equally* available for use by all bacteria; then if 1500 kg C ha^{-1} are translocated below ground as estimated for a wheat crop (Martin & Puckridge 1982) the maximum potential nitrogen fixation is 15 kg N ha^{-1}.

It has been suggested that the potential for associative nitrogen fixation is much higher in tropical grasses and cereals which possess the C_4 photosynthetic pathway (Brown 1982). Estimates for the amount of carbon translocated below ground in C_4 plants are not available, and therefore similar calculations are not possible at present.

The above estimates have assumed that all of the carbon translocated below ground is available for use by soil bacteria. Since much of it is contributed to the soil as complex compounds it is likely that turnover and mineralization will govern the amount of nitrogen fixed during the growth of a crop. Root carbon may therefore be used as a substrate for nitrogen fixation over periods after that in which it is produced. In effect, rates of nitrogen fixation may be limited by the rate of breakdown of organic matter.

Similar calculations based solely on estimates of root exudation led Barber & Lynch (1977) and later Beck & Gilmour (1983) to conclude that the potential for fixation of significant amounts of nitrogen in the rhizosphere of cereals was very low.

MEASUREMENTS OF RHIZOSPHERE-ASSOCIATED NITROGEN FIXATION

Estimates based on the acetylene reduction assay

Apart from the evidence gained in nitrogen balance studies (Table 1), most estimates of rhizosphere-associated fixation have been based on the acetylene reduction assay (Hardy, Burns & Holsten 1973). Extrapolation from these indirect, point measurements of nitrogenase activity to estimates of nitrogen fixed is fraught with problems.

The two most widely used methods of sample incubation employed for estimation of acetylene reduction activity (ARA) in the rhizosphere are incubation of soil cores (Day *et al.* 1975) and incubation of excised roots (Dobereiner & Day 1976). There are a few reports of *in situ* ARA measurements (e.g. Balandreau, Miller & Dommergues 1974). Although measurements on removed samples are intended to provide estimates of the rates of nitrogenase activity in the field, there are often long delays between the time of sample collection and exposure to acetylene, during which samples are incubated under different conditions. Bacterial multiplication has been demonstrated to occur on the excised roots in the 16-hour preincubation period commonly used (van Berkhum 1980) and, as a consequence, it has been suggested that only linear and instantaneous rates of acetylene reduction should be taken to indicate nitrogen fixation in the field (van Berkhum & Sloger 1979).

Estimates of ARA with excised root or soil core incubations have indicated very different amounts of nitrogen fixation, the excised root assay generally showing higher amounts (van Berkhum & Bohlool 1980). David & Fay (1977) found that long incubations of nitrogen-fixing bacteria in the presence of acetylene led to increased rates of nitrogen fixation. Incubation under acetylene has also been shown to stimulate nitrogenase synthesis (Ela, Anderson & Brill 1982). Further problems such as gas diffusion and equilibration through soils have been encountered and these were comprehensively reviewed by van Berkhum & Bohlool (1980). These problems are equally important with assays on intact soil–plant root systems in pots (e.g. Ela *et al.* 1982) especially if long or repeated incubations are carried out.

A major problem in applying ARA to measuring rhizosphere-associated nitrogen fixation is that acetylene inhibits the oxidation of ethylene which is a normal soil process (Nohrstedt 1975/6; Witty 1979). Despite the knowledge that ethylene production in the presence of acetylene can cause large and variable overestimates of ARA (Witty 1979; Nohrstedt 1983), little research has been done to quantify the amounts of ethylene produced in different experimental systems and its effects on potential estimates of nitrogen fixation. A particular danger is that the conditions which stimulate ARA (e.g. high temperature and soil moisture) are also conducive to ethylene production (Goodlass & Smith 1978).

Consideration and confirmation of the above problems by Lethbridge *et al.* (1982) led them to recommend that ARA should be used only as a qualitative indication of nitrogen fixation and then only with appropriate controls. This view has been supported by other workers (Rennie & Rennie 1983). It is now obvious that many of the early ARA measurements of nitrogen fixation in the rhizosphere gave rise to gross overestimates. These were based on a 3:1 or 4:1 conversion ratio for C_2H_2 reduced to N_2 fixed. Figures derived using this ratio as high as $2\cdot4$ kg N ha^{-1} day^{-1} fixed had been suggested (von Bulow & Dobereiner 1975). It is evident that estimates of nitrogen fixation in soils and plant–soil systems obtained by ARA must be interpreted with extreme caution. In the absence of proper controls ARA estimates cannot even be assumed to indicate the presence of active nitrogen-fixing bacteria.

Estimates based on isotope dilution methods

Over 99% of the nitrogen in air is $^{14}N_2$. If a plant with no associated nitrogen fixation is grown solely on ^{15}N-labelled fertilizer, there will be the same enrichment of ^{15}N in the plant tissues at harvest as in that fed as fertilizer. A plant which had taken up biologically-fixed nitrogen would have a lower tissue enrichment of ^{15}N at harvest than one fed exclusively on fertilizer. If plants with and without associated nitrogen fixation are grown in soil amended with ^{15}N labelled fertilizer they both will have a tissue enrichment of ^{15}N below that present in the fertilizer. When appreciable amounts of biologically-fixed nitrogen are taken up by the test plant the final ^{15}N enrichment will be lower than that in the control plant. The amount of nitrogen fixed in the test plant can then be calculated *providing that the proportion of soil and fertilizer nitrogen taken up is the same* in control and test plants (Rennie, Rennie & Fried 1978).

Few studies have been carried out using these techniques to date. Rennie (1980) demonstrated significant isotope dilution with N-deficient maize seedlings grown in test tubes using plants not inoculated with nitrogen-fixing bacteria as controls. A small decrease in ^{15}N enrichment was detected in plants without added carbon, suggesting that the bacteria were using root C as a substrate.

In a series of carefully controlled glasshouse experiments Lethbridge & Davison (1983a) investigated the contribution of several bacterial species to the nitrogen nutrition of wheat. In one experiment, wheat inoculated either with viable or heat-killed inoculum was grown in sand fertilized with $Ca(^{15}NO_3)_2$ (99% ^{15}N). Bacterial inocula and seed were also labelled with ^{15}N (>99% ^{15}N) and clover inoculated with *Rhizobium* was included to check the method. Nitrogen fixation was only detected with the clover plants (Table 3). Further experiments carried out in ^{15}N-labelled soil indicated that root-associated nitrogen fixation was neglible. Only when carbohydrate was added to plants grown under monoxenic sand culture was significant isotope dilution detected in wheat. Lethbridge & Davison (1983)

TABLE 3. Comparison of the response to monoxenic inoculation of wheat with free-living nitrogen-fixing bacteria and white clover with *R. trifolii* as measured by ^{15}N dilution after 6 weeks growth. (Adapted from Lethbridge & Davison 1983a)

	%[15]N Shoots	
Plant and inoculum	Dead inoculum	Live inoculum
Wheat and *Azotobacter beijerinckii*	93·8 ± 3.1	97·6 ± 3·1
Wheat and *Azospirillum brasilense* sp. 107	94·3 ± 1·0	95·0 ± 0·8
Wheat and *Bacillus polymyxa*	95·8 ± 1·2	95·7 ± 1·4
Clover and *Rhizobium trifolii*	95·0 ± 0·6	81·4 ± 0·6***

Values are means ± standard errors of four or five replicates.
*** Significantly different ($P < 0.001$) from the heat-killed control.

concluded that the method they employed would detect a contribution from fixed nitrogen if it were in excess of 5% of total plant nitrogen. Isotope dilution estimates indicating a significant contribution of root-associated nitrogen fixation in wheat have been reported (Rennie *et al.* 1983). Uninoculated controls, rather than controls receiving heat-killed inocula, were used, although the nitrogen content of the inoculum quoted was insufficient to account for the differences in isotope dilution. However, in one cultivar the uninoculated control had a significantly lower enrichment than the inoculated treatments, a result which is difficult to understand unless the control was highly contaminated with nitrogen-fixing bacteria or the ^{15}N fertilizer nitrogen added and other ^{14}N in the medium was not evenly available in all treatments.

Recent reviews on the methodology of measurements of nitrogen fixation associated with non-legumes (Rennie & Rennie 1983; Fried, Danso & Zapata 1983) agree that isotope dilution methods are suitable for field measurements. Fried, Danso & Zapata (1983) suggest that the method can only be used to demonstrate amounts of fixed N taken up above 10–15% of the total plant N. Both of these reviews acknowledge the methodological problems involved in selection of suitable control plants for isotope dilution experiments in the field, which have been elegantly explained by Witty (1983) using legume systems. Any difference in the rooting pattern of the control and test plants can lead to differences in enrichment of the plant tissues at harvest if the soil on which they are grown is not evenly labelled, both with depth and in time. It is difficult to see how these constraints can be adequately satisfied without a thorough knowledge of the distribution of actively absorbing roots throughout the growth period of both control and test plants. Chalk, Douglas & Buchanan (1983) suggest that a control plant is unnecessary if measurement of the ^{15}N enrichment of mineralized soil nitrogen are made but this would only be true if (i) all nitrogen fixation was very close to the roots and had no effect on the ^{15}N enrichment of mineralized N; (ii) the ^{15}N-labelling of the soil was constant with depth and in time; (iii) mineralization results in an even distribution of ^{15}N with depth (i.e., mineralization rates are equal in all soil horizons). It would appear that it is no easier to satisfy these criteria than it is to select an appropriate non-fixing control plant.

Estimates have been made of the fixed nitrogen absorbed by *Paspalum notatum* grown in large concrete cylinders and enclosed at the base (Boddey *et al.* 1983). ^{15}N-enriched fertilizer was added in split doses throughout the year in an attempt to reduce variation in the amount of fertilizer available with time. Benefits from nitrogen fixation in *P. notatum* cv. batatais were calculated with reference to two controls with low rhizosphere ARA: *P. notatum* cv. pensacola and *P. maritimum*. It was estimated that *P. notatum* cv. batatais derived approximately 20 kg N ha^{-1} year^{-1} (11% of total plant N accumulated) from uptake of plant-associated fixed nitrogen. Differences in results obtained between the two controls, the differential penetration of roots to lower soil horizons and more importantly the different soil enrichments and patterns of root distribution of the control and test plants in the upper soil horizons made interpretation of these results difficult. However, isotope dilution and total N difference estimates were in broad agreement in this study. It is likely that isotope dilution methods cannot accurately measure the low level of nitrogen fixation in the rhizosphere in the field.

Measurements based on ^{15}N$_2$ incorporation

Detection of ^{15}N$_2$ incorporated into plant tissues provides clear proof of uptake of fixed nitrogen by the plant. Unfortunately, due to the expense of isotopically enriched nitrogen, the low rates of nitrogen fixation under examination and the problems of maintenance of ambient gas concentrations in enclosed systems, it is only practical to use ^{15}N$_2$ for short-term assays and under controlled conditions. An attempt to investigate *in situ* nitrogen fixation associated with sugar-cane (*Saccharum* sp.) resulted in an increase in soil enrichment, but no enrichment of ^{15}N in the plant (Matsui *et al.* 1981).

Porter & Grable (1969) found ^{15}N in leaves of *Juncus* and *Carex* spp. from wet mountain meadows following incubation of complete sodmats in an atmosphere of ^{15}N$_2$ in the dark. However, the soil became anaerobic in the course of the experiment and this may have led to increased rates of nitrogen fixation by anaerobic bacteria. Other investigations have used incubation vessels with a varying degree of control of gas concentrations. Incorporation of fixed ^{15}N$_2$ has been demonstrated in the tropical grasses *Digitaria decumbens* and *Paspalum notatum* (De-Polli *et al.* 1977), sugar-cane *Saccharum* sp. (Ruschel, Henis & Salati 1975; Ruschel *et al.* 1981) rice *Oryza sativa* (Yoshida & Yoneyama 1980; Ito, Cabrera & Watanabe 1980; Eskew, Eaglesham & App 1981) and sorghum *Sorghum bicolor* (Giller *et al.* 1984). Whilst these experiments demonstrate the presence of active N-fixing organisms in the rooting medium and absorption by the plant, a close rhizosphere association of nitrogen fixation is not demonstrated. However, Ela *et al.* (1982) demonstrated ^{15}N$_2$ incorporation on or within excised root segments of maize. Nitrogen fixed by bacteria may be made available for plant uptake by active excretion or passive loss of NH$_4$ by the bacteria or by lysis and degradation of dead cells. In all of these cases the root must compete for available nitrogen with micro-organisms in the rhizosphere and therefore rates of uptake will depend on the activity of the microflora, in the third case particularly on mineralization rates.

In some experiments incorporation of fixed nitrogen into plant tops was very rapid (e.g. Giller *et al.* 1984) but this did not necessarily indicate that the bacteria were excreting N, because cell death and degradation could also be rapid. Indeed the small amounts of labelled nitrogen absorbed may not have been derived solely from nitrogen fixed in the rhizosphere. Yoshida & Yoneyama (1980) detected that up to 25% of the fixed ^{15}N was in rice plants at the end of a 13-day incubation but other evidence has stongly suggested that uptake of fixed nitrogen by rice was slow (Ito *et al.* 1980; Eskew *et al.* 1981). Studies of mineralization using rewetted rice soil which had been incubated in ^{15}N$_2$ indicate that the availability of fixed nitrogen was similar to that of immobilized fertilizer nitrogen (Ito & Watanabe 1981). Similarly bacterial inocula were utilized as a nitrogen source when added to wheat plants grown in soil (Lethbridge & Davison 1983b).

The length of time involved in attaining maximum uptake of fixed nitrogen makes comparisons with acetylene reduction assays difficult to interpret. Okon, Heytler & Hardy (1983) estimated the amount of nitrogen fixed in association with *Setaria italica* grown in pots using both ARA and ^{15}N$_2$ incorporation methods. Only 5% of the nitrogen estimated by ARA to have been fixed during a 3-day exposure to ^{15}N$_2$ was found in the plant tops at harvest. This apparent low uptake of fixed nitrogen may be due to both slow release of fixed nitrogen and to overestimation of nitrogen fixation rates by ARA.

GROWTH RESPONSES TO INOCULATION WITH NITROGEN-FIXING BACTERIA

Inoculation responses and nitrogen fixation

Inoculation with N-fixing bacteria has increased growth in many field experiments; for example, in tropical grasses and cereals in North America (e.g. Smith *et al.* 1976) and wheat in Israel (Kapulnik *et al.* 1981) and in Belgium (Reynders & Vlassak 1982). Responses usually, but not always, resulted in increased total nitrogen uptake in the inoculated plants (Smith *et al.* 1976). Recently the number of reports of inoculation experiments has increased dramatically, especially from India (Rai & Gaur 1982; Meshram & Schende 1982).

Whilst in some studies higher rates of acetylene reduction have been found in inoculated plants, in others no nitrogenase activity could be detected (Venkateswarlu & Rao 1983). Inoculation experiments have concentrated on using *Azospirillum* spp., but as van Berkhum & Bohlool (1980) emphasized, there is no evidence that members of this genus are responsible for *in situ* nitrogenase activity or that nitrogenase activity in *Azospirillum* is closely coupled to nitrogen assimilation. Often responses to inoculation have been shown when fertilizer N was added, in spite of the fact that added N has been shown to depress nitrogenase activity (van Berkhum & Sloger 1983). It is possible that the reported inoculation responses are not primarily caused by increased rhizosphere nitrogen fixation.

Inoculation responses and plant growth hormones

Gaskins & Hubbell (1979) found both live and heat-killed inocula of *Azospirillum brasilense* increased growth of *Pennisetum americanum* without increasing acetylene reduction activity. Mutants of *Azospirillum lipoferum* which were unable to fix nitrogen increased both shoot dry weight and nitrogen content when inoculated onto maize in the field (O'Hara, Davey & Lucas 1981).

The ability of free-living nitrogen-fixing bacteria to produce plant growth hormones has been demonstrated (Brown & Burlingham 1968; Tien, Gaskins & Hubbell 1979). Barea & Brown (1974) investigated growth responses to inoculation of *Paspalum* with *Azotobacter paspali* in glasshouse experiments. They concluded that plant growth regulators were of primary importance in determining the growth responses. Solutions from *Azotobacter paspali* cultures contained auxins, gibberellins and cytokinins (Brown 1976). Increased numbers of lateral roots and root hairs were induced on millet (*Pennisetum americanum*) plants by inoculation with *Azospirillum* which was attributed to production of auxins, cytokinins and gibberellins (Tien *et al.* 1979). Production of indole acetic acid by *A. brasilense* was demonstrated by Reynders & Vlassak (1979). Stimulation of root development occurred in *Setaria italica* inoculated with *Azospirillum* (Kapulnik *et al.* 1981). Enhanced uptake of nitrate, potassium and phosphorus was found in inoculated plants of *Setaria* (Lin, Okon & Hardy 1983). It has been shown also that field-grown sorghum accumulated N, P and K at faster rates when inoculated with *Azospirillum* (Sarig *et al.* 1984).

Attempts to follow the population sizes of *Azospirillum* in the rhizospheres of inoculated plants have indicated a rapid fall in bacterial numbers within a week of inoculation (Schank *et al.* 1979). These results of field experiments contrast sharply with those of pot trials where the highest populations of nitrogen-fixing bacteria in the rhizosphere have been found at early flowering stages (Okon *et al.* 1983).

Accumulating evidence suggests that the responses to inoculation occur during the early stages of seedling establishment and growth. These effects are compatible with the hypothesis that inoculation responses are principally due to production of plant growth promoting substances by the bacteria, which in turn stimulate development of a more active and extensive root system. Other processes may be important: inoculation of several crops with species of *Pseudomonas*, which lacked the ability to fix nitrogen, have resulted in significant yield increases (e.g. Kloepper, Schroth & Miller 1980). These increases have been attributed to suppression and exclusion of pathogens in the rhizosphere (Schroth & Hancock 1981).

CONCLUSION

The discovery of high rates of acetylene reduction activity associated with roots of *Paspalum notatum* (Dobereiner *et al.* 1972) stimulated much interest and research. However, despite the profusion of literature concerning nitrogen fixation in the rhizosphere, knowledge of the amounts of nitrogen contributed to plants or soil has advanced little. Much is now known of the distribution and physiology of the bacteria involved (Klingmuller 1982). Remarkably little is understood of the bacteria–plant interactions, but the evidence cited above tends to suggest a loose, casual association, although this is a matter of controversy (Dobereiner 1982; van Berkhum, McClung & Sloger 1982). In order to understand the importance of nitrogen fixation in the rhizosphere, it is crucial to know the amounts involved, but methods capable of yielding irrefutable evidence of nitrogen fixation in these systems have not been developed. Perhaps the best evidence for a significant contribution from free-living nitrogen-fixing bacteria still comes from nitrogen balance studies (Table 1) but such figures represent the cumulative total of numerous minor contributary sources. Amounts of rhizosphere nitrogen fixation calculated in the literature are almost certainly gross overestimates (Paul 1978; van Berkhum & Bohlool 1980).

The most useful evidence available for assessment of the importance of nitrogen fixation in the rhizosphere is indirect. Early suggestions that increases in crop yields due to inoculation with nitrogen-fixing bacteria were mainly due to increased nitrogen fixation are now in question. Similar responses to bacterial mutants lacking active nitrogenase (O'Hara *et al.* 1981) and the effects of growth promoting substances produced by the bacteria (Tien *et al.* 1979) indicate that responses are mainly due to enhanced mineral uptake (Lin *et al.* 1983). Measurements of carbon availability in the rhizosphere indicate that the potential for nitrogen fixation in the rhizosphere is very limited (see above, Whipps & Lynch 1983). The low proportion of plant photosynthate lost as soluble root exudates tends to suggest that dead roots may be a more important carbon source for nitrogen fixation.

Contribution of rhizosphere-fixed nitrogen in different environments

Lichens and cyanobacteria are mainly responsible for the small quantities of nitrogen fixed in polar climatic regions (Paul 1978). In temperate regions, current estimates, even from the acetylene reduction assay which tends to overestimate the amounts of nitrogen fixed, suggest a small contribution (<5 kg N ha^{-1} year^{-1}) from nitrogen fixation in the rhizosphere of cereal crops (Lethbridge, Davison & Sparling 1982; Lethbridge & Davison 1983a). Albrecht & Gaskins (1982) calculated that a diversion of photosynthate in wheat sufficient to support rhizosphere nitrogen fixation of 40 kg N ha^{-1}, a minimal crop requirement, would decrease yields by 10–20%. Witty *et al.* (1979) calculated that most (28 kg N ha^{-1}) of the positive nitrogen balance (33 kg N ha^{-1}) on the Broadbalk continuous wheat experiment could be attributed to nitrogen fixation by free-living cyanobacteria. Similarly estimates in natural herbaceous vegetation tend to suggest that nitrogen fixation by bacteria in the rhizosphere contributes less than a few kg ha^{-1} year^{-1} (Skeffington & Bradshaw 1980; Rosswall & Granhall 1980).

Granhall & Linberg (1978) suggested that free-living nitrogen-fixing bacteria utilizing decaying wood and litter make the main contribution of fixed nitrogen, and that fixation in the rhizosphere is negligible. This may contribute to the positive nitrogen balance of 34 kg N ha^{-1} year^{-1} in the Broadbalk wilderness where the activity of nitrogen-fixing cyanobacteria is low (Day *et al.* 1975). Gains from dust deposition and ammonia absorption may be greater in woodlands which have a more extensive canopy than herbaceous plant communities.

In tropical climatic regions less is known of the potential for nitrogen fixation in the rhizosphere. Calculations of the amount of carbon available for utilization by nitrogen-fixing bacteria in the rhizosphere suggest that the contribution is limited. Where the water supply is abundant (e.g. in rice paddies) the amount of nitrogen fixed by cyanobacteria may be high. In dryland farming water may limit nitrogen fixation by both phototrophic bacteria and also fixation by heterotrophic bacteria, the latter by affecting rates of mineralization of organic matter. In the absence of accurate measurements it is not possible to attribute the maintenance of fertility under continuous cropping regimes (or positive nitrogen balances) to nitrogen fixation in the rhizosphere.

The contribution of nitrogen from nitrogen fixation in the rhizosphere may only be in the order of a few kg N ha^{-1} year^{-1}, yet this may be a significant contribution if nitrogen is limiting plant production. It is reasonable to assume that if nitrogen-fixing bacteria are present they will fix atmospheric nitrogen into the soil at least during some seasons, or during some phases of their growth. In ecosystems where there is efficient internal cycling of nutrients and low loss rates, the input of nitrogen from heterotrophic nitrogen-fixing bacteria may be sufficient to meet the requirements for maintenance of productivity.

The absence of a fuller understanding of the contribution from free-living nitrogen-fixing bacteria in the rhizosphere to the nitrogen economy of plants is undoubtedly due to the absence of methods which can measure and differentiate clearly between sources of nitrogen in the field. However, past conclusions and overestimation of potentials may not be solely due to faulty techniques; Clark (1981) correctly suggested with respect to the acetylene reduction assay that 'the ready acceptance of erroneous values is due to our enthusiasms'.

ACKNOWLEDGMENTS

We thank Dr M. E. Brown and Dr D. S. Jenkinson for helpful comments and the Overseas Development Administration (Research Scheme R3648) for financial support.

Albrecht, S.L. & Gaskins, M.H. (1982). The bioenergetics of asymbiotic nitrogen fixation. *Energy, Conservation and Use of Renewable Energies in the Bio-Industries 2* (Ed. by F. Vogt), pp. 24–38. Pergamon Press, Oxford.

Balandreau, J. P. (1983). Microbiology of the association. *Canadian Journal of Microbiology*, **29**, 851–859.

Balandreau, J.P., Millier, C.R. & Dommergues, Y.R. (1974). Diurnal variations of nitrogenase activity in the field. *Applied Microbiology*, **27**, 662–665.

Baldani, V.L.D. & Dobereiner, J. (1980). Host plant specificity in the infection of cereals with *Azospirillum* spp. *Soil Biology and Biochemistry*, **12**, 433–440.

Barber, D.A. & Lynch, J.M. (1977). Microbial growth in the rhizosphere. *Soil Biology and Biochemistry*, **9**, 305–308.

Barber, D.A. & Martin, J.K. (1976). Release of organic substances by cereal roots into soil. *New Phytologist*, **76**, 69–80.

Barea, J.M. & Brown, M.E. (1974). Effects on plant growth produced by *Azotobacter paspali* related to synthesis of plant growth regulating substances. *Journal of Applied Bacteriology*, **37**, 583–593.

Beck, S.M. & Gilmour, C.M. (1983). Role of wheat exudates in associative nitrogen fixation. *Soil Biology and Biochemistry*, **15**, 33–38.

Boddey, R.H., Chalk, P.M., Victoria, R.L., Matsui, E. & Dobereiner, J. (1983). The use of the ^{15}N isotope dilution technique to estimate the contribution of associated biological nitrogen fixation to the nutrition of *Paspalum notatum* cv. batatais. *Canadian Journal of Microbiology*, **29**, 1036–1045.

Brown, J.C. & Bell, W.D. (1969). Iron uptake dependent upon genotype of corn. *Proceedings of the Soil Science Society of America*, **83**, 99–101.

Brown, M.E. (1976). Role of *Azotobacter paspali* in association with *Paspalum notatum*. *Journal of Applied Bacteriology*, **40**, 341–348.

Brown, M.E. (1982). Nitrogen fixation by free-living bacteria associated with plants – fact or fiction? *Bacteria and Plants* (Ed. by M. E. Rhodes-Roberts & F. A. Skinner), pp. 25–41. Academic Press, London.

Brown, M.E. & Burlingham, S.K. (1968). Production of plant growth substances by *Azotobacter chroococcum*. *Journal of General Microbiology*, **53**, 135–144.

Brown, M.E., Burlingham, S.K. & Jackson, R.M. (1962). Studies on *Azotobacter* species in soil, II. Populations of *Azotobacter* in the rhizosphere and effects of artificial inoculation. *Plant and Soil*, **17**, 320–332.

Bulow, J.F.W. von & Dobereiner, J. (1975). Potential for nitrogen fixation in maize genotypes in Brazil. *Proceedings of the National Academy of Sciences of the United States of America*, **72**, 2389–2393.

Chalk, P.M., Douglas, L.A. & Buchanan, S.A. (1983). Use of ^{15}N enrichment of soil mineralisable N as a reference for isotope dilution measurements of biologically fixed nitrogen. *Canadian Journal of Microbiology*, **29**, 1046–1052.

Clark, F.E. (1981). The nitrogen cycle, viewed with poetic licence. *Terrestrial Nitrogen Cycles* (Ed. by F.E. Clark & T. Rosswall). *Ecological Bulletins (Stockholm)*, **33**, 13–24.

Clement, C.R. & Williams, T.E. (1967). Leys and soil organic matter, II. The accumulation of nitrogen in soils under different leys. *Journal of Agricultural Science (Cambridge)*, **69**, 133–138.

Dalton, H. & Postgate, J.R. (1969). Growth and physiology of *Azotobacter chroococcum* in continuous culture. *Journal of General Microbiology*, **56**, 307–319.

David, K.A.V. & Fay, P. (1977). Effects of long term treatment with acetylene on nitrogen fixing microorganisms. *Applied and Environmental Microbiology*, **34**, 640–646.

Day, J.M., Harris, D., Dart, P.J. & van Berkum, P. (1975). The Broadbalk experiment. An investigation of nitrogen gains from non-symbiotic nitrogen fixation. *Nitrogen Fixation by Free-living Micro-organisms* (Ed. by W.D.P. Stewart), pp. 71–84. Cambridge University Press, Cambridge.

De-Polli, H., Matsui, E., Dobereiner, J. & Salati, E. (1977). Confirmation of nitrogen fixation in two tropical grasses by [15]N_2 incorporation. *Soil Biology and Biochemistry*, **9**, 119–123.

Dobereiner, J. (1961). Nitrogen fixing bacteria of the genus *Beijerinckia* Derx. in the rhizosphere of sugar cane. *Plant and Soil*, **15**, 211–217.

Dobereiner, J. (1966). *Azotobacter paspali* sp. n. una bacteria fixadora de nitrogenio na rizosfera de *Paspalum*. *Pesquisa Agro pecueria, Brasil*, **1**, 357–365.

Dobereiner, J. (1982). Emerging technology based on biological nitrogen fixation by associative N_2-fixing organisms. *Biological Nitrogen Fixation Technology for Tropical Agriculture*: Papers presented at a workshop held at CIAT, 9–13 March, 1981 (Ed. by P. H. Graham & S. C. Harris), pp. 469–484. Centro Internacional de Agricultura Tropical, Cali, Colombia.

Dobereiner, J. & Day, J.M. (1976). Associative symbioses and free-living systems. *Proceedings of the 1st International Symposium on Nitrogen Fixation* (Ed. by W. E. Newton & C. J. Nyman), pp. 518–538. Washington State University Press, Pullman.

Dobereiner, J., Day, J.M. & Dart, P.J. (1972). Nitrogenase activity and oxygen sensitivity of the *Paspalum notatum - Azotobacter paspali* association. *Journal of General Microbiology*, **71**, 103–116.

Dobereiner, J., Marriel, I.E. & Nery, M. (1976). Ecological distribution of *Spirillum lipoferum* Beijerinck. *Canadian Journal of Microbiology*, **22**, 1464–1473.

Ela, S.W., Anderson, M.A. & Brill, W.J. (1982). Screening and selection of maize to enhance associative bacterial nitrogen fixation. *Plant Physiology*, **70**, 1564–1567.

Eskew, D.L., Eaglesham, A.R.J. & App, A.A. (1981). Heterotrophic $^{15}N_2$ fixation and distribution of newly fixed nitrogen in a rice–flooded soil system. *Plant Physiology*, **68**, 48–52.

Firth, P., Thilipoca, H., Suthipradit, S., Wetselaar, R. & Beech, D.F. (1973). Nitrogen balance studies in the central plain of Thailand. *Soil Biology and Biochemistry*, **5**, 41–46.

Fried, M., Danso, S.K.A. & Zapata, F. (1983). The methodology of measurement of N_2 fixation by non–legumes as inferred from field experiments with legumes. *Canadian Journal of Microbiology*, **29**, 1053–1062.

Gaskins, M.H. & Hubbell, D.H. (1979). Response of non-leguminous plants to root inoculation with free-living diazotrophic bacteria. *The Soil–Root Interface* (Ed. by J.L. Harley & R. Scott-Russell), pp. 176–192. Academic Press, London and New York.

Giller, K.E., Day, J.M., Dart, P.J. & Wani, S.P. (1984). A method for measuring the transfer of fixed nitrogen from free-living bacteria to higher plants using $^{15}N_2$. *Journal of Microbiological Methods* **2**, 307–316.

Goodlass, G. & Smith, K.A. (1978). Effects of organic ammendments on evolution of ethylene and other hydrocarbons from soil. *Soil Biology and Biochemistry*, **10**, 201–206.

Gorham, E., Vitousek, P.H. & Reiners, W.A. (1979). The regulation of chemical budgets over the course of terrestrial ecosystem succession. *Annual Review of Ecology and Systematics*, **10**, 53–84.

Gordon, J.K. (1981). Introduction to the nitrogen fixing prokaryotes. *Prokaryotes: a Handbook on Habitats, Isolation and Identification of Bacteria* (Ed. by M.P. Starr *et al.*), pp. 783–792. 2 Vols. Springer–Verlag, New York.

Granhall, U. & Linberg, T. (1978). Nitrogen fixation in some coniferous forest ecosystems. *Environmental Role of Nitrogen Fixing Blue-green Algae and Asymbiotic Bacteria* (Ed. by U. Granhall) *Ecological Bulletins (Stockholm)*, **22**, 178–192.

Greenland, D.J. & Nye, P.H. (1959). Increases in the carbon and nitrogen contents of tropical soils under natural fallows. *Journal of Soil Science*, **10**, 284–299.

Hardy, R.W.F., Burns, R.C. & Holsten, R.D. (1973). Applications of the acetylene reduction assay for measurement of nitrogen fixation. *Soil Biology and Biochemistry*, **5**, 47–81.

Helal, H.M. & Sauerbeck, D.R. (1983). Method to study turnover processes in soil layers of different proximity to roots. *Soil Biology and Biochemistry*, **15**, 223–226.

Ito, O. & Watanabe, I. (1981). Immobilisation mineralisation and availability to rice plants of nitrogen derived from heterotrophic nitrogen fixation in flooded soil. *Soil Science and Plant Nutrition*, **27**, 169–176.

Ito, O., Cabrera, D. & Watanabe, I. (1980). Fixation of dinitrogen-15 associated with rice plants. *Applied and Environmental Microbiology*, **39**, 554–558.

Jaiyebo, E.O. & Moore, A.W. (1963). Soil nitrogen accretion under different covers in a tropical rain forest environment. *Nature*, **197**, 317–318.

Jenkinson, D.S. (1971). The accumulation of organic matter in soil left uncultivated. *Annual Report of Rothamsted Experimental Station for 1970, Part 2*, pp. 113–137.

Jenkinson, D.S. (1977). The nitrogen economy of the Broadbalk Experiments, I. Nitrogen balance in the experiments. *Annual Report of Rothmastad Experimental Station for 1970, Part 2*, pp. 103–109.

Jensen, H.L., Petersen, E.J., Bhattacharya, P.K. de & Bhattacharya, R. (1960). A new nitrogen-fixing bacterium: *Derxia gummosa* nov. gen. nov. spec. *Archiv für Mikrobiologie*, **36**, 182–195.

Jensen, V. (1956). Nitrogen fixation by strains of *Aerobacter aerogenes*. *Physiologia Plantarum*, **9**, 130–136.

Kapulnik, Y., Kigel, J., Okon, Y., Nur, I. & Henis, Y. (1981). Effect of *Azospirillum* inoculation on some growth parameters and N-content of wheat. *Sorghum* and *Panicum*. *Plant and Soil*, **61**, 65–70.

Katznelson, H., Rouatt, J.W. & Payne, T.M.B. (1956). Recent studies on the microflora of the rhizosphere, *6th Congress International de la Science du Sols* Paris, Commission III, Biologie du sol. (Rapports) pp. 151–156.

Klingmuller, W. (ed.) (1982). *Azospirillum Genetics, Physiology and Ecology*. Proceedings of a workshop held at the University of Bayreuth, Germany, July 1981. Birkhauser Verlag. Basel, Boston, Stuttgart.

Kloepper, J.W., Schroth, M.N. & Miller, T.D. (1980). Effects of rhizosphere colonisation by plant growth promoting rhizobacteria on potato plant development and yield. *Phytopathology*, **70**, 1078–1082.

Klossak, R.H. & Bohlool, B.B. (1983). Prevalence of *Azospirillum* spp. in the rhizosphere of tropical plants. *Canadian Journal of Microbiology*, **29**, 649–652.

Lee, J.A., Harmer, R. & Ignaciuk, R. (1982). Nitrogen as a limiting factor in plant communities *Nitrogen as an Ecological Factor*. (Ed. by J.A. Lee, S. McNeill & I.H. Rorison), pp. 95–112. British Ecological Society Symposium 22. Blackwell Scientific Publications, Oxford.

Lethbridge, G. & Davison, M.S. (1983a). Root-associated nitrogen-fixing bacteria and their role in the nitrogen nutrition of wheat estimated by ^{15}N isotope dilution. *Soil Biology and Biochemistry*, **15**, 365–374.

Lethbridge, G. & Davison, M.S. (1983b). Microbial biomass as a source of nitrogen for cereals. *Soil Biology and Biochemistry*, **15**, 375–376.

Lethbridge, G., Davison, M.S. & Sparling, G.P. (1982). Critical evaluation of the acetylene reduction test for estimating the activity of nitrogen-fixing bacteria associated with the roots of wheat and barley. *Soil Biology and Biochemistry*, **14**, 27–35.

Lin, W., Okon, Y. & Hardy, R,W.F. (1983). Enhanced mineral uptake by *Zea mays* and *Sorghum bicolor* roots inoculated with *Azospirillum brasilense*. *Applied and Environmental Microbiology*, **45**, 1775–1779.

Macdonald, R.M. (1980). Cytochemical demonstrtion of catabolism in soil micro-organisms. *Soil Biology and Biochemistry*, **12**, 419–423.

Marrs, R.H., Roberts, R.D., Skeffington, R.A. & Bradshaw, A.D. (1982). Nitrogen and the development of ecosystems. *Nitrogen as an Ecological Factor* (Ed by J.A. Lee, S. McNeill & I.. Rorison), pp. 113–136. British Ecological Society Symposium 22, Blackwell Scientific Publications, Oxford.

Martin, J.R. & Puckridge, D.W. (1982). Carbon flow through the rhizosphere of wheat crops in south Australia. *The Cycling of Carbon, Nitrogen, Sulphur and Phosphate in Terrestrial and Aquatic Ecosystems* (Ed. by I.E. Galbally & J.R. Freney), pp. 77–82. Australian Academy of Science, Canberra.

Matsui, E., Vose, P.B., Rodrigues, N.S. & Ruschel, A.P. (1981). Use of ^{15}N$_2$ enriched gas to determine N$_2$ fixation by undisturbed sugar cane plant in the field. *Associative Nitrogen Fixation* (Ed. by P.B. Vose & A. P. Ruschel), pp. 153–161. CRC Press, Boca Raton, Florida.

Matthews, S.W., Schank, S.C., Aldrich, A.C. & Smith R.L. (1983). Peroxidase-antiperoxidase labeling of *Azospirillum brasilense* in field-grown pearl millet. *Soil Biology and Biochemistry*, **15**, 699–703.

McBride, J. (1983). Soil bacteria: another source of nitrogen? *Agricultural Research*, **31**, 10–12.

McClung, C.R. & Patriquin, D. (1980). Isolation of a nitrogen fixing *Campylobacter* species from the roots of *Spartina alterniflora* Loisel. *Canadian Journal of Microbiology*, **26**, 881–886.

McClung, C.R., van Berkhum, P., Davis, R.E. & Sloger, C. (1983). Enumeration and localisation of N$_2$-fixing bacteria associated with roots of *Spartina alterniflora* Loisel. *Applied and Environmental Microbiology*, **45**, 1914–1920.

Meshram, S.U. & Schende, S.T. (1982). Total nitrogen uptake by maize with *Azotobacter* inoculation. *Plant and Soil*, **69**, 275–280.

Moore, A.W. (1966). Non-symbiotic nitorgen fixation in soil and soil–plant systems. *Soils and Fertilisers*, **29**, 113–128.

Neal, J.L. & Larson, R.L. (1976). Acetylene reduction by bacteria isolated from rhizosphere of wheat. *Soil Biology and Biochemistry*, **8**, 151–155.

Nohrstedt, H.O. (1975/6). Decomposition of ethylene in soils and its relevance in the measurement of nitrogenase activity with the acetylene reduction method. *Grundforbatting*, **27**, 171–178.

Nohrstedt, H.O. (1983). Natural formation of ethylene in forest soils and methods to correct results given by the acetylene-reduction assay. *Soil Biology and Biochemistry*, **15**, 281–286.

O'Hara, G.W., Davey, M.R. & Lucas, J.A. (1981). Effect of inoculation of *Zea mays* with *Azospirillum brasilense* strains under temperate conditions. *Canadian Journal of Microbiology*, **27**, 871–877.

Okon, Y. (1982). *Azospirillum*: Physiological properties, mode of association with roots and its application for the benefit of cereal and forage grass crops. *Israel Journal of Botany*, **31**, 214–220.

Okon, Y., Heytler, P.G. & Hardy, R.W.F. (1983). N_2 fixation by *Azospirillum brasilense* and its incorporation into host *Setaria italica*. *Applied and Environmental Microbiology*, **46**, 694–697.

Okon, Y., Houchins, J.P., Albrecht, S.L. & Burris, R.H. (1977). Growth of *Spirillum lipoferum* at constant partial pressures of oxygen, and the properties of its nitrogenase in cell free extracts. *Journal of General Microbiology*, **98**, 87–93.

Ovington, J.D. (1951). The afforestation of Tentsmuir sands. *Journal of Ecology*, **39**, 363–375.

Ovington, J.D. (1956). Studies on the development of woodland conditions under different trees, IV. The ignition loss, water, carbon and nitrogen content of the mineral soil. *Journal of Ecology*, **44**, 171–179.

Parker, C.A. (1957). Non-symbiotic nitrogen fixing bacteria in soil, II. Studies on *Azotobacter*. *Australian Journal of Agricultural Research*, **6**, 388–397.

Patriquin, D.G. & Dobereiner, J. (1978). Light microscopy observations of tetrazolium-reducing bacteria in the endorhizosphere of maize and other grasses in Brasil. *Canadian Journal of Microbiology*, **24**, 734–742.

Paul, E.A. (1978). Contribution of nitrogen-fixation to ecosystem functioning and nitrogen fluxes on a global basis. *Environmental Role of Nitrogen fixing Blue Green Algae and Asymbiotic Bacteria* (Ed. by U. Granhall). *Ecological Bulletins (Stockholm)*, **22**, 157–167.

Pederson, W.L., Chakabarty, K., Klucas, R.V. & Vidaver, A.K. (1978). Nitrogen fixation (acetylene reduction) associated with roots of winter wheat and sorghum in Nebraska. *Applied and Environmental Microbiology*, **35**, 129–135.

Porter, L.K. & Grable, A.R. (1969). Fixation of atmospheric nitrogen by non-legumes in wet mountain meadows. *Agronomy Journal*, **61**, 521–523.

Rai, S.N. & Gaur, A.C. (1982). Nitrogen fixation by *Azospirillum* spp. and the effect of *Azospirillum lipoferum* on the yield and N-uptake of wheat crop. *Plant and Soil*, **69**,233–238.

Rao, V.R. (1978). Effect of carbon sources on asymbiotic nitrogen fixation in a paddy soil. *Soil Biology and Biochemistry*, **10**, 319–321.

Reiners, W.A. (1981). Nitrogen cycling in relation to ecosystem succession. *Terrestrial Nitrogen Cycles* (Ed. by F.E. Clark & T. Rosswall). *Ecological Bulletins (Stockholm)*, **33**, 507–528.

Rennie, R.J. (1980). Nitrogen-15 isotope dilution as a measure of dinitrogen fixation by *Azospirillum brasilense* associated with maize. *Canadian Journal of Botany*, **58**, 21–24.

Rennie, R.J. & Rennie, D.A. (1983). Techniques for quantifying N_2 fixation in association with non-legumes under field and greenhouse conditions. *Canadian Journal of Microbiology*, **29**, 1022–1035.

Rennie, R.J., De Freitas, J.R., Ruschel, A.P. & Vose, P.B. (1983). [15]N isotope dilution to quantify dinitrogen (N_2) fixation associated with Canadian and Brazilian wheat. *Canadian Journal of Botany*, **61**, 1667–1671.

Rennie, R.J., Rennie, D.A. & Fried, M. (1978). Concepts of ^{15}N usage in dinitrogen fixation studies. *Isotopes in Biological Dinitrogen Fixation*. Proceedings of an advisory group meeting organised by the joint FAO/IAEA Division of Atomic Energy in Food and Agriculture and held in Vienna, 21–25 November 1977, pp. 107–131. International Atomic Energy Agency, Vienna.

Reynders, L. & Vlassak, K. (1979). Conversion of tryptophan to indoleacetic acid by *Azospirillum brasilense*. *Soil Biology and Biochemistry*, **11**, 547–548.

Reynders, L. & Vlassak, K. (1982). Use of *Azospirillum brasilense* as a biofertiliser in intensive wheat cropping. *Plant and Soil*, **66**, 217–273.

Rosswall, T. & Granhall, U. (1980). Nitrogen cycling a subartic ombrotrophic mire, *Ecology of a Sub-arctic Mire* (Ed. by M. Sonesson). *Ecological Bulletins (Stockholm)*, **30**, 209–234.

Ruschel, A.P., Henis, Y. & Salati, E. (1975). Nitrogen-15 tracing of N-fixation with soil-grown sugar cane seedlings. *Soil Biology and Biochemistry*, **7**, 181–182.

Ruschel, A.P., Matsui, E., Salati, E. & Vose, P.B. (1981). Potential N_2 fixation by sugar cane (*Saccharum* sp) in solution culture. II. Effect of inoculation and dinitrogen fixation as measured directly by $^{15}N_2$. *Associative Nitrogen Fixation* (Ed. by P.B. Vose & A.P. Ruschel), pp. 127–132. Vol.II. CRC Press, Boca Raton, Florida.

Sarig, S., Kapulnik, Y., Nur, I. & Okon, Y. (1984). Response of non-irrigated *Sorghum bicolor* to *Azospirillum* inoculation. *Experimental Agriculture*, **20**, 59–66.

Schank, S.C., Smith, R.L., Weiser, G.C., Zuberer, D.A., Bouton, J.H., Quesenbury, K.H., Tyler, M.E., Milam, J.R. & Littell, R.C. (1979). Fluorescent antibody technique to identify *Azospirillum brasilense* associated with roots of grasses. *Soil Biology and Biochemistry*, **11**, 287–296.

Schroth, M.N. & Hancock, J.G. (1981). Selected topics in Biological Control. *Annual Review of Microbiology*, **35**, 453–476.

Skeffington, R.A. & Bradshaw, A.D. (1980). Nitrogen fixation by plants grown in reclaimed china clay waste. *Journal of Applied Ecology*, **17**, 469–477.

Smith, R.L., Bouton, J.H., Schank, S.C., Quesenbury, K.H., Tyler, M.E., Milam, J.R., Gaskins, M.H. & Littell, R.C. (1976). Nitrogen fixation in grasses inoculated with *Spirillum lipoferum*. *Science*, **193**, 1003–1005.

Stevenson, F.J. (1982). Origin and distribution of nitrogen in soils. *Nitrogen in Agricultural Soils* (Ed. by F.J. Stevenson), pp. 1–42. American Society of Agronomy Monograph 22, Madison.

Stotzky, G. & Norman, A.G. (1961). Factors limiting microbial agriculture in soil, I The level of substrate, nitrogen and phosphorus. *Archiv für Mikrobiologie*, **40**, 341–369.

Tarrand, J.J., Krieg, N.R. & Dobereiner, J. (1978). A taxonomic study of the *Spirillum lipoferum* group, with descriptions of a new genus *Azospirillum* gen. nov. and two species, *Azospirillum lipoferum* (Beijerinck) comb.nov. and *Azospirillum brasilense* sp. nov. *Canadian Journal of Microbiology*, **24**, 967–980.

Tien, T.H., Gaskins, M.H. & Hubbell, D.H. (1979). Plant growth substances produced by *Azospirillum brasilense* and their effect on the growth of pearl millet (*Pennisetum americanum* L.). *Applied and Environmental Microbiology*, **37**, 1016–1024.

Umali-Garcia, M., Hubbell, D.H., Gaskins, M.H. & Dazzo, F.B. (1980). Association of *Azospirillum* with grass roots. *Applied and Environmental Microbiology*, **39**, 219–226.

van Berkhum, P. (1980). Evaluation of acetylene reduction by excised roots for the determination of nitrogen fixation in grasses. *Soil Biology and Biochemistry*, **12**, 141–146.

van Berkhum, P. and Bohlool, B.B. (1980). Evaluation of nitrogen fixation by bacteria in association with roots of tropical grasses. *Microbiological Reviews*, **44**, 491–517.

van Berkhum, P. & Sloger, C. (1979). Immediate acetylene reduction by excised grass roots not previously preincubated at low oxygen tensions (and maize roots inoculated with nitrogen fixing bacterium *Spirillum lipoferum*). *Plant Physiology*, **64**, 739–743.

van Berkhum, P. & Sloger, C. (1983). Interaction of combined nitrogen with the expression of root associated nitrogenase activity in grasses and the development of N_2 fixation in soybean (*Glycine max* L. Merr). *Plant Physiology*, **72**, 741–745.

van Berkhum, P., McClung, C.R. & Sloger, C. (1982). Some pertinent remarks on N_2-fixation associated with the roots of grasses. *Biological Nitrogen Fixation Technology for Tropical Agriculture*: Papers presented at a workshop held at CIAT, 9–13 March, 1981 (Ed. by P.H. Graham & S.C. Harris), pp. 513–526. Centro Internacional de Agricultura Tropical, Cali, Colombia.

Venkateswarlu, B. & Rao, A.V. (1983). Response of pearl millet to inoculation with different strains of *Azospirillum brasilense*. *Plant & Soil*, **74**, 379–386.

Vlassak, K. & Reynders, L. (1978). Associative dinitrogen fixation in temperate regions. *Isotopes in Biological Nitrogen Fixation*. Proceedings of an advisory group meeting organised by the joint FAO/IAEA Division of Atomic Energy in Food and Agriculture, Vienna 21–25 November, 1977, pp. 71–85. International Atomic Energy Agency, Vienna.

Warembourg, F.R. & Paul, E.A. (1977). Seasonal transfers of assimilated ^{14}C in grassland: plant production and turnover, soil and plant respiration. *Soil Biology and Biochemistry*, **9**, 295–301.

Whipps, J.M. & Lynch, J.M. (1983). Substrate flow and utilisation in the rhizosphere of cereals. *New Phytologist*, **95**, 605–623.

White, J.W., Holben, F.J. & Richer, A.C. (1945). Maintenance level of nitrogen and organic matter in grassland and cultured soils over periods of 54 and 72 years. *Journal of the American Society of Agronomy*, **37**, 21–31.

Whitt, D.M. (1941). The role of blue grass in the conservation of the soil and its fertility. *Soil Science Society of America Proceedings*, **6**, 309–311.

Witty, J.F. (1979). Acetylene reduction assay can over-estimate nitrogen fixation in soil. *Soil Biology. and Biochemistry*, **11**, 209–210.

Witty, J.F. (1983). Estimating N fixation in the field using ^{15}N-labelled fertiliser: some problems and solutions. *Soil Biology and Biochemistry*, **15**, 631–639.

Witty, J.F., Keay, P.J., Frogatt, P.J. & Dart, P.J. (1979). Algal nitrogen fixation on temperate arable fields. *Plant and Soil*, **52**, 151–164.

Yoshida, T. & Yoneyama, T. (1980). Atmospheric dinitrogen fixation in the flooded rice rhizosphere as determined by the N-15 isotope technique. *Soil Science and Plant Nutrition*, **26**, 551–559.

Interactive responses of associative diazotrophs from a Nebraska Sand Hills grassland

LAWRENCE A. KAPUSTKA, PAUL. T. ARNOLD AND PAUL T. LATTIMORE

Department of Botany, Miami University, Oxford, Ohio, USA 45056

INTRODUCTION

Associative diazotrophs are bacteria which exist asymbiotically in the root zone of plants and are capable of reducing atmospheric nitrogen. Considerable taxonomic and ecological variability exists among the known asymbiotic diazotrophs. Rennie (1980) evaluated several isolates from a wide range of domesticated cereal crops as well as native grasses. Of the potential diazotrophs isolated, 4–100% of the isolates of a given taxon reduced acetylene *in vitro*.

Rangeland and other temperate grasslands receive 1–10 g N ha^{-1} annually in the form of nitrous oxides and ammonium in precipitation (Woodmansee 1979). Biological dinitrogen fixation, principally by asymbiotic bacteria, contributes <1–8 kg N ha^{-1} annually (DuBois & Kapustka 1983; Kapustka 1982).

Here we present a portion of a study undertaken to characterize the role of asymbiotic diazotrophs in the native grasslands of the Nebraska Sand Hills.

MATERIALS AND METHODS

Site description

The Nebraska Sand Hills, which extend across 52 000 km^2 of north- and west-central Nebraska, support a unique vegetation dominated by representatives of the short grass and tall grass prairies. Arapaho Prairie, a 526 ha tract owned by The Nature Conservancy and managed by the University of Nebraska-Lincoln, is typical of the drier upland prairie of the western region. The mean annual precipitation is 40 cm. Some 200 species of flowering plants occur among the ridge, slope, valley and disturbance habitats (Keeler, Harrison & Vescio 1980; for a detailed description see Barnes & Harrison 1982).

The soils of Arapaho Prairie grade from coarse sand on the dune crests to fine sand or loamy fine sand in the valleys. Ranges and means of selected chemical characteristics of the soils are: pH, 4·4–(6·4)–7·5; nitrate, <0·1–(4·9)–45·6 ppm; ammonium, <0·2–(2·9)–8·5 ppm; and phosphate 0·2–(1·8)–6·7 ppm. Residual soils enzyme activities (amylase, maltase, cellulase, and dehydrogenase) exhibited a 2–5 fold range among the several soils tested. Consistently, the highest values for all parameters except pH were in swales dominated by *Agropyron smithii*. Rhizospheres of *Muhlenbergia pungens*, a colonizer of disturbance areas (blow outs), had the lowest values for the enzyme activities and most of the nutrient values. Rhizospheres of species occupying the ridges and slopes had variable residual enzyme activities and nutrient levels (L. A. Kapustka, unpubl.).

Acetylene reduction assay (ARA)

One-hour ARAs were performed on isolated rhizosphere soil and root segments following the precautions of van Berkhum & Bohlool (1980). Approximately 100 g soil plus root tissue was placed in 1 litre ziploc bags fitted with a rubber serum stopper.

Amended nitrogenase activity was determined during a 96-hour incubation period (DuBois & Kapustka 1983). Aliquots of soil (2·0 g) were placed in 16×120 mm culture tube containing 5 ml combined carbon medium (CCM) (Rennie 1981). One-hour ARAs were performed at time 0, 24, 48, 72, and 96 hours. The tubes were vented after each assay.

Isolation and identification of diazotrophs

Soils from the rhizospheres of the predominant grasses were suspended in sterile CCM for 48 hours. Subsequently, a 1 ml aliquot was plated onto solid CCM. As colonies developed, they were transferred to fresh plates to achieve purified isolates. Only those isolates positive for ARA were retained. Several alternative media were employed to check for contamination. The purified cultures were diagnosed with the API-20E identification kit (Rennie 1980).

Bioassays of rhizosphere soils

Aqueous soil extracts were prepared from the rhizosphere soils: 200 g soil was suspended in 200 ml sterile distilled water in a Waring blender. The suspension was centrifuged for 15 min at $7500\,g$. The supernatant was passed through a sterile 0·45 μm Gelman filter. Sterility was verified by plating aliquots of the extract on various media. Twelve soils were tested against twelve diazotrophs including four ATCC cultures (see Results section for listing). Each bioassay consisted of three treatments. Treatment A contained 2·0 ml soil extract and 2·0 ml sterile distilled water. Treatment B contained 2·0 ml soil extract and 2·0 ml CCM. Treatment C contained 2·0 ml CCM and 2·0 ml sterile distilled water. These were inoculated with 0·1 ml of an aqueous suspension of the cultures. After 7 days growth was assessed by measuring the turbidity (absorbance at 650 nm) and the protein content according to the Lowry technique (Lowry *et al.* 1951). Initial absorbancy and protein content from the 0·1 ml inoculum was subtracted from the respective values. The mean initial absorbancy was 0·004 and the mean initial protein content was 8 mg.

Plant response to diazotrophs

Seeds were surface sterilized in an aqueous solution of chlorox (1%), ethanol (50%), Tween 20 (trace) for 15 min under vacuum. The seeds were rinsed with sterile distilled water and germinated on sterile, moist filter pads. Following germination, seedlings were transferred by aseptic technique to a presterilized container and inoculated with a suspension of selected diazotrophs. *Lolium perenne* (a C_3 grass) and *Bouteloua gracilis* (a C_4 grass) were grown in sterile, acid-washed quartz sand. *Schizachyrium scoparium* was grown in sterile growth pouches. Five plants in each diazotroph treatment received one-half strength N-free Hoagland's solution, five received one-half strength complete Hoagland's solution. At the conclusion of the 3-week growth period, shoot height and weight, and root length and weight were recorded.

The influence of diazotroph inocula on growth of plants and on VAM fungal infectivity was examined with non-sterile prairie soil collected from the Arapaho Prairie. Sieved soil (2 mm screen) was placed in pine-cell containers (Ray Leach Conetainers Inc., Corvallis, Oregon). Following germination, 1 ml aliquots of selected diazotroph inocula were introduced to the soil surface. Inoculated treatments received a daily watering of 1/5 strength $-N$, $-P$ Hoagland's solution. Uninoculated controls received daily watering of 1/5 strength $-P$ Hoagland's solution. After 6 weeks, plants were harvested. The measurements in the previously mentioned experiment were performed. In addition, the percentage infectivity by VAM fungi was scored utilizing a uniform, ten-class scale (Biermann & Linderman 1981). Analysis was by the non-parametric Wilcoxson statistic.

Roots were stained with a europium-calcoflor differential fluorescence stain (Johnen & Drew 1979) and observed with a Leitz Orthoplan Epifluorescence microscope.

RESULTS

Acetylene reduction assays

The rates of nitrogenase activity of rhizosphere soils were highly variable and generally low. Maximum rates were 25–30 μg N g^{-1} soil hour^{-1}. Activity occurred sporadically during the early spring when soils were near field capacity. By mid-June nitrogenase activity was below detectable levels and remained so throughout the summer. Experimental manipulation of water potential has shown a linear decline of ARA to $-2\cdot0$ MPa (Kapustaka 1982). Consistently, the highest rates of ARA were observed on rhizosphere soils from *Agropyron smithii* and *Bouteloua gracilis*.

The N addition calculated from the amended ARA measurements were consistent with the non-amended rates. *A. smithii* and *B. gracilis* had $1\cdot5$–2 fold higher rates than the other species tested. In general, the amended ARAs followed the pattern of the residual soil enzyme rates with *Muhlenbergia pungens* and *Andropogon hallii* exhibiting low rates.

Isolates

The most common heterotrophic diazotroph isolated from the rhizospheres was *Enterobacter cloacae*, with *Bacillus polymyxa* and *Azotobacter vinelandii* less prominent. *Klebsiella pneumoniae* and three unknowns were found but at low frequency. In addition, four species of the purple nonsulphur genus *Rhodopseudomonas* were isolated from the rhizosphere soils (L. R. Finke, pers comm.).

Bioassays of rhizosphere soil extracts

The growth of diazotrophs in culture expressed as absorbancy and as protein increases was greatest in treatment B (soil extract plus CCM; Table 1). Half-strength CCM (treatment C) supported slightly more growth than the half-strength soil extract (treatment A). It is particularly interesting to note the 3-fold range of variability among the different *Enterobacter cloacae* isolates. Among all isolates for a given treatment, the range in protein yield is approximately 4-fold. Among the soil extracts the protein yield range is approximately 2-fold.

Plant response to isolates

Comparison of the influence of the diazotrophs within the same nutrient condition reveals substantial differences among the three species of grasses (Table 2). *L. perenne* growth in the N-free condition was reduced by all diazotrophs tested. However, with N, there was no deleterious effect by five of the isolates, and one (*E. cloacae* from *C. longifolia*) increased growth by 35%. *B. gracilis* growth was reduced by all isolates. Growth of *S. scoparium* was enhanced in all cases in the N-free condition and in all but two cases in the N medium. None of the diazotrophs, including those from the ATCC, were capable of supplying sufficient N to *L. perenne* or *B. gracilis* grown in N-free nutrient solution. Total dry weight of gnotobiotically grown plants varied from 60% reductions to 370% increases over controls in N-free medium. With N present, the growth response varied from 15% reductions to 35% increases in plant growth. It is especially interesting to note the effect of the diazotrophs on shoot:root ratio. There is a general tendency to increase the amount of root tissue when diazotrophs are present, whereas shoot production remained constant or was reduced.

   B. gracilis growth in non-sterile Arapaho Prairie soil inoculated with diazotrophs ranged from 88 to 390% of controls (Table 3). Although the aerial growth was enhanced in several treatments, the major increases in growth was due to root production. This is illustrated by the significantly lower W_S/W_R ratios in many of the treatments. *S. scoparium* growth ranged from 81 to 150% of the control. In most instances the relative responses to the various treatments was the opposite of *B. gracilis*. VAM fungal infectivity was highly variable among treatments. Furthermore, there was a marked difference in levels of infection between larger roots ($>1\cdot0$ mm diameter) and smaller roots ($<1\cdot0$ mm diameter). Despite the absence

TABLE 1. Bioassays of rhizosphere soil extracts with diazotrophs (mg protein)

	Treatment					
	(a) Azospirillum brasilense ATCC inoculum.			(b) Isolate from Bouteloua hirsuta.		
Rhizosphere	A	B	C	A	B	C
Andropogon hallii	143	299	146	71	202	119
Bouteloua hirsuta	44	206	181	72	186	122
Calamovilfa longifolia	44	293	64	102	50	39
Koeleria pyramidata	98	200	108	79	180	21
Muhlenbergia pungens	89	168	124	74	145	73
Panicum virgatum	68	217	164	76	169	85
Barren disturbance	85	177	83	79	139	55
Sporobolus cryptandrus	87	114	112	81	180	115
Schizachyrium scoparium	88	229	159	78	253	143
Stipa comata	72	157	171	79	197	169

	(c) Soil extract from Agropyron smithii.			(d) Soil extract from Bouteloua gracilis.		
Diazotroph, source	A	B	C	A	B	C
Azospirillum brasilense, ATCC	30	165	71	97	135	63
Azotobacter chroococcum, ATCC	17	198	70	63	182	69
Klebsiella pneumoniae, ATCC	65	171	142	90	154	59
Bacillus polymyxa, ATCC	51	169	47	86	162	91
Enterobacter cloacae, S. scop.	40	146	50	67	161	57
Unknown, B. hirsuta	43	111	49	73	159	55
E. cloacae, B. gracilis	35	102	51	71	83	70
K. pneumoniae, Panicum virgatum	61	209	119	102	172	123
E. cloacae, Stipa comata	20	181	48	93	196	72
B. polymyxa, K. pyramidata	60	262	60	190	141	79
E. cloacae, Carex sp.	86	205	55	138	196	130
E. cloacae, C. longifolia	27	165	45	83	212	38

TABLE 2. Mean height, weight, and W_S/w_R ratios of gnotobiotically-grown plants in response to diazotroph inocula

Diazotroph, source		Lolium perenne			Bouteloua gracilis		Schizachyrium scoparium	
		Height (mm)	DW (mg)	W_S/W_R	DW (mg)	W_S/W_R	DW (mg)	W_S/W_R
Klebsiella pneumoniae, ATCC 13883	−N	nd	nd	nd	nd	nd	63[a]	0·37[a]
	+N	nd	nd	nd	nd	nd	48	0·83[a]
Azotobacter chroococcum, ATCC 480	−N	nd	nd	nd	nd	nd	43[a]	0·50[a]
	+N	nd	nd	nd	nd	nd	31	0·68
Bacillus polymyxa, ATCC 842	−N	nd	nd	nd	nd	nd	26	0·55[a]
	+N	nd	nd	nd	nd	nd	33	0·90[a]
Azospirillum brasilense, ATCC 19245	−N	nd	nd	nd	nd	nd	42[a]	0·61[a]
	+N	nd	nd	nd	nd	nd	67[a]	0·46
Enterobacter cloacae, Calamovilfa longifolia	−N	81	78	0·22	nd	nd	31	0·68[a]
	+N	150[a]	945[a]	0·46[a]	83[a]	0·26[a]	60[a]	0·32[a]
Klebsiella pneumoniae, Panicum virgatum	−N	71	76	0·21	23	0·51	32	0·37[a]
	+N	147[a]	736	0·39	84[a]	0·55	41	0·59
Unknown, Bouteloua hirsuta	−N	55[a]	65[a]	0·25	20	0·49	nd	nd
	+N	154	602	0·40	87	0·94	nd	nd
Enterobacter cloacae, Schizachyrium scoparium	−N	nd	nd	nd	nd	nd	39[a]	0·56[a]
	+N	nd	nd	nd	nd	nd	43	0·56
Bacillus polymyxa, Koeleria pyramidata	−N	nd	nd	nd	nd	nd	38[a]	0·56[a]
	+N	nd	nd	nd	nd	nd	59[a]	0·45
Enterobacter cloacae, Stipa comata	−N	59[a]	42[a]	0·28	21	0·37[a]	nd	nd
	+N	166	361[a]	0·42	73[a]	0·39	nd	nd
Unknown, Sporobolus cryptandrus	−N	71	71	0·16	20	0·52	nd	nd
	+N	170	661	0·52	65[a]	0·30[a]	nd	nd
Enterobacter cloacae, Carex sp.	−N	69	47[a]	0·30[a]	17[a]	0·30[a]	nd	nd
	+N	191[a]	610	0·52	87	0·65	nd	nd
Control	−N	74	99	0·22	24	0·57	17	1·37
	+N	164	699	0·46	115	0·48	38	0·53

DW, dry weight of shoots and roots.
W_S/W_R, dry weight of the shoot over the dry weight of the root.
Controls received heat-killed inoculum and 1/2 strength Hoagland's Solution with (+N) and without (−N) N.
Values marked with a are significantly different from the respective control, $P < 0.05$; nd, not determined

TABLE 3. Plant response to diazotroph inoculum added to prairie soil

Treatment Diazotroph, source	Bouteloua gracilis					Schizachyrium scoparium				
	Height (mm)	DW (mg)	W_S/W_R	VAM percentage infectivity		Height (mm)	DW (mg)	W_S/W_R	VAM percentage infectivity	
				L	S				L	S
Klebsiella pneumoniae, ATCC 13883	126[a]	16	0·93[a]	3	13[a]	92[a]	27	1·04	20	26
Azotobacter chroococcum, ATCC 480	233[a]	67[a]	0·57[a]	14	33	76	20	1·01	6	18
Bacillus polymyxa, ATCC 842	234[a]	53[a]	0·95[a]	8	33	61	30	0·91	10	15
Azospirillum brasilense, ATCC 19245	172	26	0·97[a]	8	16	68	18	1·20	19	29[a]
Enterobacter cloacae, Calamovilfa longifolia	182	25	2·42[a]	8	30	86[a]	32[a]	1·09[a]	5	27
Klebsiella pneumoniae, Panicum virgatum	233[a]	38[a]	3·05	6	24	79	19[a]	2·63[a]	10	13
Unknown, Bouteloua hirsuta	229[a]	31[a]	3·21	18	19	78	17	2·20[a]	1[a]	15
Enterobacter cloacae, Schizachyrium scoparium	170	25	2·79	8	16	84	19	2·59[a]	20	13
Bacillus polymyxa, Koeleria pyramidata	166	15	3·31	18	27	77	17	2·93[a]	5	18
Control (−N)	166	17	3·70	10	34	66	22	1·42	12	8
Control (+N)	213[a]	16	4·73[a]	11	31	64	12	4·43[a]	21	21

DW, dry weight of shoots and roots.
W_S/W_R, dry weight of the shoot over the dry weight of the root.
L, large (>1 mm) diameter roots. S, small (<1 mm diameter roots.
Control (−N) received heat-killed inoculum and 1/5 strength −N Hoagland's Solution. Control (+N) received 1/5 strength Hoagland's Solution.
Values marked with a are significantly different from Control (−N), $P < 0·05$.

of statistically significant differences, there was a trend for generally higher levels of VAM infection in *S. scoparium* and generally lower levels of VAM infection in *B. gracilis* in diazotroph inoculated treatments. This last observation is consistent with the presumed relatively lesser importance of VAMs for fibrous root sytems. Finally, there appeared to be a tendency for the treatments exhibiting reduced growth to have a higher incidence of pathogenic root fungal infections.

Localization of bacteria

The bacteria tend to occupy two areas in considerable numbers. One location is the protrusion site of branch roots which are rimmed by active bacterial cells. Another is the middle lamella region along the epidermal tissue. Both sites presumably provide abundant supplies of organic nutrition for the bacteria. Often there are scattered groups of cells along the root which fluoresce red, an indication of high metabolic activity. Where such regions occur, bacterial cells are present in large numbers along the middle lamellae of all the active cells but occur only sparsely along the middle lamellae of less active cells. To a lesser extent, bacteria are associated with root hairs.

DISCUSSION

Rates of dinitrogen fixation associated with the roots of temperate and subtropical grassland systems range from <1 to 18 kg N ha^{-1} year^{-1} (Evans & Barber 1977; also see several entries in Stewart 1975). The numerous factors affecting measurement of dinitrogen fixation make it difficult to interpret the significance of fixed N input (van Berkhum & Bohlool 1980). Rates of dinitrogen fixation vary with temperature and soil moisture and exhibit large fluctuations which may be due to pulses of nutrient availability. In the Arapaho Prairie site, the associative bacteria contribute very little N (<1 kg ha^{-1} annually), whereas cyanolichen crusts contribute 2–3 kg ha^{-1} annually at Arapaho (L. A. Kapustka, unpubl.).

   Plants inoculated with various associative diazotrophs often respond with increased growth, increased yield, and altered root morphology (Smith *et al.* 1976; Baltensperger *et al.* 1979; Abdel Wahab & Wareing 1980; Kapulnik *et al.* 1981; Avivi & Feldman 1982; Kapulnik *et al.* 1982). Brown (1974) indicated that *Azotobacter* inoculation of corn in eastern European countries caused significant yield increases. In Israel, corn yield has been enhanced $8\cdot3$–$16\cdot0\%$ by field inoculations with *Azospirillum*. There are conflicting reports regarding the role of dinitrogen fixation in mediating these responses. Some diazotrophic bacteria are known to exude pectolytic substances (Umali-Garcia *et al.* 1980) and plant hormones (Tien, Gaskins & Hubbell 1979; Inbal & Feldman 1982). These bacterial products may have a major role in causing the observed plant responses. The collective effect of these plant responses to associative diazotrophs may aid the plant in establishing an extensive root system early after germination resulting in obvious ecological advantages in the acquisition of water and nutrients.

ACKNOWLEDGMENTS

This work was supported by the National Science Foundation, Grant Number DEB-8004172.

REFERENCES

Abdel Wahab, A.M. & Wareing, P.F. (1980). Nitrogenase activity associated with the rhizosphere of *Ammophila arenaria* L. and effect of inoculation of seedlings with *Azotobacter. New Phytologist*, **84**, 711–721.

Avivi, Y. & Feldman, M. (1982).The response of wheat to the genus *Azospirillum. Israel Journal of Botany*, **31**, 237–246.

Baltensperger, A.A., Schank, S.C., Smith, R.C., Littell, R.C., Bouton, J.H. & Dudeck, A.E. (1979). Effect of innoculation with *Azospirillum* and *Azotobacter* on turf-type bermuda genotypes. *Crop Science*, **18**, 1043–1045.

Barnes, P.W. & Harrison, A.T. (1982). Species distribution and community organization in a Nebraska Sandhills mixed prairie as influenced by plant/soil-water relationships. *Oecologia*, **52**, 192–201.

Biermann, B. & Lindermann, R.G. (1981). Quantifying vesicular–arbuscular mycorrhizae: a proposed method towards standardization. *New Phytologist*, **87**, 63–67.

Brown, M.E. (1974). Seed and root bacterization. *Annual Review of Phytopathology*, **12**, 181–197.

DuBois, J.D. & Kapustka, L.A. (1983). Biological nitrogen influx in an Ohio relict prairie. *American Journal of Botany*, **70**, 8–16.

Evans, H.J. & Barber, L.E. (1977). Biological nitrogen fixation for food and fiber production. *Science*, **197**, 332–339.

Inbal, E. & Feldman, M. (1982). The response of a hormonal mutant of common wheat to bacteria of the genus *Azospirillum. Israel Journal of Botany*, **31**, 257–263.

Johnen, B.G. & Drew, E.A. (1979). Morphology of microorganisms stained with europium chelate and fluorescent brightener. *Soil Biology and Biochemistry*, **10**, 487–494.

Kapulnik, Y., Okon, Y., Kigel, J., Nur, I. & Henis, Y. (1981). Effects of temperature, nitrogen fertilization, and plant age on nitrogen fixation by *Setaria italica* inoculated with *Azospirillum brasilense* (strain cd). *Plant Physiology*, **68**, 340–343.

Kapulnik, Y., Sarig, S., Nur, I., Okon, Y. & Henis, Y. (1982).The effect of *Azospirillum* inoculum on growth and yield of corn. *Israel Journal of Botany*, **31**, 247–255.

Kapustka, L.A. (1982). The significance of asymbiotic dinitrogen fixation in grasslands. *Proceedings of the 7th North American Prairie Conference* (Ed. by C.L. Kucera). Southwest Missouri State University, Springfield, MO. University of Missouri Press, Columbia.

Keeler, K.H., Harrison, A.T. & Vescio, L.S. (1980). The flora and sandhills prairie communities of Arapaho Prairie, Arthur County, Nebraska. *Prairie Naturalist*, **12**, 65–78.

Lowry, O.H., Rosenbough, N.J., Farr, A. & Randall, R.J. (1951). Protein measurement with the folin phenol reagent. *Journal of Biological Chemistry*, **193**, 265–275.

Rennie, R.J. (1980). Dinotrogen-fixing bacteria: computer assisted identification of soil isolates. *Canadian Journal of Microbiology*, **26**, 1275–1283.

Rennie, R.J. (1981). A single medium for the isolation of acetylene-reducing (dinitrogen fixing) bacteria from soils. *Canadian Journal of Microbiology*, **27**, 8–14.

Smith, R.L., Bouton, J.H., Schank, S.H., Quesenberry, K.H., Tyler, M.E., Gaskins, M.H. & Littell, R.C. (1976). Nitrogen fixation in grasses inoculated with *Spirillum lipoferum. Science*, **193**, 1003–1005.

Stewart, W.D.P. (1975). *Nitrogen fixation by free-living microorganisms.* IBP Synthesis Vol. No. 6. Cambridge University Press, London.

Stewart, W.D.P. (1982) Nitrogen fixation—its current relevance and future potential. *Israel Journal of Botany*, **31**, 5–44

Tien, T.M., Gaskins, M.H. & Hubbell, D.H. (1979). Plant growth substances produced by *Azospirillum brasilense* and their effect on the growth of pearl millet (*Pennisetum americanum* L.). *Applied Environmental Microbiology*, **37**, 1016–1024.

Umali-Garcia, M., Hubell, D.H., Gaskins, M.H., & Dazzo, F.B. (1980). Association of *Azospirillum* with grass roots. *Applied and Environmental Microbiology*, **39**, 219–226.

van Berkhum, P. & Bohlool, B.B. (1980). Evaluation of nitrogen fixation by bacteria in association with roots of tropical grasses. *Bacteriology Review*, **44**, 491–517.

Woodmansee, R.G. (1979). Factors influencing input and output of nitrogen in grasslands. In *Perspectives in Grassland Ecology* (Ed. by N.R. French) Springer-Verlag, New York.

Influence of fungal growth pattern on decomposition and nitrogen mineralization in a model system

K. PAUSTIAN

*Department of Ecology and Environmental Research, Swedish University of Agricultural Sciences,
S-750 07 Uppsala, Sweden*

SUMMARY

1 A model of fungal growth in relation to organic matter decomposition and nitrogen mineralization is presented. The model considers fungal biomass in terms of a potentially active cytoplasmic fraction and an inactive fraction consisting of cell walls lacking cytoplasm.

2 The fungal growth pattern proposed by the model assumes that growth resources will be differentially allocated to cell wall synthesis and cytoplasm synthesis, depending on substrate availability. As a result of this growth pattern, the model predicts that the relative nitrogen requirement of the fungal biomass decreases as decomposition proceeds.

3 Model simulations are compared to data on fungal growth in submerged culture. Data on FDA-active biomass support model predictions of the effects of substrate availability on fungal biomass dynamics.

4 Simulations of the effects of fungivore grazing on decomposition rates and nitrogen mineralization are presented. Selection by fungal consumers of active or inactive parts of the biomass is proposed as an important determinant of the effects of grazing on fungal activity.

INTRODUCTION

Organic matter decomposition and nitrogen mineralization–immobilization are biologically mediated processes. Consequently, biomass dynamics and activity of soil micro-organisms and soil fauna are important elements in models describing these processes. Many models of decomposition and N mineralization (*cf.* van Veen *et al.* 1981) do not include soil organisms as explicitly defined components. Such models can have considerable predictive value for a specific set of conditions but they are less appropriate for examining process-controlling mechanisms and for predicting effects of system perturbations (van Veen *et al.* 1981).

The majority of carbon and nitrogen cycling models explicitly including micro-organisms generally represent them as a single component (Parnas 1975; Smith 1979; van Veen & Frissel 1981) or more infrequently they divide micro-organisms into subgroups such as inactive and active microbes (Hunt 1977), or bacteria and fungi (McGill *et al.* 1981).

159

Fungi have received relatively little attention by modellers interested in decomposition and N cycling, although several models specific to fungal growth (Prosser 1979) have been developed for laboratory growth studies. Existing models of nitrogen cycling (*cf.* Frissel & van Veen 1981) that include a micro-organism component typically employ growth representations, such as Monod-type kinetics, which were originally developed for describing growth in bacterial cultures. Fungi, because of their hyphal form, possess characteristics which may make such a treatment of growth processes insufficient. These characteristics include biomass increase through hyphal extension and branching, the active penetration of hyphae into the substrate matrix and the ability to recycle materials within the mycelium via cytoplasm translocation and self-lysis. The present model attempts to consider some of the special growth characteristics of fungi in order to investigate process mechanisms in decomposition and nitrogen mineralization–immobilization.

Model predictions of fungal biomass dynamics are compared to laboratory data on fungal growth in submerged culture. Simulation experiments of fungus-grazer interactions and the effects of grazer food selection on fungal activity, decomposition and nitrogen mineralization are presented.

MODEL THEORY

Growth and carbon utilization

A complete mathematical description of the model (K. Paustian & J. Schnürer, unpubl.) is being prepared, but the central concepts of the model will be outlined here as a background for the simulation experiments discussed later. The model (Fig. 1) considers both carbon and nitrogen flows. Litter is divided into two functional components, labile and structural, as proposed by McGill *et al.* (1981). The labile fraction represents readily decomposable material associated with cytoplasmic materials of plant and microbial cells, while the structural component represents more recalcitrant materials having a lower nitrogen content. The fungal biomass is divided into two compartments: cytoplasmic hyphae and evacuated hyphae. This division is supported by observations that many hyphae in soil appear to consist of cell walls lacking cytoplasm with the remaining hyphal lengths being partially or completely filled with cytoplasm (Frankland 1975). The grazer component (Fig. 1) is presented later.

Metabolic processes are modelled as a function of the hyphae containing cytoplasm; the evacuated hyphae do not contribute to uptake or growth processes. The cytoplasmic fraction in the model is not necessarily 'active' biomass but is conceptualized as potentially active biomass with activity dependent on growth-determining factors such as substrate availability, temperature and moisture.

A central feature of the model is that hyphal spread (expressed by cell wall synthesis) and biomass increase are modelled as separate but inter-related processes. The model formulations for growth and uptake are illustrated for the simple case of a single limiting-substrate.

(a)   (b)

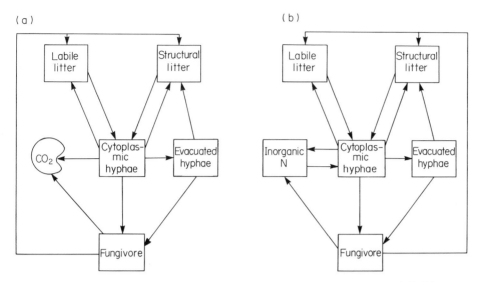

FIG. 1. Model structure and material flows of (a) carbon and (b) nitrogen. Litter is divided into two components, labile and structural. Fungal biomass is divided into hyphae containing cytoplasm and hyphae lacking cytoplasm.

Cell wall synthesis (H) is modelled as a function of a potential specific growth rate (a), cytoplasmic fungal biomass (B), and the fraction of a unit of hyphal biomass (expressed as carbon) that constitutes the cell wall (f).

$$hH = a \cdot B \cdot f$$

where:

$$a = v_m \cdot e \tag{1}$$

The specific growth parameter (a) is equivalent to the product of the maximum specific uptake rate or the biomass (v_m) and the potential growth yield efficiency (e). Maintenance respiration is treated in the model but is omitted here for the sake of simplicity.

The amount of substrate available for biomass synthesis depends on the specific uptake parameter (v_m), the amount of cytoplasmic hyphae (B) and a function describing the relative substrate availability (S). The substrate availability function depends on the type of substrate system being modelled, but it varies from 1 (when substrate availability is not limiting) to 0 (when substrate is exhausted).

The synthesis of cytoplasm (P) is expressed as the difference between total non-respired assimilate and the amount allocated to cell wall synthesis.

$$P = v_m \cdot e \cdot B \cdot S - a \cdot B \cdot f \tag{2}$$
$$= a \cdot B(S - f)$$

Biosynthates (C and N) are thus allocated to cell wall and cytoplasm synthesis depending on relative substrate availability. At decreasing substrate availabilities, proportionally more biosynthate is allocated to cell wall synthesis (hyphal spread) and additional cytoplasm is made available to the new growth through translocation. Translocation appears to occur in a wide variety of fungi (Jennings *et al.* 1974), although it is not a universal phenomenon among fungi and the rates and extent to which it occurs, particularly in soil, are poorly known (Jennings 1976). It has been suggested that translocation of cytoplasm occurs extensively in rapidly growing 'exploitative' species (Harley 1972).

Translocation of cytoplasm (T) to new growth is calculated as the difference between the amount of cytoplasm required to complement hyphal wall synthesis (H) and the actual synthesis rate of cytoplasm (P).

$$T = [a \cdot B(1 - f) - a \cdot B(S - f)]$$
$$= a \cdot B(1 - S) \tag{3}$$

The rate of formation of evacuated hyphae (E) is then,

$$E = \frac{f}{(1 - f)} \cdot a \cdot B(1-S) \tag{4}$$

and the change in the cytoplasmic biomass is equal to the difference between total (non-respired) carbon assimilation and the hyphal evacuation rate,

$$\frac{dB}{dt} = a \cdot B \cdot S - E$$
$$= a \cdot B \frac{(S - f)}{(1 - f)} \tag{5}$$

The model assumes that maximum hyphal outgrowth, expressed by cell wall synthesis, will be maintained over a wide range of growth conditions. However, when substrate availability and consequently uptake rates become low, the specific rate of cell wall synthesis decreases.

Nitrogen dynamics

Biomass and substrate nitrogen as well as inorganic nitrogen pools are included in the model (Fig. 1). Uptake of organic nitrogen follows carbon uptake, according to the C/N ratio present in the substrate pools. Inorganic nitrogen may be taken up or released depending on the nitrogen requirement for synthesis of new biomass. The nitrogen requirement for growth depends on the rate of synthesis of cell walls and cytoplasm and the C/N ratios of the two fractions. The nitrogen content of fungal cell walls varies widely between species and in response to nutritional conditions

(Cochrane 1958). Published C/N ratios for fungal cell walls range from 8 to 107, although most reported values are between 15 and 30 (K. Paustian & J. Schnürer, unpubl.). Cell cytoplasm N contents also vary with the nutritional status of the fungi and the formation of storage products in the cell (Cochrane 1958). However, the cytoplasm will have generally a higher N content than the cell wall and an average C/N of 3 has been suggested by McGill *et al.* (1981).

Due to the differential N content of the cell wall and cytoplasm fractions, the fungal biomass N requirement varies depending on the growth allocation described earlier. If substrate availability is high, for example in the early colonization of fresh litter, the allocation of growth resources to cytoplasm synthesis will be high and consequently the biomass will have a high N requirement relative to biomass increase. As substrate availability drops, the relative rate of cytoplasm synthesis decreases, accompanied by translocation from older hyphae, and the N requirement for synthesis of new biomass decreases. To use a numerical example, assume C/N ratios of 5 and 25 for cytoplasm and cell walls, respectively, a 40% yield efficiency and that 60% of hyphal carbon is associated with the cell wall. Then the nitrogen content of the substrate required to meet growth needs varies from 1·9% N (C/N = 25), where growth allocation to cytoplasm synthesis is highest, to 0·7 N (C/N = 63) where translocation is maximized.

When the supply of substrate nitrogen is insufficient to meet growth requirements, available inorganic nitrogen is immobilized. If total N uptake is insufficient to supply current growth requirements, N is first allocated to cell wall synthesis and cytoplasm synthesis rates are reduced, with an accompanying increase in cytoplasm translocation. With even greater N limitation the rate of cell wall formation is decreased and new cytoplasm synthesis ceases. Such an allocation pattern is supported by growth patterns of fungi at high carbon supply rates in liquid culture, where the final growth phase may depend on mobilization of nitrogen from older hyphae for use in new growth (Cochrane 1958). Levi, Merrill & Cowling (1968) found evidence that wood-decaying fungi maintained growth in extremely nitrogen-limited conditions by using nitrogen translocated or autolysed from older hyphae.

MODEL SIMULATIONS

Simulation of fungal growth in submerged culture

A central hypothesis in the model is that limiting growth resources are preferentially allocated to cell wall synthesis to maintain high rates of hyphal spread. As a result, part of the existing biomass loses cytoplasm and becomes inactive in terms of contributing to growth. It is therefore desirable to compare predicted model behaviour to growth experiments where total biomass as well as the metabolically active portion of the biomass have been measured.

One method proposed for measuring the metabolically active part of total fungal biomass is the fluorescein-diacetate (FDA) staining procedure (Söderström 1977). Values for FDA-stained biomass measured in soil are often quite low

(5–10%) and the method has been criticized for underestimating the active fraction of the biomass (Söderström 1979). However, in laboratory conditions with pure cultures and less disruptive sample preparation, FDA-stained hyphae may be highly correlated with metabolically active biomass. Ingham &Klein (1982) measured FDA-stained biomass and total biomass as well as O_2 and glucose utilization for two species of fungi grown in liquid culture. One of the fungi, *Penicillium citrinum*, commonly grows in pellet form in liquid culture, while the other, *Rhizoctonia solani*, grows in a more diffuse form. Incubation series of 1 and 3 g glucose l^{-1} for *P. citrinum* and 1, 3, 10 and 30 g glucose l^{-1} for *R. solani* (Ingham 1981) are compared here to model simulations.

Model parameterization

The model simulations were intended to test whether the growth-controlling factors postulated in the model could reflect the observed behaviour of two fungi with different growth characteristics over a range of substrate concentrations. To provide a robust test of the model, parameter values derived independently from the comparison data sets were used. No attempt, with one exception, was made to adjust model parameters or to create additional parameters to optimize the model fit to the data.

Maximum specific growth rates of $0·13$ h^{-1} for *P. citrinum* and $0·1$ h^{-1} for *R. solani* were estimated from hyphal outgrowth rates measured in a preliminary experiment (Ingham & Klein 1982). Maintenance coefficient ($0·02$ h^{-1}) and maximum potential yield ($0·5$) for *Penicillium* were taken from Righelato (1975) Literature data on the growth efficiency of *Rhizoctonia* was not available. Therefore, the potential yield coefficient ($0·3$) was estimated from observed yields during early growth phase (Ingham 1981) where the maintenance portion of yield should be low. The maintenance coefficient used was the same as for *Penicillium*. Substrate availability (S) was expressed as a hyperbolic function of the form,

$$S = \frac{Q}{(kB + Q)} \tag{6}$$

where Q is the solution concentration (g C l^{-1}), B is cytoplasmic fungal biomass and k a constant. This modification of the Monod equation was used by Contois (1959), who found a strong relationship between the specific uptake rate, biomass density and substrate concentration in both batch and continuous culture. Ingham's (1981) results indicated that glucose utilization rates were much lower than would be expected if substrate affinity were independent of biomass concentration. In this case the equation was used simply as an empirical relationship and a value for k was selected which gave reasonable simulated endpoints for substrate concentration. A single k value for each fungi was used for all the incubation series.

Simulation results

Simulated total fungal biomass and carbon substrate concentration for *P. citrinum* are shown in Fig. 2. There was a period of accelerating biomass increase during the early growth phase with the period of growth rate increase being slightly longer at the higher substrate concentration (Fig. 2b). After 50–60 hours the model showed a fairly linear increase in total biomass for the remainder of the simulation. This apparent linearity in biomass increase is frequently observed in liquid culture (Cochrane 1958). Although the model assumes that growth of active hyphae is inherently autocatalytic (i.e. exponential), the decline in the relative amount of active to total biomass resulted in the linear trend in total biomass increase. The model showed fairly good agreement with measured biomass and glucose carbon concentration for both incubations. In the 1 g l^{-1} glucose incubation the model underestimated total biomass at the end of the experiment and predicted substrate concentrations were somewhat below the measured values (Fig. 2a). In this case the observed yield of biomass produced per unit glucose consumed (64%) was unusually high and exceeded the maximum yield coefficient (50%) in the model. However, the potential yield coefficient is generally assumed to be independent of substrate concentration; therefore, the same value was used at both concentrations.

The fraction of cytoplasmic fungi in the model exhibited dynamics similar to the measured FDA-stained fraction of the biomass (Fig. 3). Simulated cytoplasmic biomass exceeded the FDA-stained fraction during the early growth phase, but was nearly equal at the end of incubation. It is reasonable that the cytoplasmic fraction should exceed the FDA-staining portion of the biomass, since some active hyphae may be ruptured in the staining procedure (Söderström 1979) and some cytoplasm-containing hyphal lengths may not hydrolyse FDA (E. Ingham, pers.

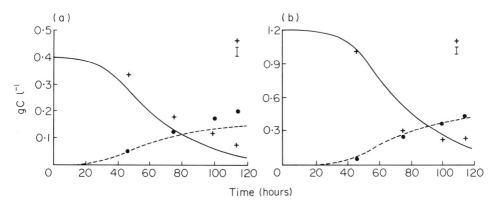

FIG. 2. Simulation of total fungal biomass and substrate concentration for *P. citrinum* with intial glucose concentrations of (a) 1 g glucose l^{-1}, (b) 3 g l^{-1}. Simulated (——) labile C and measured (+) glucose and simulated (– – –) and measured (●) fungal biomass. Bracket in upper right shows least significant difference, $P<0.05$. Measured glucose and fungal biomass dry weight from Ingham (1981) converted to carbon mass assuming 40% and 45% carbon content, respectively.

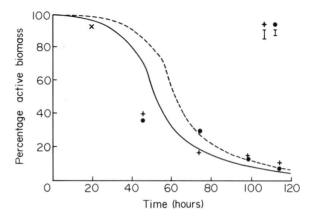

FIG. 3. Simulated percentage cytoplasmic biomass for *P. citrinum* for 1 g l^{-1} incubation (——) and 3 g l^{-1} glucose incubation (– – –). Measured percentage FDA-staining biomass (●) and (+), for 1 and 3 g glucose l^{-1} incubations, respectively. FDA-staining biomass in the early growth phase (×) was measured in a separate experiment, 18 hours after inoculation. Brackets in upper right show least significant difference, $P < 0.05$. Experimental data from Ingham (1981).

comm.). The differences in cytoplasmic hyphae predicted by the model as a function of substrate concentration were not clearly reflected in the data. The FDA-fractions in the 3 g l^{-1} incubation were slightly higher at both the early and late phases of growth (Fig. 3), but the differences were not significant.

Simulated biomass dynamics for *R. solani* in the 1 and 3 g l^{-1} series gave reasonable fits to the data and showed approximately linear increases in total hyphae for most of the incubation period (Fig. 4). At the lowest glucose concentra-

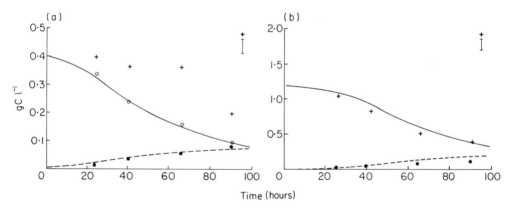

FIG. 4. Simulation of total fungal biomass and substrate concentration for *R. solani* with initial glucose concentrations of, (a) 1 g glucose l^{-1} and (b) 3 g glucose l^{-1}. Simulated (——) labile C measured (+) glucose and simulated (– – –) and measured (●) fungal biomass. Bracket in upper right shows least significant difference, $P < 0.5$. Carbon substrate concentrations (○) were calculated from respiration and biomass data where glucose measurements were unrealistic. Experimental data from Ingham (1981).

tion there was a large discrepancy between measured glucose values and the labile pool simulated by the model (Fig. 4a). In this case the measured glucose values were unreasonably high during the first half of the incubation. Considerable growth had occurred by 70 hours while glucose, which was the sole carbon source, remained almost unchanged from the initial concentration. Another estimate of substrate carbon concentration can be made using reported data on respiration rates (Ingham 1981). The difference between the initial glucose carbon addition and cumulation CO_2-C losses (estimated from O_2 utilization rates) plus measured biomass C was calculated for the 1 g l^- incubation (Fig. 4a). An important assumption is that the point estimates of respiration rate were taken at times that reasonably define the cumulative respiration curve (calculated by integrating straight-line segments between measurement points). In this case the calculated substrate concentration is more appropriate to the observed biomass increases than the measured glucose values (Fig. 4a).

Simulations of fungal biomass and substrate utilization for *R. solani* at high glucose concentrations did not fit the data well. In the 10 g l^{-1} incubation total biomass increases were overestimated during the latter half of the experiment (Fig. 5). In the data, total biomass increased only slightly after 50 hours, although glucose uptake rates were high, and the measured biomass yield based on glucose utilization, after 124 hours, was very low (7%). Nitrogen and other mineral nutrients were supplied in excess and the data indicates that carbon availability did not limit growth. The accumulation of toxic metabolites may retard or stop fungal growth where very concentrated media is used (Cochrane 1958) and this may have caused the low biomass yield observed. In that case it would be necessary to include this factor to model successfully growth in culture at high substrate concentrations. However, this aspect of growth would probably not be of concern in modelling fungal growth and litter decomposition in soil where substrate concentrations are generally low.

Simulated cytoplasmic hyphae exceeded the FDA-stained hyphal fraction dur-

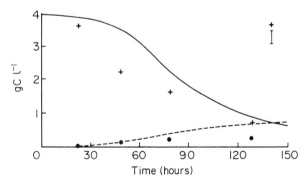

FIG. 5. Simulation of total fungal biomass and glucose for *R. solani* with an initial glucose concentration of 10 g l^{-1}. Simulated (—) and measured (+) glucose and simulated (---) and measured (●) fungal biomass. Bracket in upper right shows least significant difference, $P < 0.05$. Experimental data from Ingham (1981).

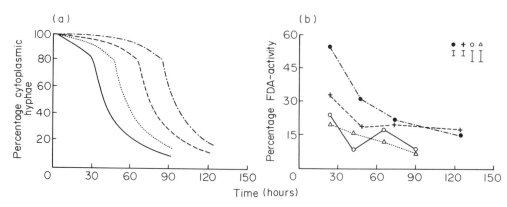

FIG. 6. Simulated percentage cytoplasmic hyphal biomass (a) and measured values of FDA-staining biomass (b) for *R. solani*. Simulated values are for 1 (——), 3 (......), 10 (– – –), 30 (–·–·) g glucose 1^{-1} initial concentration. Measured values for 1 (O), 3 (△), 10 (+) and 30 (●) g glucose 1^{-1} (Ingham 1981). Brackets in upper right show least significant difference, $P < 0.05$. Note the different scaling of the y-axis.

ing most of the incubation (Fig. 6). It is difficult to evaluate quantitatively model predictions of active biomass amounts, due to the characteristics of the FDA technique discussed above. However, the model shows the same qualitative relationship between active biomass dynamics and substrate availability as the data. Particularly in early growth phase, percentage active biomass was greater with increasing initial glucose concentration (Fig. 6). Although *R. solani* exhibited unbalanced growth patterns during late growth phase at high glucose concentrations, substrate utilization rates increased during the early growth phase with increasing initial glucose concentration (Ingham & Klein 1982). Thus, the data appear to support qualitatively the relationship between substrate utilization and the dynamics of the cytoplasmic fraction of the biomass proposed by the model.

Grazing effects on decomposition and N mineralization

It is generally accepted that soil fauna play an important role in the decomposition processes of most terrestrial ecosystems. A number of studies have focused on the effect of microbivores on microbial populations and decomposition and nitrogen mineralization rates. Several microcosm studies of fungus-grazer systems have shown increased carbon loss rates in the presence of grazers (Hanlon & Anderson 1979; van der Drift & Jansen 1977) while other experiments have reported no significant effects (O. Andrén, unpubl.; Hassal, Parkinson & Visser 1983). One proposed mechanism by which grazing may enhance microbial activity and decomposition is the recycling of growth-limiting nutrients to the microbes and the enrichment of substrate pools by faeces and other rejecta (van der Drift & Jansen 1977). To explore the hypothesis that grazing stimulates fungal activity by removing degenerate biomass and increasing inorganic nitrogen turnover, a number of simulations were performed. Since the model considers both active and inactive fractions of the biomass, the possible effects of different grazer food selection patterns on fungal activity could be examined.

Three types of grazer selection were modelled: (i) selective feeding only on cytoplasmic biomass, (ii) selective feeding on evacuated hyphae, and (iii) non-selective feeding, where consumption of active and inactive fractions is determined by their amount relative to total fungal biomass.

Model parameterization

An additional component was incorporated into the model to represent a fungal grazer capable of different food selection habits (Fig. 1). Consumption of hyphae was dependent on grazer biomass, a consumption rate constant and a linear function of fungal biomass density. In addition, a density-dependent factor was used to limit the maximum grazer biomass. Carbon assimilation by the grazer was calculated as 60% of consumption and 60% of assimilated carbon was assumed to be respired, with the remaining carbon representing production of grazer biomass. Assimilation efficiency of consumed nitrogen was 85% and nitrogen not required for production (grazer C/N = 6) was released as inorganic nitrogen. The carbon energetics quotients, representative for microbivorous nematodes (Sohlenius 1980), and the assumption about nitrogen assimilation were chosen to maximize the release of nitrogen in order to stimulate fungal growth in nitrogen-limited conditions.

To simulate plant litter decomposition, the structural litter component was included in the simulations. The simulations of pure culture growth discussed earlier employ only the labile substrate component (for glucose). Two litter types were simulated, with carbon to nitrogen ratios of 30 and 70, representing non-nitrogen-limited and nitrogen-limited conditions, respectively. In defining their functional litter fractions, McGill et al. (1981) assume both the structural and labile fraction have a constant C/N and the overall litter C/N determines the relative amount of each fraction. In the present study the C/N of the structural component was set to 100 in both cases and the labile fraction C/N was set to 10 in the case where nitrogen is not limiting. The C/N of the labile fraction for the nitrogen-limited case was increased to 30. A higher C/N was used to provoke a greater fungal growth response to nitrogen release by grazers and to give a more equal split between labile and structural fractions than if a C/N of 10, for the labile fraction, was used in both cases. This minimized differences in initial decomposition rates, due to the greater recalcitrance of the structural component, which masks the effect of nitrogen limitation on growth. Given litter C/N ratios of 70 and 30, the initial labile fractions were 18% and 26% and the structural fractions were 82% and 74% of total litter carbon, respectively.

The labile substrate uptake rate constant was set to 0.05 h^{-1}. Assuming a yield coefficient of 40%, this is equivalent to a relative growth rate of 0.02 h^{-1}, which falls in the mid-range of microbial growth rates in field studies where easily-decomposable carbon was added to soil (McGill et al. 1981, p. 71). The uptake rate constant for the structural component was given a lower value (0.02 h^{-1})

to reflect the slower decomposition of this material. Substrate availability was expressed as the amount of each fraction relative to total carbon, with a half-saturation constant included for the labile component. The initial litter quantity was 10 g carbon and grazing was initiated 48 hours from the start of the simulation to allow for an initial fungal colonization phase.

Simulation results

Total respiration was higher in the low C/N litter than in high C/N litter in all treatments (Table 1). This was expected given the lack of N limitation to growth and the higher proportion of labile material in the initial litter quantity. Grazing resulted in reduced weight loss and CO_2 evolution in all treatments, except where only evacuated hyphae were grazed (Table 1). The negative effect of grazing on respiration was greatest in the low C/N litter where cytoplasmic biomass was selectively grazed. In this case there was a rapid build-up of the grazer population due to the initial high rate of fungal production. The non-selective grazing had a slightly negative effect (<10%) on CO_2 production in both litter types. Total carbon and total fungal biomass development for the high C/N litter simulations are shown in Fig. 7.

Nitrogen mineralization was increased by grazing in all treatments (Table 1). Grazing reduced fungal N immobilization by reducing fungal production and N was released directly by the grazer. Also the potential for mineralization increased as decomposition proceeded, due to the decline in the relative N requirement of the fungi.

The percentage of cytoplasmic hyphae was generally greater in the low than in the high C/N litter, due to the growth response of the fungi to nitrogen limitation (Table 1). Grazing decreased the percentage of active mycelia when food selection included cytoplasmic hyphae. The percentage active biomass was highest where only evacuated hyphae were grazed (Table 1).

In other simulation trials of the high C/N treatment, grazing intensity was manipulated by varying the time at which grazing was initiated and the density-

TABLE 1. Effects of fungivore food selection on carbon and nitrogen mineralization in litter at two different C/N ratios after a 30-day simulation period

Food selection	Cumulative CO_2 – C (% of initial substrate C)		Net N mineralized (% of initial substrate N)		% Cytoplasmic hyphae (at 720 hours)	
	C/N = 70	C/N = 30	C/N = 70	C/N = 30	C/N = 70	C/N = 30
No grazing	36	58	0	44	19	25
Non-selective grazing	33	54	7	54	15	19
Selective grazing on cytoplasmic hyphae	27	39	17	65	3	1
Selective grazing on evacuated hyphae	38	66	17	61	49	53

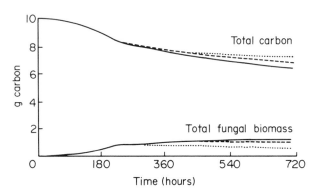

Fig. 7. Simulated total carbon (fungi + litter + grazer) and total biomass (cytoplasmic + evacuated) with litter C/N = 70: (——) no grazing, (– – –) non-selective grazing and (......) selective grazing of cytoplasmic biomass.

dependent factor controlling the maximum grazer biomass. No combination, involving non-selective grazing or selective grazing on cytoplasmic hyphae, was found in which fungal activity and CO_2 production was increased over the no-grazing treatment. The positive effect of additional N mineralization was not enough to offset reduced growth caused by grazer removal of active fungal biomass. Where only evacuated hyphae were grazed decomposition rates were increased over the no-grazing control and nitrogen mineralization was high. However, this type of feeding strategy seems unlikely in nature, barring the presence of some toxic or inhibitory substance in the cytoplasmic hyphae. A possible alternative reason for a high degree of selection of inactive hyphae would be due to a difference in availability, for example, where inactive hyphae predominated on litter particle surfaces. In that case the model would predict results intermediate to those of the non-selective grazing and the selection of evacuated hyphae.

Other effects of grazing may be more important in the possible stimulation of microbial activity than the recycling of nitrogen and grazer faeces to fungi that was simulated in the model. Enhancement of colonization rates by spore transport, stimulation of bacterial activity by suppression of the fungal community and increasing litter surface-area are other possible mechanisms (van der Drift & Jansen 1977). Knapp, Elliott & Campbell (1983) found that decomposition of straw with a C/N of 50 was not nitrogen-limited after the first few days of incubation. The model suggests that nitrogen is not an important limiting factor for fungi in most litter types, largely due to the internal recycling of nitrogen made possible by the fungal growth form.

CONCLUSIONS

The original development of the model came from a need to express biomass dynamics in the context of a larger integrated model of N processes in soil. How microbial biomass should be coupled to process rate equations and how these equations should be related to field measurements of process rates and microbial biomass are fundamental problems.

It is widely accepted that a large part of the microbial biomass in soil is inactive most of the time (Clark 1967). Therefore, if process rates, determined in the laboratory for highly active microbial populations, are extrapolated to simulate microbial processes in nature, the results will be meaningless.

The products of microbial turnover, necromass and microbial metabolites, must be considered to determine true decomposition rates of soil litter (Paul & van Veen 1978). The amount and formation rates of microbial necromass and metabolites are important in determining potential nitrogen mineralization rates, since they may constitute a major portion of the most active soil organic matter fraction (Juma & Paul 1981; Parton, Persson & Anderson 1983).

The model presented here addresses both the potentially active biomass and the formation of non-functional biomass or necromass, for soil fungi. The proposed growth mechanism on which the model is based is the differential allocation of growth resources depending on substrate availability and hyphal growth rates. This growth pattern is made possible by the transport of cellular constituents from existing hyphae to the growing tips. Working from these assumptions, analysis of the model generated the hypothesis that the relative nitrogen requirement of the fungal biomass and, therefore, the immobilization potential of the biomass, decreases as decomposition proceeds. Such a change in biomass N requirements may be part of the explanation for the wide range of critical C/N ratios observed in litter decomposition studies (Berg & Staaf 1981). This pattern of fungal growth provides a mechanism by which biomass nitrogen can be conserved, allowing growth in nitrogen-poor environments.

ACKNOWLEDGMENTS

I wish to thank Thomas Rosswall and Johan Schnürer for their helpful comments and discussion and Elaine Ingham for the use of data from her doctoral dissertation. This research was supported by grants from the Swedish Council for Planning and Coordination of Research, the Swedish Council for Forestry and Agricultural Research, the Swedish Natural Science Research Council and the Swedish Environmental Protection Board to the Project 'Ecology of Arable Land, Role of Organisms in Nitrogen Cycling'.

REFERENCES

Berg, B. & Staaf, H. (1981). Leaching, accumulation and release of nitrogen in decomposing litter. *Terrestrial Nitrogen Cycles. Processes, Ecosystem Stategies and Management Impacts* (Ed. by F.E. Clark and T. Rosswall), pp. 163–178, Ecological Bulletin, **33**, Stockholm.
Clark F.E. (1967). Bacteria in soil. *Soil Biology* (Ed. by A. Burges & F. Raw), pp. 15–49, Academic Press, London.
Cochrane, V.M. (1958). *Physiology of Fungi.* John Wiley & Sons, New York.
Contois, D.E. (1959). Kinetics of bacterial growth: Relationship between population density and specific growth rate of continuous cultures. *Journal of General Microbiology*, **21**, 40–50.

Frankland, J.C. (1975). Estimation of live fungal biomas. *Soil Biology and Biochemistry*, **7**, 339–40.

Frissel, M.J. & van Veen, J.A. (Eds) (1981). *Simulation of Nitrogen Behaviour of Soil—Plant Systems.* Pudoc, Wageningen.

Hanlon, R.D.G. & Anderson, J.M. (1979). The effects of Collembola grazing on microbial activity in decomposing leaf litter. *Oecologia*, **38**, 93–99.

Harley, J.L. (1972). Fungi in ecosystems. *Journal of Animal Ecology*, **41**, 1–16.

Hassall, M., Parkinson, D. & Visser, S. (1983). The effects of *Onychiurus subtenuis* on the microbial decomposition of *Populus tremuloides* leaf litter. *New Trends in Soil Biology* (Ed. by P. Lebrun, H.M. André, A. De Medts, C. Grégoire–Wibo & G. Wauthy), pp. 613. proceedings of VIII International Colloquium of Soil Zoology. Louvain-la-Neuve.

Hunt, H.W. (1977). A simulation model for decomposition in grasslands. *Ecology*, **58**, 469–484.

Ingham, E.R. (1981). *Use of fluorescein diacetate for assessing functional fungal biomass in soil.* Ph.D. dissertation, Colorado State University.

Ingham, E.R. & Klein, D.A. (1982). Relationship between fluorescein diacetate-stained hyphae and oxygen utilization, glucose utilization, and biomass of submerged fungal batch cultures. *Applied and Environmental Microbiology*. **44**, 363–370.

Jennings, D.H., Thornton, J.D., Galpin, M.F.J. & Coggins, C.R. (1974). Translocation in fungi. *Transport at the cellular level* (Ed. by M.A. Sleigh and D.H. Jennings), pp. 139–156. Symposium of the Society for Experimental Biology no. 28. Cambridge University Press.

Jennings, D.H. (1976). Transport and translocation in filamentous fungi. *The Filamentous Fungi. Vol. 2. Biosynthesis and Metabolism* (Ed. by J.E. Smith & D.R. Berry), pp. 32–64. Edward Arnold Ltd., London.

Juma, N.G. & Paul, E.A. (1981). Use of tracers and computer simulation techniques to assess mineralization and immobilization of soil nitrogen. *Simulation of Nitrogen Behaviour of Soil–Plant Systems* (Ed. by M.J. Frissel & J.A. van Veen), pp. 145–154, Pudoc, Wageningen.

Knapp, E.B., Elliott, L.F. & Cambell, G.S. (1983). Microbial respiration and growth during the decomposition of wheat straw. *Soil Biology and Biochemistry*, **15**, 319–323.

Levi, M.P., Merrill, W. & Cowling, E.B. (1968). Role of nitrogen in wood deterioration. VI. Mycelial fractions and model nitrogen compounds as substrates for growth of *Polyporus versicolor* and other wood–destroying and wood–inhabiting fungi. *Phytopathology*, **58**, 627–634.

McGill, W.B., Hunt, H.W., Woodmansee, R.W. & Reuss, J.O. (1981). PHOENIX, a model of the dynamics of carbon and nitrogen in grassland soils. *Terrestrial Nitrogen Cycles. Processes, Ecosystem Strategies and Management Impacts* (Ed. by F. E. Clark & T. Rosswall), pp. 49–115, Ecological Bulletins 33, Stockholm.

Parnas, H. (1975). Model for decomposition of organic material by microorganisms. *Soil Biology and Biochemistry*, **7**, 161–169.

Parton, W.J., Persson, J. & Anderson, J.M. (1983). Simulation of organic matter changes in Swedish soils. *Analysis of Ecological Systems: State-of-the-art in Ecological Modelling* (Ed. by W.K. Lauenroth, G.V. Skogerboe & M. Flug), pp. 511–516. Elsevier Scientific Publishing Co., Amsterdam

Paul, E.A. & van Veen, J.A. (1978). *The use of tracers to determine the dynamic nature of organic matter.* Transactions of 11th International Congress of Soil Science, Edmonton, Vol. 3, pp. 61–102.

Prosser, J.I. (1979). Mathematical modelling of mycelial growth. *Fungal Walls and Hyphal Growth* (Ed. by N.H. Burnett & A.P.J. Trinci), pp. 359–384. British Mycological Society Symposium, 2.

Righelato, R.C. (1975). Growth kinetics of mycelial fungi. *The Filamentous Fungi Vol.1. Industrial Mycology* (Ed. by J.E. Smith & D.R. Berry), pp. 79–103, Edward Arnold Ltd., London.

Smith, O.L. (1979). An analytical model of the decomposition of soil organic matter. *Soil Biology and Biochemistry*, **11**, 585–606.

Sohlenius, B. (1980). Abundance, biomass and contribution to energy flow by soil nematodes in terrestrial ecosystems. *Oikos*, **34**, 186–194.

Söderström, B.E. (1977). Vital staining of fungi in pure cultures and in soil with fluorescein diacetate. *Soil Biology and Biochemistry*, **9**, 59–63.

Söderström, B.E. (1979). Some problems in assessing the fluorescein diacetate-active fungal biomass in the soil. *Soil Biology and Biochemistry*, **11**, 144–148.

van der Drift, J. & Jansen, E. (1977). Grazing of springtails on hyphal mats and its influence on fungal growth and respiration. *Soil Organisms as Components of Ecosystems* (Ed. by U. Lohm & T. Persson), pp. 203–209, Ecological Bulletins **25**, Stockholm.

van Veen, J.A. & Frissel, M.J. (1981). Simulation model of behaviour of N in soil. *Simulation of Nitrogen Behaviour of Soil—Plant Systems* (Ed. by M.J. Frissel & J.A. van Veen), pp. 126–144, Pudoc, Wageningen.

van Veen, J.A., McGill, W.B., Hunt, H.W., Frissel, M.J. & Cole, C.V. (1981). Simulation models of the terrestrial nitrogen cycle. *Terrestrial Nitrogen Cycles. Processes, Ecosystem Strategies and Management Impacts.* (Ed. by F.E. Clark & T. Rosswell), pp. 25–48, Ecological Bulletins **33**, Stockholm.

Persistence under field conditions of excised fine roots and mycorrhizas of spruce

R. C. FERRIER AND I. J. ALEXANDER

Department of Botany, University of Aberdeen, Aberdeen, AB9 2UD

INTRODUCTION

About 25% of total net primary production of the trees in temperate forests is below ground in the fine root system (Fogel & Hunt 1983). This value varies with site productivity (Keyes & Grier 1981). Turnover rates are rapid and the fine roots are of considerable importance to carbon and nutrient cycles in the forest ecosystem (Persson 1979, 1980; McClaugherty, Aber & Melillo 1982; Vogt *et al.* 1982, 1983b). Estimates of production and turnover depend on the classification of excavated fine root and mycorrhizal material, usually <1 mm diameter, into such categories as 'live' and 'dead' or 'active' and 'inactive' (Marks, Ditchburne & Foster 1968; Harvey, Larsen & Jurgensen 1976; Santantonio 1978). The value of such estimates depends on the validity of these distinctions and, ultimately, on the criteria used to make them.

In previous work (Alexander & Fairley 1983; Fairley 1983) we have identified 'living' mycorrhizas and fine root tips as having an intact apex, turgid and lighter in colour than the proximal tissue, with a turgid cortex over at least part of their length. 'Dead' tips were uniformly darker in colour, were fissured or wrinkled throughout their length, were friable when manipulated, and often lacked an apex. These definitions are retained in this paper in which we described three separate field investigations involving excised fine roots (80–95% mycorrhizal), where we have found that root tips persist in a 'living' state for at least 9 months after connection with the parent tree has been severed.

EXPERIMENTAL

During an investigation of nitrogen uptake by fine roots nine passive flow lysimeter trays (Jordan 1968) were installed during October 1981 between the F and H layers of the forest floor in a 35-year old Sitka spruce (*Picea sitchensis* (Bong.) Carr) plantation. To exclude roots the edges of four lysimeters were cut and recut every 2 weeks. After 5 months five cores were taken from the forest floor in each lysimeter and the number of living root tips recorded. Living tips were still present in the root-excluded lysimeters although their number (21 ± 9; mean $\pm$ standard error) was significantly lower ($t = 10.876$; $n = 7$; $P < 0.001$) than in the rooted lysimeters (138 ± 11).

In a study of fine root decomposition a number of 10×10 cm nylon litterbags (mesh size 1 mm^2) were placed in the F horizon of the forest floor of a 38-year old

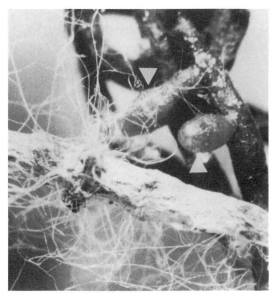

FIG. 1. Excised fine root material after field incubation in a litter bag for 4 months. The indicated root tips would currently be classified as 'active' or 'living'.

Sitka spruce plantation during April 1983. Each bag contained *c*. 1 g freshly excised fine root material consisting of fragments <5 cm in length bearing many turgid live tips. Bags were collected after 2, 4 and 8 months. At the first two sampling times many live tips were still present (Fig. 1) and considerable amounts of mycelium, associated with these tips, had proliferated within the bags. After 8 months the number of live tips had declined and many were turgid only at the extreme tip. Radiating hyphae were still evident.

In the same plantation trenched, root-excluded, areas approximately 2–3 m^2 were established in March 1983 (Harmer & Alexander, p. 267). In December 1983 it was still possible to find living fine roots within the trenched areas. Each month since trenching samples of forest floor from twenty-seven trenched and twenty-seven untrenched areas have been taken for estimation of fluorescein diacetate (FDA)-stained hyphal lengths (Söderström 1977, 1979). Fragments of FDA-stained ectomycorrhizal sheath material are readily recognizable in this procedure by their characteristic surface patterns (Chilvers 1968; Alexander 1981) and were still present in samples taken 9 months after trenching. The average frequency (57%) with which they were encountered in samples during the study period did not differ between trenched and untrenched areas.

DISCUSSION

Our observations show that excised fine root material can persist under field conditions for at least 9 months and that associated mycorrhizal fungi also remain metabolically active. This confirms the observations of Persson (1982) who demonstrated living fine root persistence for 18 months after clear-felling a mature stand of *Pinus sylvestris* L. We cannot tell from this investigation whether this phenomenon represents only the persistence of live roots present at the start of the period or whether normal processes of root extension and short root production also take place after excision.

The most likely explanation for excised root persistence is the existence of carbohydrate reserves in the tissue. The extent of these reserves will depend on the size of the excised fragments: on this basis large roots would last longest, as in Persson's (1982) clear-cut area or our trenched plots, where roots up to 20 cm diameter were severed. However, even small fine root fragments (<5 cm in length) can survive for 4 or even 8 months. Root carbohydrates also show seasonal fluctuations (Van den Driessche 1978; Ericsson & Persson 1980) so the date of excision is likely to influence persistence. In our experiments roots were excised in late autumn or early spring when reserves are thought to be high.

The proliferation of hyphae on excised roots in litter bags was striking, and our observations indicated that much of the mycelium belonged to ectomycorrhizal fungi. Favourable environmental conditions in the bag may be responsible (Lousier & Parkinson 1976) or the exclusion of fungal grazers (Gilbert & Bocock 1962), though this latter cause seems unlikely as in acid forest ecosystems most of the invertebrates would not be excluded by a 1 mm^2 mesh (Vogt *et al.* 1983a). An alternative hypothesis is that the balance between host and mycorrhizal fungus is altered after excision resulting in increased fungal growth or even that eventually the excised root could become a sink for external carbohydrate supplied *via* the mycorrhizal fungus. Field incubation of excised mycorrhizas could be a powerful tool for examining the relationship between host and fungus.

As long as excised mycorrhizal roots remain alive due to the presence of carbohydrate reserves they may continue to act as a sink for mineral nutrients, especially those, such as nitrogen and particularly phosphorus, whose uptake in the transpiration stream is unimportant (Prenzel 1979). The nitrogen and phosphorus storage capacity of the fungal component, even when excised, is well documented (Harley & Smith 1983). This has important implications for studies of field net nitrogen mineralization where uptake into excised roots may result in the depression of apparent rates of mineralization (Harmer & Alexander, p. 267). Trenching experiments will also be affected (Gadgil & Gadgil 1975; Berg & Lindberg 1980).

Under natural conditions the majority of fine roots are not 'excised' but go through a period of senescence (Orlov 1957, 1960; Marks *et al.* 1968) characterized by changes in morphology and anatomy. Some of the material in our litter bags appeared to follow this pattern between 4 and 8 months after excision, when the meristem was the last portion to die. Some natural 'excision' can however be expected due to wind throw, or wind movement of stems, or to other small-scale disturbance of the forest floor. In commercial plantations thinning will have a similar effect. Such events introduce further variability into the estimation of fine root production and mortality from sequential biomass samples.

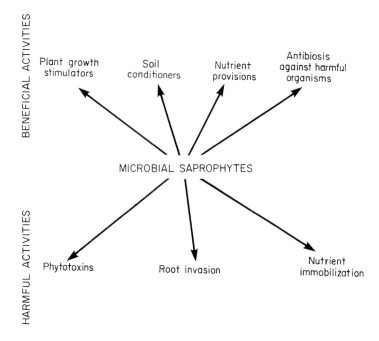

FIG. 1. Effects of microbial saprophytes on plants.

increased greatly in recent years. This may be considered as somewhat surprising because stubble burning has always been an agricultural practice. Virgil in *The Georgics* was puzzled as to 'whether the earth conceives a mysterious strength or sustenance thereby . . .' and it has not been until recent years that the value of stubble burning has been investigated scientifically. The stimulation for this has been the introduction of minimal cultivation systems, resulting in a greater concentration of substrates and hence saprophytic activity in the surface layers of soil. During the seeding of the crop there is a greater chance that the establishing seedling will come under the influence of this saprophytic activity with minimal tillage than under conventional methods of cultivation. The effect of the microbial saprophytes on plant growth has been particularly harmful when the soil is wet and relatively anaerobic. The poor seedling establishment and subsequent crop growth commonly results in a yield reduction of cereals of about 20%, although the effects can vary from smaller yield reductions to total crop failure (Lynch & Elliott 1984). The contrasting agricultural ecosystems to which these observations relate include direct-drilling of cereals in the UK, the stubble-mulch practice for corn grown in the Mid-West of the US and conservation tillage of wheat in the Pacific Northwest of the US. In the UK, it is recommended that straw should normally be removed by burning because there are no obvious advantages in retaining it. By contrast, in stubble-mulch the residues are used to retain soil moisture and in conservation tillage the residues reduce the risks of water erosion on the sloping hillsides which are characteristic of the region. The presence of crop residues can mechanically interfere with drilling (Lynch & Elliott 1984) and in many situations this contributes to loss in yield. However, even when these problems are eliminated, yield reductions still occur.

Crop residues other than straw

Straw, of all crops, is more lignocellulosic than most plant residues which enter the soil. Most straw has about 80% w/w cellulose and hemicellulose with about 14% w/w lignin (Harper & Lynch 1981a). By contrast, grass and leaves have a much smaller lignocellulose content and a much larger proportion of water-soluble components. Between these two extremes there is a vast range of plant residues which enter agro-ecosystems and natural ecosystems. These all serve as substrates for microbial saprophytes which potentially can affect plant growth. The precise nature of the substrate and associated saprophytic population will determine whether there is a net benefit or detriment to plant growth. However, it is the detrimental effects which seem to have received more attention, especially in the agricultural context. For example, when dense infestations of couch grass (*Agropyron repens*) are successfully killed with the herbicide glyphosate, there can be poor establishment of succeeding crops (Lynch & Penn 1980). Similarly when old grass swards are treated with herbicides, there can be poor establishment of a succeeding sward (Gussin & Lynch 1981). Neither of these problems are a consequence of herbicide residues remaining in the soil but result from the microbial saprophytic activity on the plant residues. The succeeding illustrations of the effects of saprophytes are from managed ecosystems but many will be relevant to natural ecosystems. In the managed ecosystem, substrates can become available at one time and at a high density; leaf litter equates with this in natural ecosystems. Root residues of both managed and natural ecosystems are spread throughout the soil profile. When this substrate is 'diluted' it should be recognized that saprophytic activity only occurs in micro-environments and the significance of this to plants can be tested experimentally by supplying small parts of the root system with microbial metabolites (Gussin & Lynch 1982).

NUTRIENT IMMOBILIZATION

The C:N ratio of straw is high (*c.* 120:1; Harper & Lynch 1981a). Decomposer micro-organisms have a much smaller C:N ratio and therefore the tendency is for available soil or fertilizer N to be immobilized during decomposition. Certainly immobilization occurs and evidence of N-deficiency is often apparent in the establishing crop. However, the N which is immobilized is not lost from the soil, but temporarily retained in the microbial biomass. Indeed there could be a net advantage during the winter in fertilized soils because immobilization would prevent mineral N being leached by rainfall and entering water-courses. The initial N-deficiency might be compensated for, depending on the time-course of the release of N from the microbial biomass to the plant. It is likely that a major proportion of N in the microbial biomass would eventually become available directly to the plant; such transfers have been demonstrated in laboratory experiments with ^{15}N by Lethbridge & Davidson (1983). Preliminary field experiments (S.H.T. Harper & J.M. Lynch, unpubl.) indicated a net immobilization of N into straw during autumn with a net release during the early spring. Both observations support the view that straw could, on balance, be beneficial to nutrient cycling in relation to crop plants but clearly point to the need for further experimentation. Such experiments should include field experiments with applications of ^{15}N-labelled fertilizers.

PHYTOTOXIN PRODUCTION

Crop residues and root-derived carbon can serve as substrates for the production of a wide range of microbial metabolites (Lynch 1976). Anaerobic soils usually result in poor crop growth and although the decreased availability of oxygen is partly responsible for the damage, phytotoxins appear to play a major role (Drew & Lynch 1980). Under these conditions fermentative metabolites can form and their subsequent catabolism can be prevented by the lack of oxygen.

Ethylene was identified as a product of anaerobic soils and in the concentrations at which it could be detected, it was toxic to seedling roots growing in solution culture (Smith & Russell 1969). It is formed from crop residues as the general substrates (Goodlass & Smith 1978; Lynch & Harper 1980). In common with other plant growth regulators, it could stimulate root extension at very low concentration (Smith & Robertson 1971). The elimination of the ethylene-producing capacity of a soil by autoclaving indicated that micro-organisms were probably involved (Smith & Restall 1971) and subsequently a fungus, *Mucor hiemalis,* and two yeasts, *Candida vartiovaarai* and *Trichosporon cutaneum* were identified as causative organisms and methionine was identified as a substrate (Lynch 1972). The apparent paradox that a fungus, normally considered as an aerobe, forms the gas under anaerobic soil conditions was resolved by demonstrating that whereas oxygen promoted the formation of an extracellular intermediate from methionine, the breakdown of the intermediate could occur abiotically and in the absence of air (Lynch & Harper 1974). Furthermore, oxygen decreased the catabolism of ethylene (Cornforth 1975). Subsequent claims were made that ethylene generally prevented the growth of aerobes and therefore anaerobic bacteria must be responsible for the formation of the gas (Smith 1973; Smith & Cook 1974). However, the claim that the gas is a general fungistatic agent has never been substantiated. The observations that the antibiotic novobiocin (effective against bacteria) prevented ethylene production by soil treated at 80°C for 30 minutes and greatly retarded production from raw soil, and that cycloheximide (inhibitory to fungi) had no effect (Sutherland & Cook 1980), are superficially difficult to explain. However, the interpretation of results from antibiotic treatment of soils has been traditionally difficult. This is because antibiotics have activity only over a limited spectrum of species and can be degraded rapidly by soil micro-organisms. In further experiments (Lynch 1983a) cycloheximide failed to inhibit ethylene production in soil and failed to prevent growth and ethylene production by pure cultures of *M. hiemalis*. There is therefore no *prima facie* case that fungal saprophytes are not the causative agents of ethylene formation in soil. Aerobic bacteria, but no anaerobes, have been demonstrated to have the capacity to produce ethylene (Primrose 1976). However, the soil biomass is dominated by fungi (Anderson & Domsch 1973) and it seems reasonable to assume that they make the largest contribution to the soil ethylene pool.

Is ethylene formation significant in soil? It certainly can be formed by a wide range of organisms and it accumulates to a greater or lesser extent in all wet soils, often in phytotoxic concentrations (Drew & Lynch 1980). However, other metabolites can also accumulate under these conditions and it is perhaps critical to consider substrate availability for the biochemical pathways. Methionine and, in some static fungal cultures, glutamate (Chalutz & Liebermann 1978; Chou & Yang 1973) are the only clearly identified substrates for ethylene production but seldom present in large concentrations in the soil substrate pool. Even in the rhizosphere, where amino acids are amongst the substrates derived from roots (Rovira 1965), methionine release is likely to be small.

It has already been indicated that cellulosic materials form a major substrate to soil saprophytes. It is therefore appropriate to consider the potential pathways of cellulose decomposition under conditions of restricted aeration. As oxygen becomes limiting in soil, cellulose is converted with a high efficiency to acetic acid (Lynch & Gunn 1978); this was recognized by the early soil microbiologists but was not scientifically substantiated until studies were made of toxic metabolites occurring in flooded soils used to grow paddi rice (Russell 1973). The accumulation is linked to the redox potential of the system and appears maximal at around 0 mV. At negative potentials the acetate itself becomes a substrate for methanogenesis and therefore disappears from the system. However, at such low redox potentials, plant damage will occur from the absence of oxygen (Drew & Lynch 1980) unless the process is only occurring in a limited number of microsites.

That acetic acid should be a major fermentative product from lignocellulose in soil should not be surprising because this pathway is also well established in other ecosystems where lignocellulolysis occurs. These systems include the rumen, silage, anaerobic sediments and the sewage digester. Acetic acid cannot be considered as a conventional phytotoxin as it is only active in millimolar concentrations (Lynch 1977) but there is a general consensus (Tang & Waiss 1978; Wallace & Elliott 1979) that it is important as a cause of crop damage following microbial decomposition of crop residues. Grass (Gussin & Lynch 1981) and weed residues (Lynch & Penn 1980) also serve as suitable substrates for its formation. Acetic acid is only toxic to seedlings when the pH is below neutrality. This is a consequence of the pKa being 4·75 and the acid has to be associated in order to be soluble in the lipid components of root membranes (Lynch 1980). Liming soil does not necessarily minimize the toxicity of the acid because it is difficult to get the neutralizing agent to the site of action, i.e. between the decomposing residue and the establishing seedling.

Having determined that the acid can be produced in the form and concentrations which are active, the next consideration is whether the toxin can move in soil. Inevitably, it serves as a suitable growth substrate for micro-organisms and therefore it is impossible to measure true diffusion. However, there is an exponential decline in concentration with distance from the substrate, the concentration being reduced to one half 1·5 cm from the surface of the straw (Lynch, Gunn & Panting 1980). Thus, with large amounts of crop residue in the surface layers of soil, such as occur with minimal tillage, with heavy leaf litter deposits or following desiccation of grass swards, management of the residues can be approached by diluting the effect with ploughing in the residues to spread them more evenly through the soil profile.

Although acetic acid, and to a lesser extent other organic acids (Drew & Lynch 1980), appear to be important and ubiquitous phytotoxins in residue-enriched soil and they can be expected to be ubiquitous; there will certainly be other phytotoxins. There are a large number of alternative phytotoxic products of micro-organisms which could also be involved in causing crop damage (Lynch 1976). For example patulin, an antibiotic and mycotoxin produced by *Penicillium urticae* (McCalla & Norstadt 1974) was thought to be responsible for the toxicity in stubble-mulch farming. However, subsequent studies by USDA scientists (Elliott, McCalla & Waiss 1978) and ourselves have failed to find phytotoxic concentrations in soil. This is not to say that under some rather specific ecological conditions this and other antibiotics (Brian 1957) will not accumulate, merely that the proof remains to be established. Indeed with the recent excitements in the biological control of plant pathogens (Cook & Baker 1983), it is likely that some antagonists will act effectively by producing antibiotics.

INFECTIONS OF SEEDS AND ROOTS

Pathogens and symbionts are traditionally considered as the major agents of infection of plant roots. However, saprophytes can also become closely associated with seeds and roots and it then becomes difficult to distinguish them from symbionts.

As seeds imbibe, membrane reorganization takes place in the embryo resulting in the release of soluble carbon compounds which are the primary substrates to the spermosphere microflora. A lesser but continuous release of carbon compounds subsequently occurs in this region. In the barley seed, these substrates are released to microbial saprophytes from the embryo end of the seed, which is also the route of water and oxygen entry to the seed (Lynch & Pryn 1977). In conditions of limiting oxygen supply, the seed and microbial saprophytes compete for the available oxygen and micro-organisms are better placed to intercept the oxygen supply (Lynch & Pryn 1977; Harper & Lynch 1981a, b). The response of the seed to the reduced oxygen is to release more substrates, favouring more microbial growth and increasing the saprophyte stress on the seed. After a few days, the seed is unable to sustain the competition and it dies.

The root itself and associated rhizosphere population may also compete for oxygen but it will be less important to the plant because of the potential for oxygen transport from the shoots. However, there are many other ways in which the rhizosphere microflora may exert a positive or negative influence on plant growth. Pathogens can, of course, be amongst the rhizosphere population and some of them have a saprophytic mode prior to becoming parasites. There are few proven cases where straw has provided a substrate base for cereal root pathogens, although the potential obviously exists. For example *Fusarium* spp., common root pathogens, usually have high cellulase activities and are good straw colonizers. Certainly *Fusarium culmorum* is an effective colonist of desiccated grass swards and couch grass rhizomes (Gussin & Lynch 1983; Penn & Lynch 1982). The presence of this pathogen appears to be a major factor in the poor establishment following herbicide treatment of grassland and soils with heavy weed infestations.

Pseudomonads are bacterial saprophytes of straw and these can enter the root cortex, influencing plant growth positively or negatively (Elliott & Lynch 1984a, b). Such colonization of the root cortex has been observed by others (Old & Nicholson 1978; M.P. Greaves, pers. comm.) and it is difficult to distinguish between a saprophytic or parasitic function. They have been variously termed subclinical pathogens, deleterious rhizobacteria and inhibitory pseudomonads but however they are designated they are certainly well placed to influence the plant.

SOIL STABILIZING MICRO-ORGANISMS

It has already been mentioned that straw is retained in North American agriculture for soil conservation purposes. There have been relatively few attempts to relate the improvements in soil structural conditions to the activity of microbial saprophytes decomposing the available substrates as opposed to the straw providing a direct physical binding without the need for decomposition. Martin (1942), Gilmour, Allen & Truog (1948) and Griffiths & Jones (1965) related the improvement in soil aggregate stability to microbial decomposition. More recently Lynch & Elliott (1983) showed the effect of the size of the biomass so produced on the soil aggregation process, and also (Elliott & Lynch 1984a, b) that the larger the N content of straw, the smaller the amount of polysaccharide (measured by viscosity) produced and the smaller the aggregating effect. Chapman & Lynch (1984) analysed the polysaccharide produced and showed it to be characteristic of micro-organisms.

In soils from many parts of the world, poor aggregate stability severely limits crop productivity and this could be improved by promoting saprophytic activity.

SAPROPHYTES AND NUTRIENT CYCLING

The fixation of dinitrogen by free-living micro-organisms in soil is severely limited by available substrates because the nitrogenase enzyme is very expensive in its ATP requirements (Postgate & Hill 1979). One of the major difficulties is that N_2-fixing bacteria are non-cellulolytic and therefore they cannot tap directly the vast pool of lignocellulosic substrates which enter soil. However, the potential of cellulolytic organisms to form suitable associations with these bacteria has been known for some time (Vartiovaara 1938; Jensen & Swaby 1941); such a phenomenon might be termed 'co-operative saprophytism'. The potential magnitude of this effect can be considered in cereal crops producing straw at about 7 Mg ha^{-1}. The co-operative associations appear to gain 6–14 mg N g^{-1} substrate consumed (Jensen & Swaby 1941; Rice & Paul 1972; Lynch & Harper 1983). Thus, for total decomposition of straw in the soil 42–98 kg N ha^{-1} would be produced, the upper limit of which is close to the amount of fertilizer N (*c.* 120 kg) which is applied to produce a crop yielding the same amount of straw in a succeeding year. The microbial biomass N is available for crop production (Lethbridge & Davidson 1983). It should be noted that the N uptake by plants is from the available N pool in soil, some of which is derived from mineralization of organic residues incorporated into soil several years previously.

Presumably N is not the only nutrient that could be made available to plants by saprophytic action. The provision of energy substrates supports large soil biomasses, including the phosphate-solubilizers. In solution culture, micro-organisms affect the uptake of P and other nutrients positively or negatively depending on the uptake period and age of the plant (Barber 1978). In soils where mycorrhizas are absent (e.g. after fumigation) or ineffective, the decreased P uptake could become quite important to the plant, although there is little evidence for the potential magnitude of the effect. There is a need for detailed evaluation of these effects in soil.

CONCLUSIONS: TOWARDS MANIPULATION

In this paper, some beneficial and harmful effects of microbial saprophytes in soil have been outlined. Presumably in most natural and managed ecosystems, there is a balance of these activities which regulate the growth of plants. Many of the effects will be interdependent. For example, the presence of toxins from micro-organisms in soils may pre-dispose a plant host to infection by pathogens (Penn & Lynch 1982; Gussin & Lynch 1983).

In managed ecosystems, some of the activities are modified deliberately by the application of agrochemicals, such as fungicides, whereas there may be inadvertent modification by other agrochemicals, such as N fertilizers, which can depress nitrogenase function.

A current prospect for soil biotechnology (Lynch 1983b) is whether the population balance in soil can be adjusted such that the beneficial organisms dominate a soil. This prospect seems unrealistic for the soil as a whole, but if inocula can be carried to the site of the action, i.e. the substrates themselves, then the prospects seem much brighter. Inoculants would have to have high titres of the necessary enzyme complement to tackle the available substrates. They should also be required to show all the necessary attributes of ecological success, probably

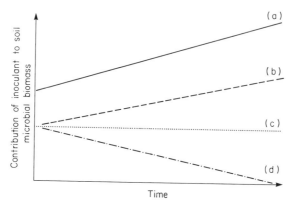

FIG. 2. Possible fates of inoculants in the short-term after their introduction to soils: (a) successful inoculants; (b) organisms only mildly successful because of small initial density; (c) poor competitors; (d) organisms die from antibiotic effect of native population.

being both autochthonous and zymogenous, or in modern terminology, r and K strategists. It is likely that they would support their high competitive abilities by producing antibiotics to minimize competition from other organisms (Fig. 2). A single organism is unlikely to have all these attributes but there is now ample evidence that well structured microbial communities can act co-operatively (Slater & Bull 1978). Such communities are only likely to be encountered by careful ecological selection and subsequent screening. Laboratory devices such as chemostats and continuous flow columns are useful in accomplishing this. We have been encouraged by our own attempts to isolate microbial communities which produce novel composts with minimum phytotoxicity while providing plant nutrient, soil conditioning and plant protection value (Lynch et al. 1984).

REFERENCES

Anderson, J.P.E. & Domsch, K.H. (1973). Quantification of bacterial and fungal contributions to soil respiration. *Archives for Microbiology*, **93**, 113–127.

Barber, D.A. (1978). Nutrient uptake. *Interactions Between Non-Pathogenic Soil Micro-Organisms and Plants* (Ed. by Y.R. Dommergues & S.V. Krupa), pp. 131–62. Elsevier, Amsterdam.

Brian, P.W. (1957). The ecological significance of antibiotic production. *Microbial Ecology* (Ed. by R.E.O. Williams & C. C. Spicer), pp. 168–88. Cambridge University Press, Cambridge.

Chalutz, E. & Liebermann, M. (1978). Inhibition of ethylene production in *Penicillium digitatum*. *Plant Physiology*, **61**, 111–114.

Chapman, S.J. & Lynch, J.M. (1984). A note on the formation of microbial polysaccharide from wheat straw decomposed in the absence of soil. *Journal of Applied Bacteriology*, **56**, 337–342.

Chou, T.W. & Yang, S.F. (1973). The biogenesis of ethylene in *Penicillium digitatum*. *Archives of Biochemistry and Biophysics*, **157**, 73–82.

Cook, R.J. & Baker, K.F. (1983). *The Nature and Practice of Biological Control of Plant Pathogens*. The American Phytopathological Society, St. Paul.

Cornforth, I.S. (1975). The persistence of ethylene in aerobic soils. *Plant & Soil*, **42**, 85–96.

Drew, M.C. & Lynch, J.M. (1980). Soil anaerobiosis, micro-organisms and root function. *Annual Review of Phytopathology*, **18**, 37–67.

Elliott, L.F. & Lynch, J.M. (1984a). Pseudomonads as a factor in the growth of winter wheat (*Triticum aestivum* L.). *Soil Biology & Biochemistry*, **16**, 69–71.

Elliott, L.F. & Lynch, J.M. (1984b). The effect of available carbon and nitrogen in straw on soil aggregation and acetic acid production. *Plant & Soil*, **78**, 335–343.

Elliott, L.F., McCalla, T.M. & Waiss, A. (1978). Phytotoxicity associated with residue management. *Crop Residue Management Systems* (Ed. by W.R. Oschwald), pp. 131–163. American Society of Agronomy, Wisconsin.

Gilmour, C.M., Allen, O.N. & Truog, E. (1948). Soil aggregation as influenced by the growth of mould species, kind of soil and organic matter. *Soil Science Society of America Proceedings*, **13**, 292–296.

Goodlass, G. & Smith, K.A. (1978). Effect of organic amendments on evolution of ethylene and other hydrocarbons from soil. *Soil Biology & Biochemistry*, **10**, 201–205.

Griffiths, E. & Jones, D. (1967). Microbiological aspects of soil structure. II. Soil aggregation by the extracellular polysaccharide of *Lipomyces starkeyi*. *Plant & Soil*, **27**, 187–200.

Gussin, E.J. & Lynch, J.M. (1981). Microbial fermentation of grass residues to organic acids as a factor in the establishment of new grass swards. *New Phytologist*, **89**, 449–457.

Gussin, E.J. & Lynch, J.M. (1982). Effect of local concentrations of acetic acid around barley roots on seedling growth. *New Phytologist*, **92**, 345–348.

Gussin, E.J. & Lynch, J.M. (1983). Root residues: substrates used by *Fusarium culmorum* to infect wheat, barley and ryegrass. *Journal of General Microbiology*, **129**, 271–275.

Harper, S.H.T. & Lynch, J.M. (1981a). The chemical components and decomposition of wheat straw leaves, internodes and nodes. *Journal of the Science of Food & Agriculture*, **32**, 1057–1062.

Harper, S.H.T. & Lynch, J.M. (1981b). Effects of fungi on barley seed germination. *Journal of General Microbiology*, **122**, 55–60.

Jensen, H.L. & Swaby, R.J. (1941). Nitrogen fixation and cellulose decomposition by soil micro-organisms, II. The association between *Azotobacter* and facultative anaerobic cellulose decomposers. *Proceedings of the Linnean Society of New South Wales*, **66**, 89–102.

Lethbridge, G. & Davidson, M.S. (1983). Root associated nitrogen-fixing bacteria and their role in the nitrogen nutrition of wheat estimated by ^{15}N isotope dilution. *Soil Biology & Biochemistry*, **15**, 365–374.

Lynch, J.M. (1972). Identification of substrates and isolation of micro-organisms responsible for ethylene production in the soil. *Nature*, **240**, 45–46.

Lynch, J.M. (1976). Products of soil micro-organisms in relation to plant growth. *CRC Critical Reviews in Microbiology*, **5**, 67–107.

Lynch, J.M. (1977). Phytotoxicity of acetic acid produced in the anaerobic decomposition of wheat straw. *Journal of Applied Bacteriology*, **42**, 81–87.

Lynch, J.M. (1980). Effects of organic acids on the germination of seeds and growth of seedlings. *Plant Cell & Environment*, **3**, 255–259.

Lynch, J.M. (1983a). Effects of antibiotics on ethylene production by soil micro-organisms. *Plant & Soil*, **70**, 415–420.

Lynch, J.M. (1983b). *Soil Biotechnology: Microbiological Factors in Crop Productivity*. Blackwell Scientific Publications, Oxford.

Lynch, J.M. & Elliott, L.F. (1983). Aggregate stabilization of volcanic ash and soil during microbial degradation of straw. *Applied Environmental Microbiology*, **45**, 1348–1401.

Lynch, J.M. & Elliott, L.F. (1984). Crop residues. *Crop Establishment: Biological Requirements and Engineering Solutions* (Ed. by M.K.V. Carr), Pitmans, London.

Lynch, J.M. & Gunn, K.B. (1978). The use of the chemostat to study the decomposition of wheat straw in soil slurries. *Journal of Soil Science*, **29**, 551–556.

Lynch, J.M., Gunn, K.B. & Panting, L.M. (1980). On the concentration of acetic acid in straw and soil. *Plant & Soil*, **56**, 93–98.

Lynch, J.M. & Harper, S.H.T. (1974). Formation of ethylene by a soil fungus. *Journal of General Microbiology*, **80**, 187–195.

Lynch J.M. & Harper, S.H.T. (1980). Role of substrates and anoxia in the accumulation of soil ethylene. *Soil Biology & Biochemistry*, **12**, 363–367.

Lynch, J.M. & Harper, S.H.T. (1983). Straw as a substrate for co-operative nitrogen fixation. *Journal of General Microbiology*, **129**, 251–253.

Lynch, J.M. & Penn, D.J. (1980). Damage to cereals caused by decaying weed residues. *Journal of the Science of Food & Agriculture*, **31**, 321–324.

Lynch, J.M., Harper, S.H.T., Chapman, S.J. & Veal, D.A. (1984). Approach to the controlled production of novel agricultural composts. *Microbiological Methods for Environmental Biotechnology* (Ed. by J. M. Grainger & J.M. Lynch), Academic Press, New York.

Lynch, J.M. & Pryn, S.J. (1977). Interaction between a soil fungus and barley seed. *Journal of General Microbiology*, **103**, 193–196.

McCalla, T.M. & Norstadt, F.A. (1974). Toxicity problems in mulch tillage. *Agriculture and Environment*, **1**, 153–174.

Martin, J.P. (1942). The effect of composts and compost materials upon the aggregation of the silt and clay particles of Collington sandy loam. *Soil Science Society of America Proceedings*, **7**, 218–222.

Old, K.M. & Nicholson, T.H. (1978). The root cortex as part of a microbial continuum. *Microbial Ecology* (Ed. by M.W. Loutit & J.A.R. Miles), pp. 291–4. Springer-Verlag, Berlin.

Penn, D.J. & Lynch, J.M. (1982). The effect of bacterial fermentation of couch grass rhizomes and *Fusarium culmorum* on the growth of barley seedlings. *Plant Pathology*, **31**, 39–43.

Postgate, J.R. & Hill, S. (1979). Nitrogen fixation. *Microbial Ecology. A Conceptual Approach* (Ed. by J.M. Lynch & N.J. Poole), pp. 191–213. Blackwell Scientific Publications, Oxford.

Primrose, S.B. (1976). Ethylene–forming bacteria from soil and water. *Journal of General Microbiology*, **97**, 343–346.

Rice, E.L. (1974). *Allelopathy*. Academic Press, New York.

Rice, W.A. & Paul, E.A. (1972). The organisms and biological processes involved in asymbiotic nitrogen fixation in waterlogged soil amended with straw. *Canadian Journal of Microbiology*, **18**, 715–723.

Russell, E.W. (1973). *Soil Conditions and Plant Growth* (10th edn.). Longman, London.

Rovira, A.D. (1965). Plant root exudates and their influence upon soil micro–organisms. *Ecology of Soil-borne Plant Pathogens—Prelude to Biological Control* (Ed. by K.F. Baker & W.C. Snyder), pp. 170–86, John Murray, London.

Slater, J.H. & Bull, A.T. (1978). Interactions between microbial populations. *Companion to Microbiology* (Ed. by A.T. Bull & P. M. Meadow), pp. 181–206. Longmans, London.

Smith, A.M. (1973). Ethylene as a cause of soil fungistasis. *Nature*, **246**, 311–313.

Smith, A.M. & Cook, R.J. (1974). Implications of ethylene production by bacterial for biological balance of soil. *Nature*, **252**, 703–705.

Smith, K.A. & Russell, R.S. (1969). Occurrence of ethylene and its significance in anaerobic soil. *Nature*, **222**, 769–771.

Smith, K.A. & Restall, S.W.F. (1971). The occurrence of ethylene in anaerobic soil. *Journal of Soil Science*, **22**, 430–443.

Smith, K.A. & Robertson, P.O. (1971). Effect of ethylene on root extension of cereals. *Nature*, **234**, 148–149.

Sutherland, J.B. & Cook, R.J. (1980). Effects of chemical and heat treatments on ethylene production in soil. *Soil Biology & Biochemistry*, **12**, 357–362.

Tang, C.S. & Waiss, A.C. (1978). Short-chain fatty acids as growth inhibitors in decomposing wheat straw. *Journal of Chemical Ecology*, **4**, 225–232.

Vartiovaara, U. (1938). The associative growth of cellulose-decomposing fungi and nitrogen-fixing bacteria. *Journal of Scientific Agricultural Society, Finland*, **10**, 241–264.

Vokou, D., Margaris, N.S. & Lynch, J.M. (1984). Effects of volatile oils from aromatic shrubs on soil micro–organisms. *Soil Biology & Biochemistry (in press)*.

Wallace, J.M. & Elliott, L.F. (1979). Phytotoxins from anaerobically decomposed wheat straw. *Soil Biology and Biochemistry*, **11**, 325–330.

Mycorrhizal mycelia and nutrient cycling in plant communities

D.J. READ, R. FRANCIS AND R.D. FINLAY

Department of Botany, University of Sheffield, Sheffield S10 2TN

SUMMARY

1 This paper describes some of the structural and functional attributes of the external mycelium of mycorrhizal roots.

2 Root chambers have been employed to investigate the development of the mycelia of both ecto- and vesicular-arbuscular mycorrhizal systems and the observations have been supplemented with simulated and actual field experiments.

3 It is demonstrated that hyphae of the external mycelium of both ecto- and VA mycorrhizal fungi can initiate mycorrhizal infection in intra- and interspecific combinations of host plants.

4 This pattern of infection development leads to the establishment of a persistent network of hyphal interconnections between plants.

5 Using $^{14}CO_2$ it is shown that carbon moves freely between plants connected by the mycorrhizal mycelium and that the movement occurs along concentration gradients which can be induced by shading.

6 It is concluded that these attributes of the mycorrhizal system will improve the vigour of the individual receiver plant, optimize the efficiency of resource distribution within the plant community and enhance the conservation of nutrients at the ecosystem level by restricting losses caused by immobilization in the general soil microflora or by leaching.

INTRODUCTION

Colonization of the terrestrial environment by early land plants would be expected to lead to an increased supply of energy-rich substrates to heterotrophic micro-organisms. Loss of materials from roots would constitute one of the most important supply processes and this would in turn lead both to increased microbial activity and to an intensification of interspecific competition in the zone which we now recognize as the rhizosphere. In these circumstances micro-organisms with the ability to enter into a closer relationship with the autotroph, either as internal or surface occupants of the root, would be at a great advantage since they would obtain direct access to assimilate supplies. Evidence that such close associations were formed by the earliest land plants comes from studies of the roots of fossil pteridophytes of the Devonian period in which vegetative structures similar to those seen in present day VA mycorrhizas can be seen (Nicolson 1975). Since the roots of these early land plants were very poorly developed, exploitation of the soil

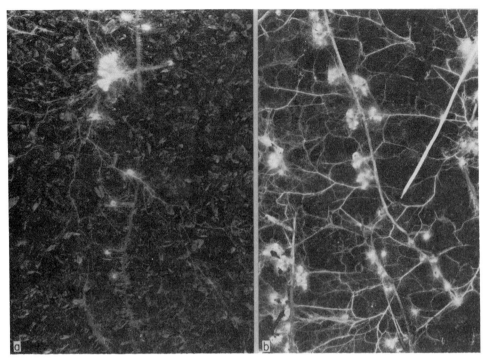

FIG. 3. Levels of strand formation in soil. (a) Strands developing from an established mycorrhizal root cluster act as sources of inoculum on emerging laterals further along the same root system. All mycorrhizal laterals thus become interconnected by strands ($\times$ 5). (b) Mature region of root chamber showing complete exploitation of the root environment by anastomosing network of strands ($\times$ 10).

mycelia have been harvested after such a feeding experiment and the major labelled product has been shown to be trehalose (D.J. Read, unpubl.). Current assimilates are thus rapidly translocated over large distances from roots into soils through the mycelial network. Where interconnections have formed between roots it is clear that the newly-formed mycorrhizas of the 'receiver' plant are major sinks for the carbon (Fig. 1d). Quantitative determination of the radioactivity (Table 1)

TABLE 1. The distribution of radioactivity (d.p.m. per mg dry wt) in roots and shoots of mycorrhizal (M) and non-mycorrhizal (NM) *Pinus contorta* plants grown in association with 'donor plants' of the same species

	Shoot	Root	Activity in whole receiver plant as % of that in donor
NM	98·2	304·3	0·028
M (*Suillus bovinus*)	304·9 (P <0·05)	3640·7 (P <0·01)	0·234
M (*Suillus granulatus*)	145·4 (NS)	1569·4 (P <0·01)	0·216

Donor plants were fed with 50 μCi of NaH^{14}CO$_3$ for 72 hours.

Significance levels refer to differences between NM and M treatments and are based on analysis of variance of 1n-transformed d.p.m. data.

TABLE 2. Host and fungus combinations in which interplant mycorrhizal connections have been synthesized and $^{14}CO_2$ transfer demonstrated (+). Cases where mycorrhizas have not formed in the 'receiver' are indicated with zero (0). Dash (−) indicates synthesis not attempted

Donor species	Receiver species	Fungal species					
		Amanita muscaria	Rhizopogon roseolus	Paxillus involutus	Suillus granulatus	Suillus bovinus	Suillus luteus
Pinus sylvestris	Pinus sylvestris	+	+	+	+	+	+
	P. contorta	+	+	+	+	+	+
	Picea abies	+	+	+	0	0	0
	P. sitchensis	+	0	+	0	−	−
	Betula pubescens	+	+	+	0	0	0
Pinus contorta	Pinus contorta	+	+	+	+	+	+
	P. sylvestris	+	+	+	+	+	+
	Picea abies	+	+	+	0	0	0
	P. sitchensis	+	+	+	0	0	0
	Betula pubescens	+	+	+	0	0	0

shows that significantly larger amounts of label occur in both shoots and roots of M than in those of NM plants. In the cases which have so far been examined this pattern of distribution between plants appears to be repeated when the linkages occur at the interspecific level (Table 2).

In addition to the transfer of carbon from autotroph to the advancing hyphal front, it has been shown that water moves in the opposite direction (Duddridge, Malibari & Read 1980), presumably along a gradient of water potential from soil to the transpiring plant. Analysis of the structure of mature strands of *Suillus* species reveals that they are differentiated structures with central 'vessel' hyphae of large diameter which lack cytoplasm and cross walls, surrounded by dense cytoplasmic sheathing hyphae of narrower diameter (Duddridge *et al.* 1980). A similar structure has been observed in strands of *Rhizopogon* (Foster 1981). Clearly the 'vessel' hyphae have a much greater hydraulic conductivity than the sheathing hyphae and it is likely that they function as the major channels of water and solute transport. They are therefore in functional terms analogous to xylem elements. Carbon transport, occurring in the opposite direction, must be restricted to the outer elements of the strand which are therefore analogous in function to phloem. The strands are thus functional extensions of the root system to which they are attached.

Field studies

It is clearly desirable to extend such laboratory investigations to forest situations but the complexity of root distribution and the delicacy of mycorrhizal mycelia make this extremely difficult. However, an investigation of the pattern of carbon transfer between mature trees and naturally regenerating plants has recently been made in an area with 35 year old, 15 m high trees of *Pinus contorta* in which each tree was surrounded by a sward of naturally regenerating plants (D.J. Read, unpubl.). The regenerating plants were of mixed species, most being of *P. contorta* growing under the parent trees, but together with these were young plants of *Pinus sylvestris*, *Picea sitchensis*, *Betula pubescens*, *Chamaecyparis lawsoniana* and *Ilex aquifolium*. The crowns of selected large *P. contorta* trees were enclosed in purpose-built polythene sacks in July, the remaining branches of the tree were removed, and 20 mCi of $^{14}CO_2$ was released into the sack. During the period of feeding, the crowns of some of the naturally regenerating *P. contorta* plants were loosely covered with black polythene sacks to increase shade. The base of the 'donor' tree was checked for signs of radioactivity for a period after feeding using a portable Geiger-Muller radiation monitor. The first indication of the presence of activity at ground level was obtained 8 weeks after release of the isotope.

At this stage selected main roots of the regenerating trees around each 'donor' were excavated to the points at which healthy fine roots could be collected. Approximately 100 ectomycorrhizal roots of shaded and of unshaded *P. contorta* were sampled around each donor. Forty roots of each of the other species in the area were also excised. These included roots of *C. lawsoniana* and *I. aquifolium* both of which are VA mycorrhizal species. All roots were transferred to the laboratory where levels of radioactivity were determined by the method described earlier.

Not surprisingly variation between individual roots of receiver plants was very large, some showing no activity while others were highly radioactive. This variability probably arises through a combination of factors, not least of which will be the difficulty of sampling roots in such a heterogeneous environment, a random pattern of distribution of mycelial connections between the different trees and variation in the physiological ages of the roots. Because of the variability only the highest levels of activity found on a given individual receiver tree are shown (Fig. 4). Considerable quantities of carbon have been transported to some trees and the fact that highest levels are found in receivers which had been artificially shaded suggests that the process was under physiological control. The absence of activity from all roots of the VA plants *C. lawsoniana* and *I. aquifolium* further indicates that transfer was through ectomycorrhizal interconnections. Experiments of this type provide indications that the processes observed in the laboratory are found also in the field, but they also highlight the difficulties of making meaningful measurements of mycelial transfer processes in the mature forest.

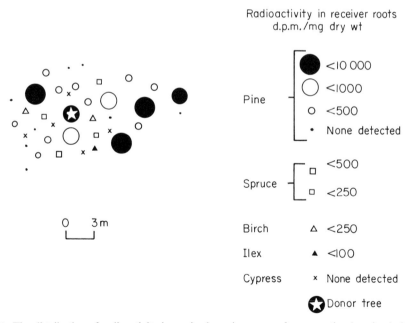

FIG. 4. The distribution of radioactivity in a mixed species group of regenerating 'receiver' plants growing around a mature *Pinus contorta* tree, the crown of which had been exposed to 20 mCi of $^{14}CO_2$ 8 weeks previously. All trees marked were sampled and the distribution of the highest levels of activity is shown.

MYCELIA OF VESICULAR-ARBUSCULAR MYCORRHIZAS

Experimental methods

Since the network of VA mycelium in soil is made up of single hyphae, methods for study of its form and function require microscopic techniques. In our studies, mycorrhizas were first synthesized by inoculating 'donor' seedlings of *Plantago lanceolata* either with surface sterilized spores, or with a mixed inoculum of VA fungi on infected root pieces collected from the field. After infection had occurred the mycorrhizal (M) plants were transferred to shallow transparent dishes containing a 1 cm layer of sterilized or unsterilized dune sand. Young non-mycorrhizal seedlings of the same species or of *Festuca ovina* were then introduced into the dishes. A parallel series of dishes was established in which the 'donor' plants were non-mycorrhizal (NM). After 6–8 weeks of incubation in a growth room the M and NM dishes were divided into two sets, one of which was used for examination of the pattern of development of the mycelium and the other for analysis of viability and function of the network using $^{14}CO_2$ as a tracer.

In order to determine the distribution of mycelium the horizontal walls of the dishes were cut away. Loose superficial layers of sand were then removed by means of a fine jet of water. Below the surface the grains are held by the meshwork of mycelium. Adhering sand grains were removed individually under a dissecting microscope using storksbill forceps. Some breakage of hyphae is inevitable, but since most of the mycelium develops on the transparent base of the dish the network remains intact and its arrangement relative to roots of the host plants and the soil can be analysed and photographed.

Shoots of donor plants of both M and NM treatments were exposed to $^{14}CO_2$ in sealed chambers. Where distribution of radioactivity was to be assessed by autoradiographic means the procedure described above for display of the mycelium was repeated. The freshly excavated system was first photographed and then exposed to autoradiographic stripping film (Kodak AR10). Film was removed and developed after 2 days of incubation and the distribution of radioactivity was analysed with reference to the photographs taken earlier.

Where radioactivity was to be determined quantitatively, pots or seed trays were used instead of transparent dishes. The surface of all pots was covered with paraffin wax before exposure of donor plants to $^{14}CO_2$ in order to eliminate transfer of gaseous $^{14}CO_2$ from the soil to the shoots. In the pot experiments shoots of some M and NM receiver plants were covered by loosely fitting caps which reduced light levels by half or to total darkness. In these experiments the donor plants was *Plantago lanceolata* and receivers were seedlings of *Festuca ovina*.

A simple sward situation was simulated by growing seedlings in a seed tray around a relatively mature plant. Groups of seedlings of *Plantago lanceolata* which were to be used as 'donor' plants were first grown in either the mycorrhizal (M) condition by inoculation with infected root pieces, or in the non-mycorrhizal (NM) condition. After the establishment of mycorrhizal infection in the M plants, both M and NM *Plantago* were transferred to a central position in a seed tray (38 × 24 cm) containing a layer of irradiated dune sand of 5 cm depth. *Plantago* seedlings were planted to form a central line through the trays and two outer rows of NM *Festuca* seedlings were placed on either side of the *Plantago* row. The distance between each seedling was 6 cm and the distance from the donor to the corner plants was 18 cm. The trays were placed in a growth chamber for 8 weeks, a period which previous studies had shown to be adequate to enable infection and spread to all M plants in a tray. At this time half of the seedlings in both M and NM trays were covered by a frame of aluminium foil for 48 hours to exclude light. During the shading period the shoot of the central *Plantago* plant in both the M and NM trays was sealed into a clear perspex box and 100 μCi of $^{14}CO_2$ was released around the shoots. After 48 hours of exposure to the gas the donor shoots were removed and all receiver shoots were harvested. Sand was carefully removed from roots and whole seedlings were air dried and prepared for autoradiography.

In both pot and tray experiments, quantitative analysis of distribution of isotope was carried out. Roots and shoots of the seedlings were processed in the same manner as ectomycorrhizal seedlings.

Results of analyses of the VA systems

A brief account of results obtained in the dish experiments has been provided elsewhere (Francis & Read 1984). Examination of dishes after removal of sand reveals that the VA mycorrhizal mycelium has grown from the donor root to form a network of hyphae covering the base of the dish in the M treatments. No differences could be detected between the patterns of mycelial development in sterilized and unsterilized sand. Hyphae making contact with receiver roots form VA infections provided that the contacts are made in regions of the root which are susceptible to infection. One of the most striking features revealed in these plant fungus associations is the distance across which hyphal connections are formed. Distances of several centimetres are commonplace (Fig. 5a, b). Some of the hyphae forming connections between roots have distinctive characteristics which have led us to call them 'arterial' hyphae. They have larger diameters, are less branched than other hyphae in the system and pass directly between roots (Fig. 5a, b). The situation is thus reminiscent of that seen in ectomycorrhizas where the formation of interconnections leads to enhanced development of the strands involved. In NM dishes virtually no mycelial development is observed, irrespective of whether the sand has been sterilized then exposed continuously to the open air, inoculated with saprophytes, or used in the field condition (Fig. 9a). This confirms the view that practically all of the mycelium observed in the mycorrhizal chambers is produced by the VA fungus.

Autoradiographic analysis of these dishes reveals that as in the case of ectomycorrhizal chambers, the tracer moves rapidly from donor plant into the mycelial network in the soil (Fig. 6a). Arterial hyphae become particularly heavily labelled (Fig. 7b). Radioactivity from these conduits accumulates first in the

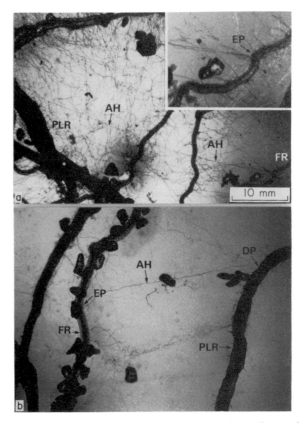

FIG. 5. Showing the development of vesicular-arbuscular mycorrhizal mycelium and the formation of inter-plant connections in interspecific associations of *Plantago lanceolata* and *Festuca ovina*.
(a) Mycorrhizal mycelium has spread from previously infected *Plantago* 'donor' (PLR) plants to colonize the sand and infect *Festuca* 'receiver' roots (FR). An arterial hypha (AH) is shown passing over 4 cm from a donor root to infect a *Festuca* root (× 15). Inset: higher magnification of infection point on receiver root (× 30). (b) Arterial hypha (AH) passing from *Plantago* 'donor' root (PLR) to infect 'receiver' *Festuca* root (FR). Hyphal departure (DP) and entry points (EP) are marked. Adherence of sand grains to *Festuca* roots is caused by prolific root hair production in this species (× 38).

internal fungal mycelium and so provides clear demarcation of infected and uninfected areas of the root in the autoradiographs (Figs 6a and 8b). Other major sinks for assimilates are spores developing on the external mycelium (Figs 6a, b and 8b). Labelled material is later transferred from fungal structures and accumulates in root apices (Fig. 6b) as well as in the shoots of receiver plants (Table 3). The fungal vesicles within the root remain heavily labelled after the major transfer of label to host tissue has occured (Fig. 6b), the radioactivity presumably having accumulated in storage compounds.

Despite the very close proximity of roots of donor and receiver plants in the non-mycorrhizal systems (Fig. 9a), no transfer of radioactivity can be detected in autoradiographs of these associations (Fig. 9b). This demonstrates that even

TABLE 3. Radioactivity in shoots and roots of receiver plants (*Festuca ovina*) growing in pots containing *Plantago* as a donor, under three light regimes. Counts are expressed as d.p.m./mg dry weight (with 95% confidence limits) and as percentage of the total activity in the donor plants at harvest ($n = 6$). (From Francis & Read 1984)

Receiver category

Treatment	*Festuca:* Mycorrhizal			*Festuca:* Non-Mycorrhizal		
	Activity in root (d.p.m./mg dry wt)	Activity in shoot (d.p.m./mg dry wt)	Activity in whole receiver plant as % of that in donor	Activity in root (d.p.m./mg dry wt)	Activity in shoot (d.p.m./mg dry wt)	Activity in whole receiver plant as % of that in donor
Full light	8908** ± 3530	363*NS ± 242	0·0151	656NS ± 403	147NS ± 194	0·0019
Half light	18072** ± 7647	479*NS ± 75	0·048	264NS ± 365	229NS ± 372	0·0010
Dark	57218*** ± 12372	51NS ± 38	0·112	117NS ± 100	ND*NS	0·0005

* Indicates significant difference between this and all other values of activity at both treatment and category levels at $P<0.05$.

** Indicates significant differences between this and other figure so marked at $P<0.01$.

NS, Figure not significantly different from other treatments in the same category.

ND, No counts above background detected.

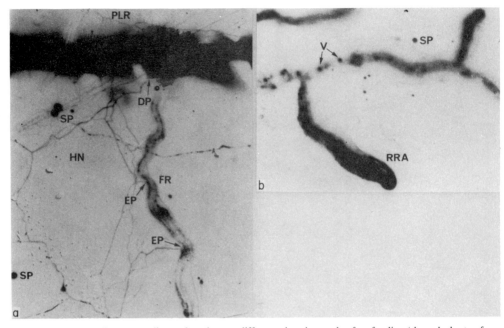

FIG. 6. Stripping film autoradiographs taken at different time intervals after feeding 'donor' plants of *Plantago lanceolata* with $^{14}CO_2$. (a) 48 hours after release of label over 'donor' shoots. Activity has moved from the heavily labelled donor root (PLR) into the hyphal network in sand (HN), and into infected regions of the *Festuca* 'receiver' root (FR). Departure (DP) and entry (EP) points are marked. Accumulation of label in spores is also seen (SP) ($\times$ 65). (b) At 72 hours after feeding of 'donor' plants, label has been transferred from infected regions of roots to receiver root apices (RRA). Activity remains high in external spores (SP) and internal vesicles (V) of the fungus ($\times$ 65).

though some leakage of carbon is to be expected from these roots, the magnitude of such loss bears little relation to the movement which is occurring into the mycorrhizal network. Facilitated uptake of leaked materials by scavenging mycorrhizal hyphae if it occurs at all, will therefore be a highly inefficient process, when compared to the direct transfer pathway revealed in the autoradiographs.

Quantitative determination of distribution of radioactivity confirms that label accumulates in roots of infected receiver plants and that it is eventually transferred to shoots (Table 3). The pattern of movement between mycorrhizal plants is greatly influenced by the light environment of the shoots, those kept in darkness during the isotope feeding period accumulating considerably more activity than those in half light, and half-light treatments in turn accumulating more than the full light.

Experiments with swards show that transfer between infected plants can occur at both intra- and interspecific levels over distances of at least 18 cm in 48 hours (Table 4a), and that in this circumstance also, shading has a major influence upon the pattern of assimilate distribution. The highest levels of activity are found in the roots of shaded mycorrhizal plants but a relatively small proportion of the activity is transferred to the shoot in this treatment. It may be that higher transpiration rates

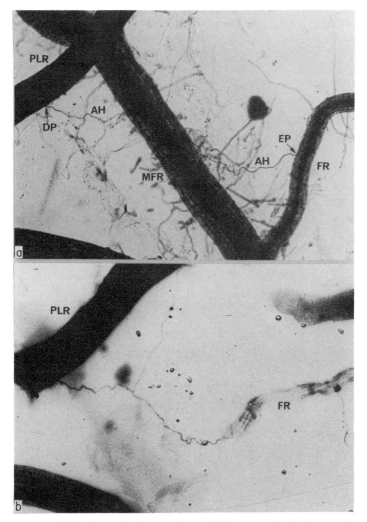

FIG. 7. Direct transfer of infection and of isotope through interconnecting hyphae in interspecific association of *Plantago* and *Festuca*. (a) Arterial hypha (AH) leaving donor root (PLR) at departure point (DP) and passing a mature *Festuca* root (MFR) which is not susceptible to infection and entering (EP) a young susceptible root (FR). Prolific root hair (RH) development is seen on *Festuca* roots (× 130). (b) Stripping film autoradiograph of (a) showing direct transfer of label form donor root (PLR) to infected area of receiver root (FR). (× 130).

in the more exposed unshaded part of the sward give rise to more rapid transfer from root to shoot. This possibility will be tested experimentally. Levels of activity in non-mycorrhizal trays are extremely low (Table 4b). Such results demonstrate that VA mycorrhizal mycelium provides a network in the soil through which all infected plants can exchange carbon, the direction of greatest net movement being determined by shade and hence physiological need.

TABLE 4(a). Distribution of plants and of radioactivity (d.p.m. mg d wt^{-1}) in simulated sward of mycorrhizal *Festuca* (square symbols) and *Plantago* (circular symbols) plants, 48 hours after feeding the central *Plantago* donor (PL. DONOR) with $^{14}CO_2$. Closed symbols represent shaded plants, open symbols fully illuminated plants. The upper figure at each plant is shoot radioactivity, the lower root radioactivity. (b) Details as for (a) but in sward of non-mycorrhizal plants.

(a) *Mycorrhizal sward*

Shoot	616	122	9	556	656
	■	■	●	□	□
Root	736	268	374	215	612
	625	366	42	565	743
	■	■	●	□	□
	15681	13750	2442	2142	1134
	412	135	PL. DONOR	426	639
	■	■	★	□	□
	45072	87555		7233	3162
	260	435	109	303	181
	■	■	○	□	□
	88997	102070	1183	7212	1659
	118	56	47	230	133
	■	■	○	□	□
	5580	4490	1370	2238	2405

(b) *Non-mycorrhizal sward*

Shoot	21	235	15	30	70
	□	□	●	■	■
Root	15	0	46	143	300
	13	5	0	3	65
	□	□	●	■	■
	19	36	7	118	50
	0	7	PL. DONOR	12	27
	□	□	★	■	■
	0	20		19	21
	32	1	49	318	12
	□	□	○	■	■
	23	42	220	81	112
	13	40	10	12	425
	□	□	○	■	■
	47	57	5	3	0

DISCUSSION

Though large numbers of studies of the structure and function of mycorrhizal roots have been carried out (see recent review by Harley & Smith 1983) relatively little attention has been given to the mycelial phase of the symbiosis in the soil. The main reason is probably that while entire mycorrhizal roots or even root systems can readily be extracted from soil with little damage, it is extremely difficult to carry out a non-destructive investigation of the mycelial system. Nevertheless, the importance of the external mycelia for the nutrition of individual plants with both ecto- and VA mycorrhizas has been demonstrated.

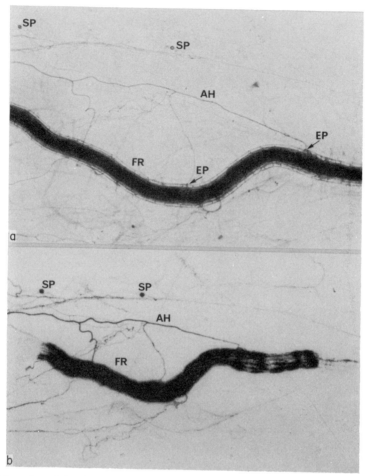

FIG. 8. Distal portion of arterial hyphae showing points of entry (EP) into receiver roots (FR). (a) Arterial hypha (AH) of relatively larger diameter branching in the vicinity of a receiver root (FR) to form two entry points (EP) ($\times$ 130). (b) Stripping film autoradiograph of this region showing heavy labelling of the arterial hypha (AH) and of spores (SP) in the external mycelium, and of the zone of infection within the receiver root (FR) ($\times$ 130).

Melin & Nilsson (1950, 1952, 1953) clearly showed that ectomycorrhizal mycelia growing in pure culture with seedlings could absorb nutrients and provide channels for their transport to roots. The occurrence of mycorrhizal strands in soil, at considerable distances from the roots to which they were attached, was later demonstrated by Schramm (1966). Bowen (1973) pointed out that such strands should enhance plant nutrient uptake largely by growing though the depletion zones which surround roots, into nutrient-rich regions beyond the rooting area. It was later shown (Skinner & Bowen 1974) that mycelial strands attached to roots of *Pinus radiata* are able to absorb phosphate ions which were then translocated to mycorrhizal roots.

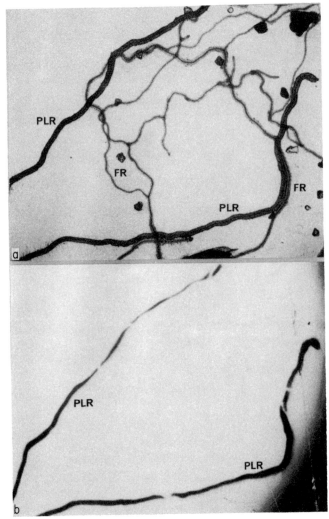

FIG. 9. Light micrograph and autoradiograph of non-mycorrhizal root chamber. (a) Closely intermingled roots of *Plantago* (PLR) and *Festuca* (FR) are shown. Very few hyphae are seen in such systems even when the plants are grown under unsterile conditions (× 13). (b) Stripping film autoradiograph of this system taken after feeding *Plantago* shoots with $^{14}CO_2$ under the same circumstances as those used for M systems. No transfer of radioactivity from heavily labelled *Plantago* roots (PLR) is detectable even though extensive physical contact between donor (PLR) and receiver (FR) roots is seen in (a) (× 13).

The major function of the ectomycorrhizal sheath appears to be that of nutrient storage (Harley & McCready 1950, 1981; Bowen & Theodorou 1967). Nutrients such as phosphate are released to the soil solution in seasonal flushes of microbial activity, P ions being captured and stored in the fungal sheath in the form of polyphosphate granules (Harley & McCready 1981). Since the capacity of mycorrhizal roots to absorb phosphate is as much as five times greater than that of their non-mycorrhizal counterparts (Harley & McCready 1950), the sheath must play a major role in the storage and conservation of such nutrients. The quantity of mycelium external to the sheath differs in different host species. Thus, while the external mycelial system of pine would be expected to play a major role in the capture of phosphate and its transport to the sheath, such a system is relatively poorly developed in beech forest (Harley 1978) where the sheath has few attachments to the soil. This difference in pattern of distribution of external mycelium, which is likely to be of considerable significance for the soil ecosystem as a whole (Harley 1978; Read 1984), may arise as a result of the different patterns of nutrient release in coniferous and deciduous tree litter. Nykvist (1963) showed that assimilates in fresh leaf litter of several deciduous tree species could be readily leached by water, but that losses from coniferous litter were slower and occurred over a longer time span. Whereas an extensive external network of mycelium might provide an energetically efficient absorptive system in the coniferous situation, the maintenance costs might be prohibitive in the case of the ephemeral nutrient release pattern asociated with the deciduous litter.

Analysis of the development of mycelial systems in root chambers shows that as well as being nutrient-absorbing structures the strands are responsible for the initiation and spread of infection. Their role in this capacity in the field has recently been demonstrated by Fleming (1983) who showed that whereas in an agricultural soil a number of non-strand-forming fungi may act as mycorrhizal colonists of seedling birch roots, in the forest the main mycorrhiza formers are strand-forming fungi which are almost certainly growing from roots of the older forest trees. There is thus indirect evidence that seedlings in the forest are quickly integrated into the common mycelial system.

The importance of ectomycorrhizal fungi in the soil ecosystem was stressed by Harley (1978) who suggested that their mycelia might be the source of the considerable quantities of CO_2 evolved in soil 'respiration' which could not be accounted for in terms of decomposition processes. Recalculating data provided by Romell (1939), Harley showed that mycorrhizal fruit body production alone could utilize carbon equivalent to as much as 10% of that required for annual timber production. Studies at the ecosystem level (Fogel & Hunt 1979; Vogt et al. 1982) have confirmed Harley's view. In Douglas fir forest, Fogel & Hunt found that up to 50% of the annual throughput of dry matter takes place in the fungal component of the forest and that 23% of this takes place through the hyphae, most of which can be assumed to be associated with mycorrhizas. Vogt et al. (1982), working in Abies amabilis forests, showed that fungal reproductive structures could account for up to 15% of net primary production of the stands. Since the fruit bodies of most mycorrhizal fungi are relatively ephemeral structures it seems likely that in most cases the vegetative mycelium will contribute an even greater sink for carbon over the full growing season. The analysis of structure and function of the external mycelium of ectomycorrhizal roots helps to explain the massive investment of carbon in this part of the system. It appears that in many members of the Pinaceae, at least, the major function of absorption from soil is fulfilled by the external

direct supplies of carbon would be expected to provide significant competitive advantages to mycorrhizal biotrophs growing through soil. Gadgil & Gadgil (1975) provided indirect evidence for the presence of such an advantage by showing that the activities of litter decomposing saprotrophs was inhibited in the presence of mycorrhizal roots. We are currently undertaking a comparative analysis of respiratory activities in intact mycorrhizal and non-mycorrhizal root chambers with a view to providing discrimination between the activities of the biotrophic and saprotrophic components of the soil microflora.

The occurrence of direct interplant transfer of carbon is likely to be of considerable significance to the physiology of the interconnected plants. It is known that seedlings of some species of both woodland and grassland habitats can survive prolonged periods of exposure to deep shade (Salisbury 1930; Chippindale, 1932). When growing in nutrient-poor soil some of these plants are even able to survive for long periods in complete darkness (Hutchinson 1967). Most of the species shown to be strongly shade-tolerant in such studies would be expected to be mycorrhizal in the field and it is evident that their survival in nature could be assisted by transfer of resources from illuminated over-storey plants. Mahmoud & Grime (1974) proposed that the remarkable shade tolerance of the grass *Deschampsia flexuosa* is attributable to its inherently low respiration rates. Clearly, again, the provision of additional respiratory substrates from nearby plants through mycorrhizal mycelia could contribute to the survival of plants in stressed circumstances of this kind. The experimental results described above suggest that carbon will move readily from well illuminated over-storey plants to shaded seedlings. The chance that seedlings will receive access to this supply even in a community of mixed species is enhanced by the lack of mycorrhizal specificity. Even if the amount of assimilate received were sufficient only to sustain the mycorrhizal association the elimination of such a respiratory drain would be a significant advantage for a shaded seedling. Autoradiographic and quantitive analyses reveal, however, that labelled assimilate is moved from regions of infection to root apices and to the shoots of infected plants so the benefits are distributed more widely.

Internal recycling of nutrients is a well known mechanism for the redistribution and conservation of resources within individual plants, but the possibility that resources may pass between individuals within the community has not been widely recognized. Resource transfer of this kind arises as a secondary but nonetheless important consequence of infection of the individual plants. Selection for the mycorrhizal habit in individuals probably arose because of the improvement of nutrient status and survival potential which infection provided. The formation of connections between individuals and species—with all of the repercussions in relation to patterns of nutrient circulation within the plant community—inevitably follows, both because of the low levels of host specificity shown by most mycorrhizal fungi, and because the normal method of infection involves growth of a hypha from a resource base on one root to a susceptible uninfected root elsewhere in the soil. Restriction of specificity in the fungus would be favoured by natural selection since it increases the chance that the heterotroph will obtain access to a suitable nutrient source.

There is normally an inverse relationship between the intensity of mycorrhizal infection and the fertility of the soil. As a result, the numbers of mycelial interconnections and, therefore, the potential for interplant transfer of nutrients will be greatest in those infertile situations where the benefits of such transfer would be most strongly felt. Many unproductive soils have high species densities. Grime (1973) has suggested that the ability of species to coexist in these environments is derived from the capacity of individuals to persist for long periods of time despite the low nutrient status of the soil environment, rather than from their competitive abilities. Transfer of nutrients from established plants to seedlings may be a crucial factor enabling young plants of a range of species to survive in these circumstances.

Direct nutrient transfer pathways have the further significant advantage that they greatly improve nutrient conservation at the ecosystem level. The current nutrient capital of the system is retained in circulation between the autotrophs, thus restricting losses which inevitably arise through leaching or microbial immobilization when such resources enter the soil system. The evidence obtained so far confirms the existence of the direct transfer pathway in both ecto- and VA mycorrhizal systems. Further studies are now required to determine the quantitative significance of the transfer process under a range of environmental circumstances.

ACKNOWLEDGMENTS

We thank Dr A. Fitter for helpful criticism of the manuscript and the NERC and the Forestry Commission for financial assistance.

REFERENCES

Ames, R.N., Reid, C.P.P., Porter, L.K. & Canbardella, C. (1983). Hyphal uptake and transport of nitrogen from two ^{15}N labelled sources by *Glomus mosseae*, a vesicular-arbuscular mycorrhizal fungus. *New Phytologist*, **95**, 381–396.

Anderson, J.P.E. & Domsch, K.H. (1973). Quantification of bacterial and fungal contributions to soil respiration. *Arkiv für Microbiologie*, **93**, 113–127.

Bethlenfalvay, G.J., Brown, M.S. & Pacovsky, R.S. (1982). Relationships between host and endophyte to development in mycorrhizal soybeans. *New Phytologist*, **90**, 537–543.

Bowen, G.D. (1973). Mineral nutrition of mycorrhizas. *Ectomycorrhizas* (Ed. by G.C. Marks & T.T. Kozlowski), pp. 151–201. Academic Press, New York.

Bowen, G.D. & Theodorou, C. (1967). Studies of phosphate uptake by mycorrhizas. *Proceedings of 14th IUFRO Congress*, **5**, 116–138.

Brownlee, C., Duddridge, J.A., Malibari, A. & Read, D.J. (1983). The structure and function of mycelial systems of ectomycorrhizal roots with special reference to their role in forming interplant connection and providing pathways for assimilate and water transport. *Plant and Soil*, **71**, 433–443.

Chiariello, N., Hickman, J.C. & Mooney, H.A. (1982). Endomycorrhizal role for interspecific transfer of phosphorus in a community of annual plants. *Science*, **217**, 941–943.

Chippindale, H.G. (1932). The operation of interspecific competition in causing delayed growth of grasses. *Annals of Applied Biology*, **19**, 221–242.

Duddridge, J.A., Malibari, A. & Read, D.J. (1980). Structure and function of mycorrhizal rhizomorphs with special reference to their role in water transport. *Nature*, **287**, 834–836.

Fleming, V. (1983). Succession of mycorrhizal fungi on birch: infection of seedling plants around mature trees. *Plant & Soil*, **71**, 263–267.

Fogel, R. & Hunt, G. (1979). Fungal and arboreal biomass in a western Oregon Douglas fir ecosystem: distribution patterns and turnover. *Canadian Journal of Forest Research*, **9**, 265–266.

Foster, R.C. (1981). Mycelial strands of *Pinus radiata*: ultrastructure and histochemistry. *New Phytologist*, **88**, 705–712.

Francis, R. & Read, D.J. (1984). Direct transfer of carbon between plants connected by vesicular-arbuscular mycorrhizal mycelium. *Nature*, **307**, 53–56.

Gadgil, R.L. & Gadgil, P.D. (1975). Suppression of litter decomposition by mycorrhizal roots of *Pinus radiata*. *New Zealand Journal of Forest Science*, **5**, 35–41.

Graham, J.H., Lindermann, R.G. & Menge, J.A. (1982). Development of internal hyphae by different isolates of mycorrhizal *Glomus* spp. in relation to root colonization and growth of troyer citrange. *New Phytologist*, **91**, 183–190.

Grime, P.G. (1973). Competition and diversity in herbaceous vegetation. *Nature*, **244**, 311–313.

Harley, J.L. (1950). Recent progress in the study of endotrophic mycorrhiza. *New Phytologist*, **49**, 213–247.

Harley, J.L. (1978). Ectomycorrhizas as nutrient absorbing organs. *Proceedings of the Royal Society of London B*, **203**, 1–21.

Harley, J.L. & McCready, C.C. (1950). Uptake of phosphate by excised mycorrhizas of beech I. *New Phytologist*, **49**, 388–397.

Harley, J.L. & McCready, C.C. (1981). Phosphate accumulation in *Fagus* mycorrhizas. *New Phytologist*, **89**, 75–80.

Harley, J.L. & Smith, S.E. (1983). *Mycorrhizal Symbiosis*. Academic Press, London.

Heap, A.J. & Newman, E.I. (1980). The influence of vesicular-arbuscular mycorrhiza on phosphorus transfer between plants. *New Phytologist*, **385**, 173–179.

Hutchinson, T.C. (1967). Comparative studies of the ability of species to withstand prolonged periods of darkness. *Journal of Ecology*, **55**, 291–299.

Janos, D. (1980). Vesicular-arbuscular mycorrhizae affect tropical rain forest plant growth. *Ecology*, **61**, 151–152.

Kramer, P.J. & Bullock, H.C. (1966). Seasonal variations in the proportions of suberised and unsuberised roots of trees in relation to the absorption of water. *American Journal of Botany*, **53**, 200–204.

Mahmoud, H. & Grime, P.G. (1974). A comparison of negative relative growth rates in shaded seedlings. *New Phytologist*, **73**, 1215–1219.

Melin, E. & Nilsson, H. (1950). Transfer of radioactive phosphorus to pine seedlings by means of mycorrhizal hyphae. *Physiologia Plantarum*, **3**, 88–92.

Melin, E. & Nilsson, H. (1952). Transfer of labelled nitrogen from an ammonium source to pine seedlings through mycorrhizal mycelium. *Svensk Botanisk Tidskrift*, **46**, 281–285.

Melin, E. & Nilsson, H. (1953). Transfer of labelled nitrogen from glutamic acid to pine seedlings through the mycelium of *Boletus variegatus* (S.W.) Fr. *Nature*, **171**, 434.

Mosse, B. (1959). Observations on the extra-matrical mycelium of a vesicular-arbuscular endophyte. *Transactions of the British Mycological Society*, **42**, 439–448.

Nicolson, T.H. (1959). Mycorrhiza in the Gramineae, I. Vesicular-arbuscular endophytes with special reference to the external phase. *Transactions of the British Mycological Society*, **42**, 421–438.

Nicolson, T.H. (1975). Evolution of vesicular arbuscular mycorrhiza. *Endomycorrhizas* (Ed. by F.E. Sanders, B. Mosse & P.B. Tinker), pp. 25–34. Academic Press, London.

Nykvist, N. (1963). Leaching and decomposition of water-soluble organic substances from different types of leaf and needle litter. *Studia Forestalia Suecica*, **3**, 1–31.

Pearson, V. & Tinker, P.B. (1975). Measurement of phosphorus fluxes in the external hyphae of endomycorrhizas. *Endomycorrhizas* (Ed. by F. E. Sanders, B. Mosse & P. B. Tinker), pp. 277–287. Academic Press, London.

Read, D.J. (1984). The structure and function of the vegetative mycelium of mycorrhizal roots. In *The Ecology and Physiology of the Fungal Mycelium* (Ed. D.H. Jennings & A.D.M. Rayner), Symposium No. 8 of the British Mycological Society, pp. 215–240. Cambridge University Press, Cambridge.

Read, D.J., Kouchecki, H.K. & Hodgson, J. (1976). Vesicular-arbuscular mycorrhiza in natural vegetation system, I. The occurrence of infection. *New Phyologist*, **77**, 641–651.

Reid, C.P.P. & Woods, F.W. (1969). Translocation of ^{14}C labelled compounds in mycorrhiza and its implications in interpreting nutrient cycling. *Ecology*, **50**, 179–181.

Romell, L.G. (1939). Barrskogens morksvampa och deras roll i skogens liv. *Svenska Skogsvardsforeningen Tidskrift*, **37**, 238–75.

Salisbury, E.J. (1930). Mortality amongst plants and its bearing on natural selection. *Nature*, **125**, 817.

Sanders, F.E. & Tinker, P.B. (1973). Phosphate flow into mycorrhizal roots. *Pesticide Science*, **4**, 384–395.

Sanders, F.E., Tinker, P.B., Black, R.L.B. & Palmerley, S.M. (1977). The development of endomycorrhizal root systems, I. Spread of infection and growth promoting effects with four species of vesicular-arbuscular endophytes. *New Phytologist*, **78**, 257–268.

Schramm, J.R. (1966). Plant colonization studies on black wastes from anthracite mining in Pennsylvania. *Transactions of the American Philosophical Society (NS)*, **56**, 1–194.

St John, T.V. (1980). Root size, root hairs and mycorrhizal infection: a re-examination of Baylis's hypothesis with tropical trees. *New Phytologist*, **84**, 483–487.

Skinner, M.F. & Bowen, G.D. (1974). The uptake and translocation of phosphate by mycelial strands of pine mycorrhizas. *Soil Biology and Biochemistry*, **6**, 53–56.

Tisdall, J.M. & Oades, J.M. (1979). Stabilization of soil aggregates by the root systems of ryegrass. *Australian Journal of Soil Research*, **17**, 429–441.

Trappe, J. (1981). Mycorrhizae and productivity of arid and semi-arid rangelands. *Advances in food-producing systems for arid and semi-arid lands*. pp. 581–599. Academic Press, New York.

Vogt, K.A., Grier, C.C., Meier, C.E. & Edmonds, R.L. (1982). Mycorrhizal role in net primary production and nutrient cycling in *Abies amabilis* (Dougl.). Forbes ecosystems in Western Washington. *Ecology*, **63**, 370&380.

Warnock, A.J., Fitter, A.H. & Usher, M.B. (1982). The influence of a springtail *Folsomia candida* (Insecta, Collembola) on the mycorrhizal association of leek *Allium porrum* and the vesicular-arbuscular mycorrhizal endophyte *Glomus fasciculatum*. *New Phytologist*, **90**, 285–292.

Whittingham, J. & Read, D.J. (1982). Vesicular-arbuscular mycorrhizas in natural vegetation systems, III. Nutrient transfer between plants with mycorrhizal interconnections. *New Phytologist*, **90**, 277–284.

Soil pH and vesicular-arbuscular mycorrhizas

GAMIN MA WANG*, D. P. STRIBLEY, P. B. TINKER

Department of Soils and Plant Nutrition, Rothamsted Experimental Station, Harpenden, Herts AL5 2JO

AND

C. WALKER

Forestry Commission, Northern Research Station, Roslin, Midlothian, EH25 9SY, Scotland

INTRODUCTION

The effects of soil pH on natural populations of vesicular-arbuscular (VA) fungi have not been studied systematically. Ecological surveys such as those of Read, Koucheki & Hodgson (1976) often confound other soil factors with soil pH. A useful opportunity for a study specifically on pH is afforded by the long-term liming experiments at Rothamsted Experimental Station where two sites, initially uniform, have been mantained at four levels of pH (*c.* 4·5, 5·5, 6·5 and 7·5 respectively) for 22 years by differential liming. We present observations on the mycorrhizal fungi of these sites, and investigations on the mechanism of the pH effects.

MATERIALS AND METHODS

Mycorrhizal fungi in the long-term liming experiment

The background and history of the two long-term liming experiments—one at Rothamsted on a flinty loam (Batcombe series), the other at Woburn on sandy loam (Cottenham series) overlying Greensand—are described by Bolton (1971, 1977).

These experiments were sown to spring oats ('Peniarth') in 1981 and 1982, and to maincrop potatoes ('Pentland Crown') in 1983, the crops being chosen for their ability to grow at a wide range of pH values. Root samples from plots of each pH were taken every year with a corer to 15 cm depth, the roots washed out, then cleared and stained. The fractional infection by mycorrhiza was measured by a grid-intersect method (Giovannetti & Mosse 1980). It is always difficult to identify the different VA endophytes in root infections (Abbott & Robson 1979). The only distinction which can be made with certainty is between 'fine' endophyte (possibly *Glomus (Rhizophagus) tenue*: Hall 1977) with hyphae of around 3 μm diameter, and others, the 'coarse' endophytes, with hyphae of around 5–10 μm diameter. The relative proportions of 'fine' to 'coarse' endophytes in our samples were measured on small subsamples by a grid-intersect method under the compound microscope.

*On leave from Brazilian National Soybean Research Center.

219

Chlamydospores (>50 μm diameter) were extracted from soils by wet-sieving and decanting.

Effects of additions of calcium carbonate

Finely-ground $CaCO_3$ was added to field soil from the most acid plots at Rothamsted, to give soils of pH 4·5, 5·5, 6·5 and 7·5 respectively. Spring oat was grown on these soils in pots in the glasshouse and the ratio of 'fine' and 'coarse' endophytes in the roots measured after 15 weeks.

Effects of aluminium and manganese in sand culture

Winter oat ('Pennal') was grown on acid-washed sand, with a basal nutrient supply, at four levels of aluminium, or four levels of manganese, respectively at pH 4·5. The inoculum was a laboratory culture of *G. caledonicum* maintained in soils of pH 7·5 with leek (*Allium porrum* L.) as host. Fractional infection was measured at 10 weeks.

Measurement of aluminium in soil solution

Soils were moistened to field capacity and incubated for 12 weeks. Soil solutions were displaced by addition of distilled water to the top of the column of soil, centrifuged, and their Al contents measured after acidification with HCl on an inductively-coupled optical emission spectrometer.

RESULTS AND DISCUSSION

pH and VA fungi in the long-term liming experiments

The fraction of the length of host root with mycorrhiza (fractional infection) at harvest was around 0·4 for both spring oat and potato and was affected little by pH in any year or crop. This is consistent with the result of Read *et al.* (1976) who found a constant level of infection in the perennial grass *Festuca ovina* L. over the pH range 4·2–7·0. Similarly, Sparling & Tinker (1978) observed little effect of pH in the range 4·9–6·2 in three grassland sites.

Table 1 shows that in the Rothamsted liming experiment roots of spring oat in 1981 were exclusively colonized by 'fine endophyte' on the most acid plots. The ratio of 'fine' to 'coarse' endophytes decreased with increasing pH, and at 7·5 coarse endophytes occurred exclusively. From the work of Wilson & Trinick (1983)

TABLE 1. Fractions of total length of mycorrhizal root occupied by fine or coarse endophytes in spring oat grown at various pH levels on the Rothamsted long-term liming experiment, 1981

	Soil pH			
	4·5	5·5	6·5	7·5
Fine endophytes	1·00	0·24	0·10	0·00
Coarse endophytes	0·00	0·76	0·90	1·00
	LSD at 5% = 0·09		CV = 17·3%	

TABLE 2. Chlamydospores of vesicular-arbuscular mycorrhizal fungi recovered from soils at various levels of pH of the Rothamsted long-term liming experiment in 1982

Fungal species	Approximate pH of soil			
	4·5	5·5	6·5	7·5
Glomus etunicatum	−	+	+	+
G. caledonicum	−	+	+	+
G. albidum	−	+	+	+
G. macrocarpum	−	−	+	+
G. fasciculatum	−	−	+	+
Glomus sp. (putative new species)	−	+	+	+
Glomus sp. (multiple-walled)	−	+	+	+
Acaulospora spp.	−	+	+	+

it seems likely that relative levels of soil inoculum were similar to these internal infections. Similar observations were made in 1982 and 1983 and also on the contrasting soil at Woburn in all years. The dominance of fine endophyte (or endophytes) at low pH suggests that it is very tolerant to acid soil. In natural communities on acid soil it has often been observed to be the dominant endophyte (Sparling & Tinker 1978), though in such cases there is usually also a different host at the different pH levels. There may also have been changes with pH in the proportions of different species of coarse endophyte but we were unable to test this.

Table 2 shows the various species of 'coarse' endophyte at Rothamsted, as determined by identifying the spores of diameter >50 μm washed out of the samples of field soil. The coarse endophyte infections observed in roots of spring oats therefore probably involved several different species. There is remarkably little information on the occurrence of species in relation to pH (Mosse, Stribley & LeTacon 1981). No spores were found at pH 4·5 whereas almost all species found were present at 5·5. Our techniques would not have separated out spores of fine endophyte, which are claimed to be around 10 μm diameter (Hall 1977). The absence of G. fasciculatum at this pH is unexpected since this species was originally described from an acid (*Sphagnum*) peat bog in Canada (Thaxter 1922). Investigations into the tolerance to pH of this species should be regarded with caution for there is very considerable doubt about the veracity of the identity of many fungi ascribed to this taxon (Walker 1983). For example, the endophyte termed 'G. fasciculatum E3', found by Tavares & Hayman (1983) to be infective at low soil pH, is *not* G. fasciculatum (C. Walker, unpubl.).

Effects of sudden change in pH

Table 3 shows that in the pot experiment, as in the field, fine endophyte occurred exclusively in roots at low pH but although the proportion of length of mycorrhizal root that this fungus occupied was gradually reduced by increasing pH, it was still present at pH 7·5. Propagules of 'coarse' endophytes must have been present, even though no spores were found, since mycorrhizas formed by coarse endophytes were

TABLE 3. Fractions of total length of mycorrhizal root occupied by fine or coarse endophytes in spring oat grown at various pH levels on soil originally taken from the acid plots from the Rothamsted long-term liming experiment, and subsequently limed

		Soil pH		
	4·5	5·5	6·5	7·5
Fractional infection				
By fine endophyte	1·00	0·35	0·31	0·11
By coarse endophytes	0·00	0·65	0·69	0·89

LSD at 5% = 0·12   CV = 16·8%

observed above pH 5·5. Their proportion relative to mycorrhizas formed by fine endophyte increased with increasing pH, as observed in the field.

Effects of aluminium and manganese

Our data imply that coarse endophytes have a broad tolerance to changes in soil acidity in the pH range 5·5–7·5, but not below pH 5·5. The presence of an inhibitory factor in acid soil is suggested by exclusion of coarse endophytes below pH 5·5 and by the relatively small sensitivity to increase in pH shown by fine endophyte in the pot experiment. Aluminium may be responsible, since its solubility in soil increases sharply below pH 5·5 (Cabrera & Talibudeen 1977; Fairbridge & Finkl 1979). It is considered to be primarily responsible for effects of soil acidity on vascular plants (Rorison 1973) and on the *Rhizobium* symbiosis (Cooper, Wood & Holding 1983).

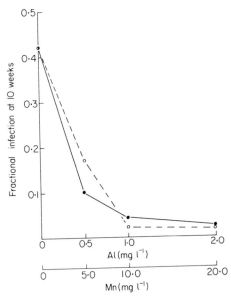

FIG. 1. Fractional infection at 10 weeks in roots of winter oat grown in sand culture at various levels of aluminium (●) and manganese (○) and inoculated with *Glomus caledonicum* (laboratory culture).

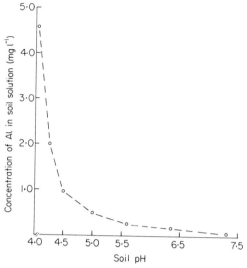

Fɪɢ. 2. Concentrations of aluminium in soil solution from plots of various pH levels of the Rothamsted long-term liming experiment.

Figure 1 shows that mycorrhizal infection by a stock isolate of *G. caledonicum* was inhibited strongly by aluminium in solution culture. Manganese was about ten times less inhibitory. The levels of aluminium used in this experiment were closely similar to those in solution in the soils from the Rothamsted experiment (Fig. 2). We were unfortunately unable to test the Al tolerance of *G. tenue* because of difficulty in isolating a pure culture of this species. This aspect of the physiology of *G. tenue* merits further study.

CONCLUSIONS

On the soils investigated here:

(i) only fine endophyte could infect at soil pH 4·5;
(ii) effects of pH were mainly on the species composition of the VA fungal flora, particularly the proportions of 'fine' to 'coarse' endophytes;
(iii) aluminium was probably responsible for inhibition of coarse endophytes below pH 5·5.

REFERENCES

Abbott, L.K. & Robson, A.D. (1979). A quantitative study of the spores and anatomy of mycorrhizas formed by a species of *Glomus*, with reference to its taxonomy. *Australian Journal of Botany*, **27**, 363–375.

Bolton, J. (1971). Long-term liming experiments at Rothamsted and Woburn. *Annual Report of Rothamsted Experimental Station for 1970, Part 2*, pp. 98–112.

Bolton, J. (1977). Changes in soil pH and exchangeable calcium in two liming experiments on contrasting soils over 12 years. *Journal of Agricultural Science (Cambridge)*, **89**, 81–86.

Cabrera, F. & Talibudeen, O. (1977). Effect of soil pH and organic matter on labile aluminium in soils under permanent grass. *Journal of Soil Science*, **28**, 259–270.

Cooper, J.E., Wood, M. & Holding, A.J. (1983). The influence of soil acidity factors on rhizobia. *Temperate Legumes: Physiology, Genetics and Nodulation* (Ed. by D.G. Jones & D.R. Davies), pp. 319–335.

Fairbridge, R.W. & Finkl, C.W. Jr. (Eds) (1979). *The Encyclopaedia of Soil Science, Part 1*. Dowden Hutchinson and Ross Inc., Pennsylvania, USA.

Giovannetti, M. & Moose, B. (1980). An evaluation of techniques for measuring vesicular-arbuscular mycorrhizal infection in roots. *New Phytologist*, **84**, 489–500.

Hall, I.R. (1977). Species and mycorrhizal infections of New Zealand Endogonaceae. *Transactions of the British Mycological Society*, **68**, 341–356.

Mosse, B., Stribley, D.P. & LeTacon, F. (1981). Ecology of mycorrhizae and mycorrhizal fungi. *Advances in Microbial Ecology*, **5**, 137–210.

Read, D.J., Koucheki, H.K. & Hodgson, J. (1976). Vesicular-arbuscular mycorrhiza in natural vegetation systems. 1. The occurrence of infection. *New Phytologist*, **77**, 641–653.

Rorison, I.H. (1973). The effect of extreme soil acidity on the nutrient uptake and physiology of plants *Acid Sulphate Soils*, Vol. 1 (Ed. by H. Dost). Proceedings of an International Symposium on Acid Sulphate Soils, Publication 18. International Institute of Land Reclamation and Improvement, Wageningen.

Sparling, G.P. & Tinker, P.B. (1978). Mycorrhizal infection in Pennine grassland, 1. Levels of infection in the field. *Journal of Applied Ecology*, **15**, 943–950.

Tavares, M. & Hayman, D.S. (1983). Soil pH preference of different VA mycorrhizal endophytes. *Annual Report of Rothamsted Experimental Station for 1982, Part 1*, p. 218.

Thaxter, R. (1922). A revision of the Endogonaceae. *Proceedings of the American Academy of Arts and Sciences*, **57**, 291–351.

Walker, C. (1983). Glomus fasciculatus: a taxon misunderstood! *Current Mycorrhizal Research* (Ed. by J. Dighton), Abstracts of communications presented at the mycorrhiza group meeting, Lancaster University, 28–30 March, 1983. Merlewood research and development paper no. 95, Institute of Terrestrial Ecology, Grange-over-Sands.

Wilson, J.M. & Trinick, M.J. (1983). Infection development and interactions between vesicular-arbuscular mycorrhizal fungi. *New Phytologist*, **93**, 543–553.

Vesicular-arbuscular mycorrhizal mediation of trace and minor element uptake in perennial grasses: relation to livestock herbage

K. KILLHAM

Department of Soil Science, University of Aberdeen, Aberdeen, AB9 2UE

SUMMARY

1 In the north-east of Scotland, trace element deficiencies in the soil and in local forage are sometimes responsible for poor performance in cattle and sheep.

2 To determine if VA mycorrhizal fungal infection mediates herbage trace (and minor) element uptake, three commonly grazed perennial grases (*Lolium perenne, Dactylis glomerata* and *Festuca rubra*) were grown with and without VAM fungal infection (*Glomus fasciculatum*: E3 Rothamsted) in soils sampled from areas of known incidence of trace element deficiencies in livestock.

3 Enhanced uptake of Cu, Co and Mg (to concentrations adequate for livestock herbage requirements) was associated with inoculation with *G. fasciculatum*, although the VAM fungal population indigenous to the soils had no such effect. This difference in mediation of plant metal uptake appears to be due to the very low degree of root infection by the indigenous VAM fungi compared to that by *G. fasciculatum*.

4 The soil conditions which may be responsible for the poor development of indigenous VAM fungal infection are discussed.

INTRODUCTION

Trace and minor element deficiencies have long been associated with poor health and reduced yields of sheep and cattle in certain areas of north-east Scotland. The trace metals Cu and Co, for example, are often applied to 'deficient' and 'borderline' soils at application rates of at least 25 kg ha^{-1} $CuSO_4$ and 2 kg ha^{-1} $CoSO_4$ to increase the concentration of these metals in herbage and to alleviate Cu and Co deficiencies in livestock (Reith 1975; Mitchell 1974). Similarly, applications in excess of 400 kg ha^{-1} of calcined magnesite have been used to alleviate Mg deficiency at certain sites (Reith 1967). Vesicular-arbuscular mycorrhizal (VAM) fungal infection is known to enhance, under certain circumstances, plant uptake of Cu and Zn (Killham & Firestones 1983; Lambert, Baker & Cole 1979; Swaminathan & Verma 1979), Ni (Killham & Firestone 1983), Ca (Rhodes & Gerdemann 1978a; Tinker 1978), S (Rhodes & Gerdemann 1978b), and Cl (Buwalda, Stribley & Tinker 1983), as well as P uptake. The exact mechanism of VA mycorrhizal enhancement of plant nutrient uptake is uncertain and may well vary from one nutrient to another, although one would expect the degree to which an element can be absorbed and translocated via VAM fungi to

225

depend on its diffusion coefficient in soil, and on its Km for uptake by the fungus compared to that of the plant root.

The aim of this experiment was to determine if VAM fungal inoculation can mediate changes in tissue concentrations of Cu, Co and Mg in host perennial grasses when they are grown in soils which are associated with herbage deficiencies of these metals, and to relate this to the function of VAM fungal populations indigeneous to these soils.

MATERIALS AND METHODS

Soil type and seed preparation

Seeds of three commonly grazed perennial grasses (*Lolium perenne, Dactylis glomerata,* and *Festuca rubra*) were surface sterilized prior to planting. Seeds were planted in three different soil types (Table 1), sampled from areas of known incidence of trace and minor element deficiency. The two 'Turriff' sites are both known to be associated with Cu and Co deficiencies, while the Glentanar site has a history of Mg deficiency. The soils were autoclaved for 1·5 h on three successive occasions prior to packing into 12·5 cm diameter, surface sterilized pots.

Mycorrhizal infection

Half of the pots were inoculated with the VAM fungus *Glomus fasciculatum* (E3 Rothamsted) by mixing into the top 2·5 cm, 10 g of the appropriate soil which had previously supported strawberries infected with *G. fasciculatum* ('inoculated treatment'). The remainder of the pots received a similar but sterile soil amendment ('non-mycorrhizal treatment'). Fifty seeds were than planted per pot. In addition, seeds were also planted into fresh soils to enable infection of the grasses by indigenous mycorrhizal fungal populations ('non-inoculated treatment').

Growth conditions

The pots (three replicates per treatment) were sprayed with simulated rain (pH 5·6, with ionic composition to reflect local rainfall) at rates equivalent to 100 cm year^{-1}.

TABLE 1. Some chemical characteristics of the soils used

Soil	Grid Ref.	pH	Organic C (%)	Total organic N (%)	Extractable* Cu (ppm)	Extractable† Co (ppm)	Extractable† Mg (ppm)
T1	NJ734506	4·4	3·3	0·25	0·50	0·01	30·6
T2	,,	4·3	3·1	0·18	0·57	0·01	27·2
G	NO485967	3·8	3·6	0·25	0·97	0·06	13·3

T1, Turriff 1 soil; T2, Turriff 2 soil; G, Glentanar soil.
* 0·5 M EDTA extraction.
† 0·5 M acetic acid extraction.

The solutions applied to the non-mycorrhizal plants also contained 50 μm P, applied as KH_2PO_4, to offset any increased uptake of this nutrient due to the presence of mycorrhizal infection. To assess the normal supply of P from the soil, a number of non-mycorrhizal plants were also grown without a phosphate supplement. Pots were sprayed twice-weekly over a 7-week period.

Analyses

At harvest, roots were examined microscopically with epifluorescent illumination to confirm the presence of autofluorescent arbuscules in VAM fungal infected roots (Ames, Ingham & Reid 1982) and to ensure that there was no infection, the non-mycorrhizal roots were also examined. To determine the degree of VAM fungal infection, the method of Phillips & Hayman (1970) for clearing and staining roots was used. After destaining in lactophenol for 20 min at 50°C, the roots were arranged on a grid of half-inch squares, according to the method of Marsh (1971). The intersections of roots and grid lines were counted with a dissecting microscope as either infected or uninfected. The degree of infection, in terms of per cent of root length infected, was then calculated. Shoots were oven-dried and ground for analysis. Acid (1 part conc. $HClO_4$:4 parts conc. HNO_3:1 part conc. H_2SO_4) digests were used for Cu, Co and Mg analyses of shoot tissue, determined by atomic absorption spectrophotometry, and for spectrophotometric determination of total phosphorus (Franson 1975). Soil pH was determined using a 1:1 distilled water:soil slurry and a glass pH electrode. Total soil organic carbon was determined by rapid titration (Walkley & Black 1934), and total N by Macro-Kjeldahl digestion (Hesse 1971). Atomic absorption spectrophotometric determination of soil Mg and Co were performed using 0·5 M acetic acid soil extracts, and those for Cu with 0·5 M EDTA soil extracts (2 h and 1 h extractions respectively). An SNK multiple-range test was used for statistical analysis of the data (Zar 1974).

RESULTS AND DISCUSSION

Concentrations of Cu and Co in shoot tissue of the non-inoculated grasses planted in the Turriff soils were considerably less than 5 μg g^{-1} dry wt and 0·08 μg g^{-1} dry wt respectively (Figs 1 and 2), concentrations recognized as being those below which Cu and Co deficiency in grazing livestock is likely (ARS 1981). Similarly, Mg concentrations in shoots of non-inoculated grasses grown in the Glentanar soil were less than 2 mg g^{-1} dry wt, the concentration below which grazing livestock tend to be affected by Mg deficiency (ARS 1981). The soils used, therefore, produced herbage likely to cause livestock metal deficiencies of the type known to occur at the selected sites.

Concentrations of Cu, Co and Mg were significantly ($P \leqslant 0·05$) higher in shoots of inoculated grasses than in both non-inoculated and non-mycorrhizal plants, while shoot metal concentrations of non-inoculated and non-mycorrhizal grasses were not significantly different (Figs 1, 2 and 3). In 'metal deficient' soils, only the

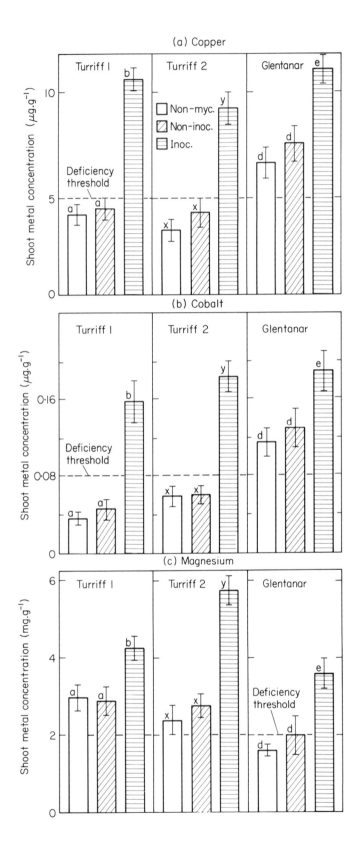

inoculated grasses contained shoot metal concentrations above the thresholds of suitability for livestock herbage (Figs 1, 2 and 3). Concentrations of Cu in inoculated grasses were 2·4 times greater than in non-inoculated grasses and 2·6 times greater than in non-mycorrhizal grasses grown in the Turriff 1 soil (Fig. 1). Shoot concentrations of Co and Mg were also increased by VAM fungal inoculation, although to a slightly lesser extent than for Cu (Figs 2 and 3).

VA mycorrhizal enhancement of plant Cu uptake has been reported in numerous studies (e.g. Killham & Firestone 1983; Timmer & Leyden 1980), but enhanced plant uptake of Co and Mg has not previously been reported. Studies of VAM enhancement of plant metal uptake have illustrated the interaction of soil P-status with mycorrhizal development, enhancement occurring in P-deficient soils, while plant metal deficiencies have been associated with reduced mycorrhizal development in soils of high P availability (Timmer & Leyden 1980). The low P-status of the soils used in this study, characterized by the low shoot P concentrations and low dry matter yields of non-mycorrhizal plants that did not receive additional P (data not shown), may well have been an important factor in the mediation of enhanced plant metal uptake by VAM fungal infection with *G. fasciculatum*.

For any one soil type, there were no major differences between the three different grass species in terms of shoot metal concentrations for corresponding non-inoculated and non-mycorrhizal plants. The data presented, therefore, are restricted to one species (*L. perenne*). Increased shoot metal concentration with mycorrhizal fungal infection occurred for all three grass species, although the magnitude of the increase varied from one species to another. Compared to non-mycorrhizal plants of the same species in Turriff 1 soil, shoot Cu concentrations for inoculated grasses were 1·5 times greater for cocksfoot (*D. glomerata*), 2·2 times greater for red fescue (*F. rubra*) and 2·6 times greater for perennial rye-grass (*L. perenne*).

Dry matter yields of the inoculated grasses were generally not significantly different from non-mycorrhizal grasses (suggesting that addition of P to the latter plants successfully offset enhanced uptake of this nutrient due to mycorrhizal infection), although considerably greater than non-inoculated grasses (Table 2). These trends in growth yield correlated closely with shoot P-status (Table 2), the inoculated grasses having shoot P concentrations similar to non-mycorrhizal grasses, but about twice as great as non-inoculated grasses.

Of fundamental interest in this study were the low concentrations of Cu, Co and Mg in the non-inoculated grasses grown in fresh soil where only the indigenous mycorrhizal fungi were able to infect the roots. It was decided to quantify the degree of VAM fungal infection of roots to determine if differences in shoot

---

FIG. 1. Concentrations of Cu, Co and Mg respectively, in shoots of non-mycorrhizal, non-inoculated and inoculated plants of *L. perenne*. For any one soil type, letters indicate significant ($P \leqslant 0.05$) difference between means, SNK multiple range test.

TABLE 2. Shoot dry matter yields, shoot P concentrations and percentages of root length which are VAM fungal-infected in plants of *L. perenne*

		Inoculated	Non-mycorrhizal	Non-inoculated
Dry matter yield (g.m^{-2})	T1	39·6 ± 6·2 (a)	31·2 ± 5·4 (a)	17·5 ± 5·0 (b)
	T2	32·7 ± 5·6 (x)	27·9 ± 4·9 (x)	11·3 ± 4·7 (y)
	G	26·6 ± 6·0 (d)	24·4 ± 5·7 (d)	7·8 ± 3·5 (e)
Shoot P (mg.g^{-1})	T1	2·2 ± 0·3 (a)	2·0 ± 0·2 (a)	1·2 ± 0·2 (b)
	T2	1·8 ± 0·2 (x)	1·8 ± 0·2 (x)	0·8 ± 0·2 (y)
	G	1·5 ± 0·3 (d)	1·3 ± 0·2 (d)	0·7 ± 0·1 (e)
Root infection (%)	T1	55·3 ± 11·0 (a)	–	19·4 ± 8·2 (b)
	T2	46·2 ± 9·7 (x)	–	16·0 ± 7·6 (y)
	G	40·1 ± 7·3 (d)	–	8·5 ± 5·2 (e)

For any one soil type, letters indicate significant ($P \leqslant 0·05$) difference between means, SNK multiple range test. T1, Turriff 1 soil; T2, Turriff 2 soil; G, Glentanar soil.

elemental composition between inoculated and non-inoculated grasses were related to level of mycorrhizal infection. Inoculated plants had over 50% of their root length infected with *G. fasciculatum,* while non-inoculated plants grown in fresh soil had less than 20% of their root length infected with VAM fungi (Table 2). This low degree of infection by the indigenous soil fungi was particularly marked in the Glentanar soil, where many of the plant roots were totally free of VAM fungal infection and where the average degree of infection was only about 10% of the root length. It would appear, therefore, that the low shoot concentrations of Cu, Co and Mg in the non-inoculated grasses grown in fresh soil were due to a low degree of VAM fungal infection. The acidity of the three soils (Table 1) may have been a factor in retarding the development of indigenous VAM associations. (In fact, the Glentanar soil which had the lowest degree of VAM fungal infection of roots, both when the soil was unamended and to a lesser extent when an inoculum was used, was also the most acid of the soils.) Different VAM fungi certainly seem to vary greatly in their preference for a particular pH range (Mosse 1972). It would be quite surprising, however, if the *G. fasciculatum* inoculum were considerably more acid tolerant than the indigenous mycorrhizal fungal population of an acid soil. It was, perhaps, the considerable inoculum density which produced such a high degree of VAM fungal infection on the inoculated roots. A second possible reason for low rates of infection by the indigenous fungal populations may have been the poorly draining nature of these soils. In the field, the VAM fungal population may well have been suppressed by frequently excessive soil moisture, a factor known to significantly reduce the growth and infection of mycorrhizal fungi (Reid & Bowen 1979). On sampling, sieving and packing, however, soil moisture characteristics may have improved to enable the *G. fasciculatum* inoculum to develop quite successfully, while the non-inoculated soils would have had a sparse, indigenous, VAM fungal population because of the previously unfavourable soil moisture conditions. Current research in our laboratory should enable us to determine which soil factors suppress development of VAM associations in these 'problem' soils.

CONCLUSIONS

In this study, enhanced uptake of Cu, Co and Mg through inoculation of perennial grasses with the VAM fungus *G. fasciculatum* has been identified. This effect is of particular significance in the soils tested since non-mycorrhizal plants had tissue Cu, Co and Mg concentrations regarded as 'deficient' for grazing livestock, while inoculated plants had adequate concentrations of these metals to ensure acceptable herbage quality. The inability of the VAM fungi indigenous to the soils to similarly enhance trace and minor element uptake would appear to be due to their very low degree of root colonization and infection.

Understanding the soil conditions and properties which prevent the success of VA mycorrhizal development may provide a solution to the problem of the livestock herbage grown on soils 'deficient' in trace and minor elements.

ACKNOWLEDGMENTS

I thank Dr J.W.S. Reith for useful discussions and Mrs L. Donald for her excellent technical assistance.

REFERENCES

ARS (1980). *Nutrient Requirements for Ruminant Livestock.* Technical Review, Agricultural Research Service, Commonwealth Agricultural Bureau.

Ames, R.N., Ingham, E.R. & Reid, C.P.P. (1982). Ultraviolet-induced autofluorescence of arbuscular mycorrhizal root infections: an alternative to clearing and staining methods for assessing infections. *Canadian Journal of Microbiology*, **28**, 350–355.

Buwalda, J.G., Stribley, D.P. & Tinker, P.B. (1983). Increased uptake of bromide and chloride by plants infected with vesicular-arbuscular mycorrhizas. *New Phytologist*, **93**, 217–225.

Franson, M.A. (Ed.) (1975). *Standard Methods for the Examination of Water and Wastewater.* (14th edn.) pp. 481–483. American Public Health Association, Washington, D.C.

Hesse, P.R. (1971). *A Textbook of Soil Chemical Analysis.* J. Murray, London.

Killham, K.S. & Firestone, M.K. (1983). Vesicular-arbuscular mycorrhizal mediation of grass response to acidic and heavy metal depositions. *Plant and Soil*, **6**, 277–284.

Lambert, D.H., Baker, D.E. & Cole, H. (1979). The role of mycorrhizae in the interactions of phosphorus with zinc, copper, and other elements. *Soil Science Society of America Journal*, **43**, 976–980.

Marsh, B. (1971). Measurement of length in random arrangement of lines. *Journal of Applied Ecology*, **8**, 265–267.

Mitchell, R.L. (1974). Trace element problems on Scottish soils. *Netherland Journal of Agricultural Science*, **22**, 295–304.

Mosse, B. (1972). Influence of soil type and *Endogone* strain on the growth of mycorrhizal plants in phosphate deficient soil. *Reviews of Ecological and Biological Society*, **9**, 529–537.

Phillips, J.M. & Hayman, D.S. (1970). Improved procedures for clearing roots and staining parasitic and vesicular-arbuscular mycorrhizal fungi for rapid assessment of infection. *Transactions of the British Mycological Society*, **55**, 158–161.

Reid, C.P.P. & Bowen, G.D. (1979). Effects of soil moisture on VA mycorrhizal formation and root development in *Medicago*. *The Soil–Root Interface* (Ed. by J.L. Harley & R.S. Russel), pp. 211–219. Academic Press, London.

Reith, J.W.S. (1967). Effects of soil magnesium levels and of magnesium dressings on crop yields and composition. *Technical Bulletin of the Ministry of Agriculture, Fisheries and Food*, **14**, 97–109.

Reith, J.W.S. (1975). Copper-deficiency in plants and effects of copper dressings on crops and herbage. *Proceedings of the Symposium on Copper in Farming*, pp. 25–37. Copper Development Association, London.

Rhodes, L.H. & Gerdemann, J.W. (1978a). Translocation of calcium and phosphate by external hyphae of vesicular-arbuscular mycorrhizae. *Soil Science*, **126**, 25–126.

Rhodes, L.H. & Gerdemann, J.W. (1978b). Hyphal translocation and uptake of sulphur by vesicular mycorrhizae of onion. *Soil Biology and Biochemistry*, **10**, 355–360.

Swaminathan, K. & Verma, B.C. (1979). Responses of three crop species to vesicular-arbuscular mycorrhizal infection on zinc-deficient Indian soils. *New Phytologist*, **82**, 481–487.

Timmer, L.W. & Leyden, R.F. (1980). The relationship of mycorrhizal infection to phosphorus-induced copper deficiency on sour orange seedlings. *New Phytologist*, **85**, 15–23.

Tinker, P.B. (1978). Effects of vesicular-arbuscular mycorrhizas on plant nutrition and plant growth. *Physiologie Végétale*, **16**, 743–751.

Walkley, A. & Black, I.A. (1934). An examination of the Degtjareff method for determining soil organic matter and a proposed modification of the chromic acid titration method. *Soil Science*, **37**, 29–38.

Zar, J.H. (1974). *Biostatistical Analysis*. Prentice Hall, New Jersey.

Use of an ecosystem model for testing ecosystem responses to inaccuracies of root and microflora productivity estimates

H. PETERSEN

Molslaboratoriet, Femmøller, DK 8400 Ebeltoft, Denmark

R. V. O'NEILL AND R. H. GARDNER

Environmental Sciences Division, Oak Ridge National Laboratory*, P.O. Box X, Oak Ridge, Tennessee 37830, USA

SUMMARY

1 A seventy-compartment model for a Danish beech forest ecosystem is described in outline.

2 The unmodified model predicts considerable accumulation of wood litter and decreasing accumulation through secondary to final decomposition products. Increment rates are similar for all components of the detritus-based food chain.

3 Modification of fine root production rate produces strong, positive response for root litter, and smaller, but still significant, responses for detritus, humus and the components of the decomposer food chain.

4 Increase of microbial biomass with adjustments of metabolism and production causes reduced accumulation of detritus and humus. The soil organisms respond according to food source.

5 Use of the model for testing the sensitivity of the ecosystem to inaccuracies of root and microflora estimates is discussed.

INTRODUCTION

The Hestehave Project was an extensive, multidisciplinary ecosystem study of a typical Danish beech forest stand conducted under the auspices of the Danish IBP-committee. Detailed analysis of specific portions of the collected data have been published (for references see DeAngelis, Gardner & Shugart 1981; Kjøller & Struwe 1982; Petersen & Luxton 1982).

Computer modelling provides an ideal technique for synthesizing such complex and extensive ecosystem data (O'Neill 1979), thus enabling a simultaneous account of all available data. A model for the Hestehave beech forest ecosystem was developed for that purpose, and for the resulting ability to extrapolate measured trends or various assumptions and hypotheses into the future in order to examine logical consequences for the whole ecosystem.

One obvious specific application of the ecosystem model is to test the extent to

*Operated by Martin Marietta Energy Systems Inc., under contract No. DE-AC05-840R21400 with the US Department of Energy. ESD Publication No. 2382.

233

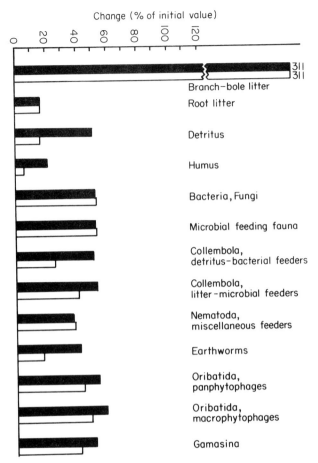

FIG. 2. Predictions of the unmodified model and responses to modification of microbial biomass, metabolism and production. The columns show percentage change after 300 years relative to the initial value. Unmodified model (solid). Microbial biomass doubled, production and respiration adjusted (unshaded).

consumption. The root litter component and exclusive or partial root feeders show relatively low increments because the fine root standing crop is kept constant.

Effects of modifying the production rate of fine roots

Figure 3 illustrates for selected components of the litter–soil stratum the relative changes, i.e. change shown by the modified model as percentage of change shown by the unmodified model, as a result of doubling or halving the production rate of fine roots. Strong response is shown for the root litter compartment while the responses of the secondary decomposition products, i.e. detritus and humus, are damped but still very significant. All organisms of the decomposer food chain

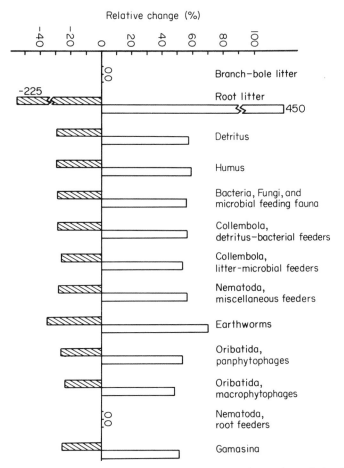

FIG. 3. Responses to modifications of fine root production rate. The columns show relative changes, i.e. change of modified model after 300 years as percentage of change shown by the unmodified model. Fine root production halved (shaded), fine root production doubled (unshaded).

including the predators (e.g. *Gamasina*) react in agreement with the most important primary food source, i.e. detritus. Note that the biomass of root feeders such as root-feeding nematodes does not change, because fine root biomass is fixed in the model. Thus, changes in root production affect litter production but have no consequence for root growth from year to year.

Halving the fine root production rate produces a relative reduction which is half the relative increments resulting from doubling the fine root production rate. This would be expected because the halving process is bounded by zero while the doubling process is unlimited. Halving the fine root production rate results in reduction of the root litter standing crop below the initial value. For all other components which show increments in the unmodified model the increment remains positive when fine root production is halved.

Effects of modifying microbial biomass production and metabolism

Figure 2 (open columns) illustrates the effect of doubling the initial microbial biomass and adjusting the production (recycling) and respiration to account for the higher maintenance cost required by the larger biomass. The litter components are unaffected whereas both detritus and humus show significantly lower increments during the simulation period than the primary model. Bacteria and fungi show the same increase relative to initial state as in the primary model. The same is true for the microbial feeding fauna, whereas faunal groups with mixed diets show different responses depending on the proportions of litter, detritus, and microbes composing the food of the individual trophic groups.

DISCUSSION

A large ecosystem model was described in outline based on the extremely simplified assumption that annual energy transfers to other components, to respiration or to biomass increment are constant fractions of the components from which they originate. It was argued and illustrated by examples that such models can be utilized to examine logical consequences of measured trends in biomass development and test effects on total ecosystem dynamics as the result of modifying parameter values.

Modifying the fine root production rate resulted in a marked change of increments of soil organic matter and most soil and litter organisms, and thus stresses the sensitivity of the ecosystem to inaccuracy of this estimate. Unfortunately, fine root production is one of the most difficult ecosystem parameters to estimate (Ulrich *et al.* 1981; Fogel 1983). The most commonly used methods based on summation of positive biomass changes measured for intervals through the year are very laborious and fail to account for continuous losses due to sloughing, exudation and rapid turnover of root hairs and mycorrhiza (Atkinson, p. 43). Edwards & Harris (1977), however, suspected such losses between their monthly samples to be within the limits of precision of the direct measurement of root biomass dynamics. The net root biomass production calculated for the Tennessee tulip poplar site of these authors was 9000 kg ha^{-1} year^{-1} ($= 18700$ kJ m^{-2} year^{-1}) which is $2 \cdot 8$ times the above-ground net wood production (Harris, Kinerson & Edwards 1977) and more than three times the root production estimate used in the present study. With the support of other recent studies (Fogel 1983) it suggests that the root production estimate used in the Hestehave model may be an underestimate. The strong influence of the root production estimate on total soil organic matter and soil organisms indicated by the model may provide a basis for testing this hypothesis.

A number of assumptions and tentative estimates for production efficiency, metabolic cost of maintenance and allocation of microbial metabolism to bacteria and fungi, respectively, make the production estimates for microflora employed in the primary model questionable. In addition, biomass estimates of bacteria and fungi may differ by orders of magnitude depending on methodology (Parkinson, Gray & Williams 1971; Holm & Jensen 1972).

Adjusting the microbial biomass to higher values implies that soil organic matter standing crop increases less during the simulation period or ultimately decreases below the initial value if microbial biomass is adjusted above a certain level. Such consequence of manipulating the microbial biomass is clearly unlikely in the Hestehave beech forest, where total litterfall is assumed to increase as a result of the growing standing crop of beech boles. Thus, our example illustrates use of the model to restrict the range of possible values for ecosystem parameters by evaluation of the effect on other components of the ecosystem.

ACKNOWLEDGMENTS

H. Petersen wishes to thank the Environmental Sciences Division, Oak Ridge National Laboratory, for providing working facilities and computer time for preparation of this work during his stay as visiting scientist at the laboratory. The stay was financed by the Danish Natural Science Research Council and the Natural History Museum, Århus. Mrs Gertrud Poulsen is thanked for typing the manuscipt.

REFERENCES

Carnahan, B., Luther, H.A. & Wilkes, J.O. (1969). *Applied Numerical Methods.* John Wiley & Sons, New York.
DeAngelis, D.L., Gardner, R.H. & Shugart, H.H. (1981). Productivity of forest ecosystems studied during the IBP: the woodland data set. *Dynamic Properties of Forest Ecosystems* (Ed. by D.E. Reichle), pp 567–672. International Biological Programme 23. Cambridge University Press, Cambridge.
Edwards, N.T. & Harris, W.F. (1977). Carbon cycling in a mixed deciduous forest floor. *Ecology*, **58**, 431–437.
Fogel, R. (1983). Root turnover and productivity of coniferous forests. *Plant and Soil*, **71**, 75–85.
Göttsche, D. (1972). *Verteilung von Feinwurzeln und Mykorrhizen in Bodenprofil eines Buchen-und Fichtenbestandes in Solling.* Kommissionsverlag Buchhandlung Max Weidebusch, Hamburg.
Harris, W.F., Kinerson, R.S. Jr & Edwards, N.T. (1977). Comparison of below-ground biomass of natural deciduous forest and loblolly pine plantations. *Pedobiologia*, **17**, 369–381.
Heal, O.W. & McLean, S.F. Jr (1975). Comparative productivity in ecosystems—secondary productivity. *Unifying Concepts in Ecology* (Ed. by W.H. van Dobben & R.H. Lowe-McConnell), pp. 89–108. W. Junk, The Hague.
Holm, E. & Jensen, V. (1972). Aerobic chemoorganotrophic bacteria of a Danish beech forest. *Oikos*, **23**, 248–260.
Holm, E. & Jensen, V. (1980). Microfungi of a Danish beech forest. *Holarctic Ecology*, **3**, 19–25.
Humphreys, W.F. (1979). Production and respiration in animal populations. *Journal of Animal Ecology*, **48**, 427–453.
Kjøller, A. & Struwe, S. (1982). Microfungi in ecosystems: fungal occurrence and activity in litter and soil. *Oikos*, **39**, 391–422.
Luxton, M. (1972). Studies on the oribatid mites of a Danish beech wood soil, I. Nutritional biology. *Pedobiologia*, **12**, 434–463.
Møller, C.Mar., Müller, D. & Nielsen, J. (1954). Graphic presentation of dry matter production of European beech. *Det forstlige Forsøgsvæsen i Danmark*, **21**, 327–335.

O'Neill, R.V. (1979). A review of linear compartmental analysis in ecosystem science. *Compartmental Analysis of Ecosystem Models* (Ed. by J.H. Matis, B.C. Patten & G.C. White), pp. 3–28. International Co-operative Publishing House, Fairland, Maryland.

Parkinson, D., Gray, T.R.G. & Williams, S.T. (Eds). (1971). *Methods for Studying the Ecology of Soil Micro-organisms.* IBP Handbook, No. 19. Blackwell Scientific Publications, Oxford.

Persson, T. & Lohm, U. (1977). Energetical significance of the soil- and litter-inhabiting annelids and arthropods in a Swedish grassland ecosystem. *Ecological Bulletins (Stockholm)*, **23**, 1–211.

Petersen, H. & Luxton, M. (1982). A comparative analysis of soil fauna populations and their role in decomposition processes. *Oikos*, **39**, 287–388.

Twinn, D.C. (1974). Nematodes. *Biology of Plant Litter Decomposition* (Ed. by C.H. Dickinson & G.J.F. Pugh), pp. 421–465. Academic Press, London.

Ulrich, B., Benecke, P., Harris, W.F., Khanna, P.K. & Mayer, R. (1981). Soil processes. *Dynamic Properties of Forest Ecosystems* (Ed. by D.E. Reichle), pp. 265–339. International Biological Programme 23. Cambridge University Press, Cambridge.

Yeates, G.W. (1971). Feeding types and feeding groups in plant and soil nematodes. *Pedobiologia*, **11**, 173–179.

Population and community dynamics in the soil ecosystem

MICHAEL B. USHER

Department of Biology, University of York, York YO1 5DD

SUMMARY

1 The review concentrates on the soil arthropods (predominantly mites and springtails) and on the effects of the environment on their population dynamics.

2 Among the many environmental factors which influence the fecundity, speed of development and mortality of the soil fauna, the species of organisms available for food, as well as their nutritional status, are important in determining such demographic parameters. Egg diapause, and the possibility of an egg bank in the soil, could also have important implications for year-to-year fluctuations in population density.

3 Competition between soil arthropods is generally asymmetrical (one species adversely affected, the other species unaffected). Predation probably plays an important role in regulating populations of the grazing species. Little is yet understood about more complex interactions in the soil ecosystem.

4 Succession is generally related to an increase in both the number of species and the density of most groups of the soil fauna. However, management effects are linked directly to modification of the physical habitat and to change in the soil organic matter content. The potential for perturbation experiments, replicated and designed to investigate interactions, is discussed.

INTRODUCTION

As the title of this review is wide, it is appropriate to focus particular attention on some aspects of population and community dynamics. Two restrictions will be made. First, the effect of the environment on the populations of soil animals will be reviewed. The word 'environment' in this context is defined as widely as possible, and includes all aspects of the physical, chemical and biological environments in which the soil fauna are living. However, which environmental factors influence the population dynamics, as opposed to the distribution, of the soil animals? The review is very largely concerned with the dynamic rather than the distributional aspects of soil communities: the latter have been reviewed by Usher (1976). Second, examples illustrating the review will be chosen from soil arthropod communities. These are comprised of some insects, mostly of the order Collembola (springtails), and all of the soil mites (Acari: especially Cryptostigmata, Mesostigmata and Prostigmata).

The review follows relatively traditional lines by examining three levels in the hierarchy of complexity of interactions in the soil environment. The first level is concerned with the dynamics of single species populations, and contains an analysis of the factors contributing to population increase or decline. Simple community types, of only two or three species, the subject of laboratory experimentation, are considered in the second level. Naturally occurring communities are looked at in the third level. The aim of the review is, therefore, to investigate the factors determining the sizes of the individual species populations, and the way in which these interact to form the community of soil animals. The review thus differs from two previous reviews on similar topics (Usher *et al.* 1979; Usher, Booth & Sparkes 1982).

DYNAMICS OF SINGLE SPECIES POPULATIONS

The basic equation in population dynamics is

$$N_{t+1} = N_t + B - D + I - E$$

where N is the number of animals, B and D are the number being born and dying respectively in the time period t to $t + 1$, and I and E are the numbers of immigrants and emigrants during the same period. Various facets of this equation will be considered in the review, which will not be concerned with the methodology of parameter estimation from field data (cf. Straalen 1983b) or with the prediction of final population densities (cf. Sager & Stelter 1981). Relatively little is known about the demography of soil animal populations, as evidenced by the final plea in Joosse's (1983) review of Collembolan ecophysiology.

Fecundity

A number of species, particularly those associated with more stable environments, breed only once (Huhta & Mikkonen 1982; Leinass & Bleken 1983) or twice (Grégoire-Wibo 1979) per year. The capacity for increase can be small: a value of $r = 0 \cdot 00008$ per degree-week was estimated for *Ceratozetes kananaskis* by Mitchell (1977). However, in less stable environments, more generations are likely to occur: Hutson (1981) showed that all Collembola colonizing reclaimed land in Northumberland had at least two generations per year and *Isotoma notabilis* reproduced continuously throughout the year.

Sexual breeding has generally been assumed. However, Petersen (1978) has shown that about 25% of species of Collembola in a beechwood breed parthenogenetically: these species are generally the most abundant since they comprise 72% of all individuals. *Folsomia candida* has frequently been used in laboratory studies because it is parthenogenetic. Arrhenotoky (haploid male, diploid female) is probably not uncommon in the Mesostigmata, and, with overlapping generations and without cannibalism or oophagy, Usher & Davis (1983) have suggested that at least one soil mite, *Hypoaspis aculeifer*, may be approaching the eusocial threshold.

Before eggs are laid it has been suggested that a protein-rich meal is required

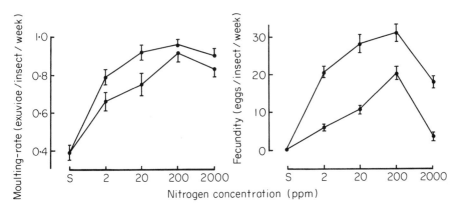

FIG. 1. The rate of development, measured as the moulting rate, and the fecundity of *Folsomia candida* when starved or feeding on two species of fungi, *Hypholoma fasciculare* (upper graphs) and *Coriolus versicolor* (lower graphs), which had been cultured in media with a variety of nitrogen concentrations. Vertical bars indicate ±1 standard error. Data from Booth (1979).

(pollen feeding by *Entomobrya socia*, Waldorf 1981), or that methionine is essential (required by *Achipteria holomensis*, Stamou *et al.* 1981). A 6-year study of a population of *Allodinychus flagelliger* by Athias-Bince (1978/79) indicated that fecundity was strongly density-dependent. Webb & Elmes (1979) showed that the number of eggs and pre-larvae of *Steganacarus magnus* was closely correlated with female size, and Grimnes & Snider (1981) showed that four electrophoretically distinct strains of the parthenogenetic *F. candida* have different fecundities. However, in an interesting study that related fecundity to food quality, Booth (1979) grew two species of fungus in liquid media with 2, 20, 200, and 2000 ppm of nitrogen, and determined the fecundity of *F. candida* when fed with these eight types of food (see Fig. 1). It is clear that fecundity was greater when *F. candida* was feeding on *Hypholoma fasciculare* than on *Coriolus versicolor*, and that fecundity increased with increasing nitrogen content up to 200 ppm. The critical experiment that Booth (1979) performed was to see if *F. candida* preferred the food that optimized fecundity. This was the case when *F. candida* were feeding on the less preferred fungus (Table 1), but there was no such relationship with the preferred

TABLE 1. The mean number of *Folsomia candida* observed feeding on fungi grown in media with different nitrogen concentrations. The data are taken from two experiments performed by Booth (1979), one experiment using *Coriolus versicolor* and the other *Hypholoma fasciculare*. Standard errors (SE) and least significant differences (LSD) are based on the analyses of variance of the data of each experiment.

Nitrogen concentration (ppm)	C. versicolor	H. fasciculare
2	6·4	4·4
20	4·6	1·4
200	9·4	2·0
2000	5·1	3·8
SE of means	1·0	0·8
LSD (at 5% level)	3·0	2·2

fungus. The extent to which both food type and food quality affect fecundity in field populations remains unknown.

It is possible that the eggs may go into diapause. In northern Europe, egg diapause by *Lepidocyrtus lignorum* is terminated after cold treatment so that there is synchronous hatching in the spring (Leinaas & Bleken 1983), and Valpas (1969) showed that hatching of eggs occurred whilst samples, particularly those that were frozen or water-logged, were being extracted in the laboratory. Some eggs of *Sphaeridia pumilis* diapause, whilst others, particularly those laid in autumn, did not (Blancquaert, Mertens & Coessens 1982). This leads to speculation that the soil may contain an 'egg bank' of arthropod eggs in much the same way as it contains a 'seed bank' of higher plant seeds. Studies of the Antarctic mite *Gamasellus racovitzai* (Usher & Bowring 1984) suggest the possibility of an egg bank, which is also indicated by Healey (1965) who says that 'recruitment of newly hatched individuals [of *Onychiurus procampatus*] occurs at all times'. He discusses some January samples which were frozen and appeared like blocks of ice: on extraction there were juveniles, but whether these had been present in the samples as juveniles or as eggs that hatched during extraction is uncertain. Valpas' (1969) finding of hatching during extraction is clearly important in understanding the dynamics of the egg bank in the soil. In turn, an egg bank would be important in population or community dynamics since it could release pulses of animals if there was synchronous hatching after unusual events, such as late summer rains in the Mediterranean region (Blancquaert *et al. 1982)*.

Speed of development

One would expect the speed of development to be closely related to temperature. Grégoire-Wibo's (1979) field study of *Folsomia quadrioculata* phenology showed that development of different cohorts, e.g. those overwintering or those hatching from eggs in the spring, was temperature-dependent. Hutson's (1978) laboratory studies of *Folsomia candida* are similar, as development from hatching to the age of first oviposition was 6 days at 25°C, 14 days at 10°C and 85 days at 5°C. The utility of a physiological scale for measuring development time is shown by Mitchell's (1977) study of development of *Ceratozetes kananaskis* which required 465 degree-weeks (approximately 2 years at field temperatures), though this assumes linearity which is not shown in Hutson's (1978) study quoted above.

There are, however, other factors which would appear to over-ride the effect of temperature. These include the effects of water (Verhoef & Selm 1983), drought (Takeda 1983), substrate (Booth 1983), and the availability of a protein-rich food, usually pollen (Allman & Zettel 1983). The fungi being grazed can also be important. Stefaniak & Seniczak (1981) attempted to culture *Oppia nitens* on twelve species of fungi. Three species remained uneaten, the mites dying without ovipositing. The rate of ingestion of three other fungi was low, and juveniles took an average of 32·7 days to develop into adults. The ingestion rate of four fungi was high, and juveniles had a greater survival and developed into adults in an average of

30·3 days. With the two other species, which were very highly ingested, large numbers of eggs were laid, juvenile survival was high, and development averaged 26·5 days. Booth's (1979) study also showed that the speed of development, measured by the rate of production of exuviae, was greater with *Folsomia candida* feeding on *Hypholoma* than on *Coriolus* (Fig. 1). The choice experiments (Table 1) could not indicate whether *F. candida* selected the form of *Coriolus* that maximized development rate or fecundity, since the reaction to both was similar: selection of *Hypholoma* was not consistent with such optimization.

Mortality

There are two types of mortality: continuous mortality and mortality associated with a particularly harsh season of the year. The latter, usually summer heat associated with desiccation or winter cold associated with freezing, can have important consequences for the dynamics of populations. Baker (1978) has described the biology of the Australian millipede, *Ommatoiulus moreletii*, in which summer mortality can be high in populations aestivating in grasslands whereas it is negligible in woodland populations. Joosse (1983) reviewed cold adaptation in the Collembola and illustrated Burn's (1982) data for *Cryptopygus antarcticus* in which population mortality is approximately proportional to the number of insects with gut contents.

There are relatively few data on continuous mortality. Mitchell (1977) illustrated a survivorship curve for *Ceratozetes kananaskis*, which showed an exponential decay in mortality rate during the development from egg to adult. Straalen (1983a) considered methods of estimating mortality rates for populations of Collembola: the rates varied between 0·7 and 0·50 per week for *Orchesella cincta* and between 0·19 and 0·41 per week for *Tomocerus minor*. Athias-Binche (1978/79), after a 6-year study of a population of *Allodinychus flagelliger*, concluded that mortality was density-dependent. Massive mortalities (greater than 50% in less than one day) have been recorded for *Onychiurus justi* by Snider & Butcher (1973) in environments with relative humidities below 100%. The experiments of Stefaniak & Seniczak (1981) also indicated that mortality is dependent on the type of food available to the soil fauna.

However, the question which appears never to have been addressed is how the various forms of mortality can be apportioned in field populations. What proportion is due to environmental extremes, what proportion to predators, what proportion to unavailability of appropriate food, and what proportion to other causes?

Immigration and emigration

There are few data on the speed of movement of the soil fauna. Berthet (1964) marked mites radioactively, and recorded their wanderings over the subsequent few weeks: he found that the rate of daily displacement was small. In the Netherlands, studies monitoring the spread of arthropod species into newly created pol-

ders (e.g. Haeck, Hengeveld & Turin 1980), indicated a reasonably fast colonization, though this was not always due to movement within the soil or litter.

Studies of individual species have tended to indicate little movement. Petersen's
(1978) study of *Tullbergia macrochaeta* in Jutland indicated that there was little
mixing of a population of sexual forms (extending to at least 36 m from the
seashore) and a population of parthenogenetic forms (occuring more than 50 m
from the shore). Allmen & Zettel (1982) documented the vertical movement of a
population of *Entomobrya nivalis* on trees: females descend to the litter to oviposit,
and juveniles ascend to feed. A second migration occurs at the beginning of winter
when the majority migrate down to hibernate under loose flakes of bark.

However, it remains uncertain in what way many of the species of soil animals
are distributed. The wide geographical distribution of many species, contrasted
with the extremely localized distributions of cave species, tends to imply efficient
dispersal. Studies of colonization of new habitats, e.g. Hutson & Luff (1978) and
Hutson (1981), indicate that some species arrive relatively quickly, but the number
of species which arrive and fail to establish themselves is unknown. Wind dispersal
(see Macfadyen in Hale 1963) is probably important, but there are no quantitative
data on numbers of species or the number of individuals distributed in this way. For
some species of Mesostigmata, phoresy is probably an alternative method of dispersal. Research into the methods and speed of dispersal would have interest both for
the dynamics of populations and communities and for the biogeography of the soil
fauna.

DYNAMICS OF COMMUNITIES OF TWO OR THREE SPECIES

Usher *et al.* (1979) reviewed the population dynamics of soil arthropods, concen-
trating on two approaches, termed 'divisive' (whereby inferences are made from
sampling field communities) and 'agglomerative' (the inferences being drawn from
putting together simple communities in the laboratory). No such approach was
taken by either Usher, Booth & Sparkes (1982) or Parkinson (1983), both of
whom selected a variety of topics of interest to them. In this section of the review a
different approach will be taken: the three most frequent interactions—interspecific competition, predator–prey, herbivore–plant—will each be considered,
and then there will be a brief review of interactions involving three trophic levels.

Interspecific competition

Competition has frequently been inferred in soil arthropod communities, but has
less frequently been demonstrated. Greenslade & Greenslade (1980) suggested
that diffuse competition (in which any one species interacts with a patchwork of
others at varying densities and in varying combinations) occurred in a community
of Isotomid Collembola from the Solomon Islands, as did Kaczmarek (1975) for a
pine forest community of Collembola in Poland. Vegter (1983), however, demonstrated that a field population of *Orchesella cincta* and *Tomocerus minor*, in the

TABLE 2. A summary of the competitive interactions between *Onychiurus ambulans* and *Isotoma viridis* (Sparkes 1982). In the table results are given in the form *x/y*, where *x* represents the effect of *O. ambulans* on the survival of *I. viridis*, and *y* the effect of *I. viridis* on the survival of *O. ambulans*. O indicates no significant effect, and − a significant depression in survival

Initial percentage abundance of		Temperature		
O. ambulans	*I. viridis*	16°C constant	8°C constant	8°C mean
67	33	− / −	0/0	0/0
50	50	0/ −	0/ −	0/0
33	67	0/ −	− / −	0/0

presence of three other species of Entomobryid Collembola, tended to avoid competition by food and microhabitat specialization. Can either competition or avoidance of competition be demonstrated in the laboratory?

Competition between two Collembola, *Isotoma viridis* and *Onychiurus ambulans*, from the community in an abandoned chalk quarry in the Yorkshire Wolds, has been investigated by Sparkes (1982). The species were cultured, either individually or mixed in varying proportions, at 8°C, 16°C, and at an average of 8°C but varying diurnally from 4°C to 12°C. At this variable temperature neither species affected the other significantly (Table 2). At constant temperature three results showed a one-way interaction (*O. ambulans* survival being depressed by the presence of *I. viridis*), and two showed genuine competition whereby the survival of both species was reduced by the presence of the other. This result is, however, over-simplistic, since *O. ambulans* often increased its proportion in the culture (Fig. 2), due to differential mortality. *I. viridis* had a greater mortality rate than *O. ambulans*, and hence in mixed species cultures *O. ambulans* often dominated eventually.

Another factor to consider in such competition experiments is the fecundity of the competing species. Sparkes' data indicated that neither temperature nor the presence of *O. ambulans* significantly affected the fecundity of *I. viridis* (approxi-

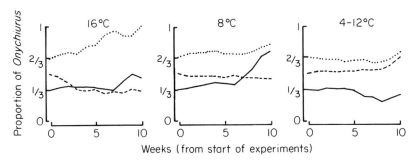

FIG. 2. The proportion of *Onychiurus ambulans* in mixed cultures with *Isotoma viridis*, over a 10-week period. The cultures were maintained in three temperatures, two constant (8° and 16°C) and one variable with a mean of 8°C but varying diurnally from 4°C to 12°C. Continuous lines represent cultures which started with 33% *O. ambulans*, dashed lines with 50% *O. ambulans*, and dotted lines with 67% *O. ambulans*. Data from Sparkes (1982).

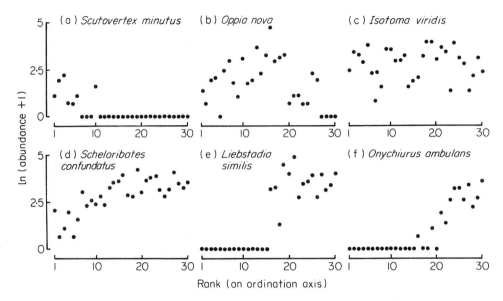

FIG. 3. The distribution of six arthropods (two Collembola, *I. viridis* and *O. ambulans*, and four Cryptostigmata) in relation to a successional gradient in Wharram Quarry Nature Reserve, Yorkshire. The successional ranking, based on a multivariate analysis as described in the text, runs from youngest (1) to oldest (30). Data from Parr (1980).

mately 0·7 eggs per female per day, assuming a 1:1 sex ratio). However, *O. ambulans* was strongly affected by both temperature and the presence of *I. viridis*: at 8°C a female laid approximately 3·7 eggs per day, whereas in the presence of *I. viridis* this rate dropped to 0·6 (again assuming a 1:1 sex ratio).

Laboratory studies of competition have tended to yield confusing results (e.g. Longstaff 1976). Few studies have demonstrated true competition, in which the fitness of both competing species is reduced, but many have indicated amensalism, whereby the fitness of only one competitor is reduced whilst that of the other is not altered significantly. Such asymmetrical competition (Lawton & Hassell 1981) is widespread in the insects. However, competition may not lead to the species with an unimpaired fitness 'winning': Fig. 2 demonstrates that *O. ambulans* often becomes proportionately more abundant in cultures, and Fig. 3c and f also indicates that it slowly becomes dominant in the field (though competition in the field has not been investigated). The actual life-history attributes of the species, such as fecundity, mortality and speed of development from egg to reproductive adult, are as important in determining the outcome of competition as the actual interaction between the species.

Predation

There are essentially five questions to answer: what are the predators, how do they locate their prey, how do they catch their prey, how many do they eat, and what are the effects of predation on the prey population?

Beetles of the family Carabidae frequently predate Collembola, e.g. *Notiophilus* sp. (Higgins 1982) and *Loricera pilicornis* (Bauer 1982). Dennison & Hodkinson's (1983) study of a community of Carabid beetles showed that 17 out of 25 species in a woodland had been predating Collembola and 17, 13, 9, 8 and 3 had been predating mites, nematodes, spiders, enchytraeid worms and isopods, respectively. Spiders are probably important predators as well: it seems inconceivable that a population of *Notiomaso australis* in a moss-turf on South Georgia, with a density of approximately $1 \cdot 5$ cm^{-2}, was predating anything other than the abundant Collembola (Usher 1984). This species subsequently moulted and laid eggs when maintained in a laboratory on the Collembola, *Setocerura georgiana*. The Mesostigmata also contains many species of small predators, and some Collembola will eat animal material (Tosi 1977).

Studies have shown that many predatory mites appear to wander randomly when searching for prey (Davis 1978; Usher & Bowring 1984), although it is possible that there are chemical stimuli deriving from prey faeces (Hislop & Prokopy 1981). Relatively little is known about the mechanisms of capture, although studies of the chelicerae of Mesostigmata (Harris 1974; Karg 1983) indicated the type of prey, and Usher & Bowring (1984) found that some species use their forelegs for holding the prey whilst the chelicerae pierce the prey's side. When attacked, some Cryptostigmata are able to withdraw into a hardened cuticle (e.g. *Hoploderma* spp.), some Collembola are able to jump, and others may possess a chemical defence (e.g. pseudocelli function in *Onychiurus*: Usher & Balogun 1966; Rusek & Weyda 1981).

Rates of predation vary according to the species of prey, the degree of satiation of the predator, and the reproductive state of an adult mite. Harris & Usher (1978) showed that the form of the functional response of *Pergamasus longicornis* changed

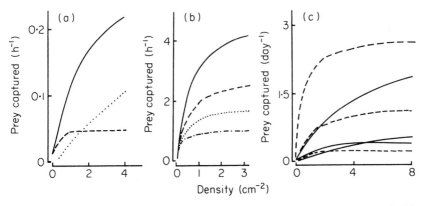

FIG. 4. Functional response curves for a variety of Mesostigmata predating Collembola in the laboratory. (a) male *Pergamasus longicornis* predating *Folsomia fimetaria* (——), *Sinella coeca* (....) and *Hypogastrura denticulata* (– – –). Data from Harris & Usher (1978). (b) male *Pergamasus crassipes* predating *Onychiurus armatus*. The four lines, from top to bottom, represent satiation of the predator (0–2 h, 2–4 h, 4–6 h and 6–8 h feeding respectively). Data from Longstaff (1980). (c) *Hypoaspis aculeifer* predating *Sinella coeca* (——) and *Hypogastrura denticulata* (– – –). The three lines (from top down) represent feeding on small, medium and large sized prey respectively. Data from Davis (1978).

from approximately a linear increase when the species was predating *Sinella coeca* to a very limited response when it was predating *Hypogastrura denticulata* (see Fig. 4a). Longstaff (1980), working with the closely related *Pergamasus crassipes*, found a similar type 2 functional response (Holling 1966), but the rate of predation decreased as the predators became satiated (Fig. 4b).

These results can be related to the possibility of predatory control of populations in the field. A type 2 functional response implies that, for all but the smallest prey populations, a doubling of prey density is not accompanied by a doubling of predation rate. This led Harris & Usher (1978) to suggest that predators were unlikely to exert a controlling influence on a species such as *H. denticulata*, which relies on chemical repulsion of the predator, but much more likely to be able to control a species such as *S. coeca*. However, Joosse (1981) has argued that the risk of predation is related to prey locomotion. The predation rate on *Tomocerus minor* by *Notiophilus biguttatus* was 26% in 8 days in dry environments, whereas in wet environments, where the prey moves less, the rate was 5% in 8 days. There is still much that is unknown about the effects of predators on the density of populations in the field.

Herbivory

The aim of this section is not to review the feeding of soil arthropods, but rather to concentrate on two questions: (i) what happens to fungi when they are grazed, and (ii) do the fungi have any defences against grazing animals? The review will concentrate on fungi rather than on bacteria since it is known that the majority of Cryptostigmata (Behan-Pelletier & Hill 1983) and Collembola (Petersen 1971; Takeda & Ichimura 1983) are fungal grazers. However, fungal grazing may not be particularly important in relation to the whole soil ecosystem: Mitchell & Parkinson (1976) estimated that only 2% of the fungal standing crop in an aspen woodland soil was consumed in one year by Cryptostigmatid mites.

Cancela da Fonseca, Kiffer & Poinsot-Balaguer (1979) described the associations of mites and fungi that they detected in the forest soil, but what might be the reason for these associations? Four effects of grazing have been noted in the literature. First, grazing may contribute to the structuring of the fungal community. Parkinson, Visser & Whittaker (1979) have demonstrated how the competitive colonizing ability of two fungi can be altered by *Onychiurus subtenuis* grazing, although they were wary of extrapolating laboratory results to a field situation. Wicklow & Yocom (1982), however, felt that grazing by larval *Lycoriella mali* (Diptera: Sciaridae) in rabbit dung reduced the number of species of coprophilous fungi, and hence was important in structuring fungal communities. Second, grazing may affect the respiration of the fungi. Grazing by *Oynchiurus armatus* on *Mortierella isabellina* (Bengtsson & Rundgren 1983) reduced respiration when grazing was continuous, but it increased respiration above that of an ungrazed culture when grazing was interrupted. Third, arthropods can distribute fungal spores in the soil, although many spores will be damaged and become non-viable with ingestion.

Behan & Will (1978) found that only 14% of faeces of seven Cryptostigmata species contained viable spores, and Ponge & Charpentie (1981) showed that spore viability decreased on passage through the gut of *Pseudosinella alba* (for *Penicillium* sp. it decreased from 69–100% to 3–11%). Fourth, it is possible that soil arthropods are associated with root pathogens, although Ulber (1983) has shown that infection of sugar-beet with *Pythium ultimum* is lower in the presence of Collembola.

Very little is known about how fungi can defend themselves against grazing. Wicklow (1979) has suggested a mechanical means whereby perithecial hairs protect the ascocarps of *Chaetomium bostrycodes*. The poisonous nature of many mycorrhizal fungi has led Shaw (p. 333) to speculate that the poisons may be anti-herbivore compounds, though *Drosophila* larvae are able to develop in the fruiting bodies of some of these poisonous species (Lacy 1984).

Three trophic levels

The majority of studies of the soil ecosystem have tended to infer relationships from field and from simple laboratory experiments. Some studies have used a wider spectrum of organisms: a producer (usually a higher plant), an associated microbe, and a grazing arthropod; or alternatively a microbe, its grazer and a predator of that grazer. Insufficient is yet known about the biology of soil organisms to know if there are 'top carnivores', and hence whether or not it would be possible to study the grazer–predator 1–predator 2 form of triple interaction.

There are a few studies of the higher plant–mycorrhiza–Collembola interaction (Warnock, Fitter & Usher 1982; Finlay 1983; Finlay, p. 319; Shaw, p. 333). Warnock *et al.* (1982) used a simple laboratory system of leeks, *Allium porrum*, infected with *Glomus fasciculatum*, and grazed by *Folsomia candida*. They found that infected plants in the presence of Collembola grew little better than uninfected plants, whilst infection in the absence of Collembola led to a significant increase in leek production and phosphate uptake. They argued that the Collembola grazed the external hyphae of this vesicular-arbuscular mycorrhiza. The presence of mycorrhizal infection also affected the populations of *F. candida*, which increased from an initial 20 per pot to an average of 53 in non-mycorrhizal treatments and to 140 in mycorrhizal treatments during a 12-week period. Analysis of gut contents indicated that the majority of Collembola with contents contained fungal material, although about 45% of the population had no visible gut contents. This figure is similar to that recorded by Anderson & Healey (1972), McMillan (1975) and Muraleedharan & Prabhoo (1978), the latter quoting ranges of 33–55%, 33–77% and 50–89% for three species of Collembola collected from the field in India. The former authors suggested that Collembola are only active for 50–60% of their adult life, whilst the latter authors attributed the lack of gut contents to a 'non-feeding phase of the inter-moult period'. What implications such a 'resting phase' may have for the population dynamics of Collembola remains unknown.

There are too few data to draw any general conclusions on such complex interactions. Some studies have shown that mycorrhizal infection can be rendered virtually inoperative in the presence of grazers, whilst Ulber (1983) shows that lethal infections of sugar-beet with *Pythium* can be reduced from 100% to 22–75% in the presence of *Onychiurus* and to 9% in the presence of *Folsomia*. The introduction of a predatory mite into such systems is an experiment that has not yet been undertaken.

DYNAMICS OF NATURAL COMMUNITIES

Number of species and diversity

There has been some discussion as to whether or not the community of soil animals is unusually diverse: Anderson (1975) and Ghilarov (1977) have argued that it is, whilst Usher *et al.* (1979) have argued that it is possibly less diverse than communities above ground. The purpose of this review is to concentrate on some aspects of diversity that will be used subsequently.

Stanton (1979) looked at the communities in Wyoming and Costa Rica, sampling pine forests, broad-leaved forests and field communities in each country. The diversity of all four forest sites was similar (12–14 species of litter-inhabiting mites per 100 g litter) and greater than the field sites (5 species per 100 g of litter). Stanton plotted species–area curves for all six sites, indicating that the usual relationship between the number of species (S) and area (A),

$$S = CA^z$$

where C and z are constants, holds for the soil fauna as well as for many other organisms (see, for example, the review by Connor & McCoy 1979). The implications of this relationship seem not to have been realized by soil zoologists when writing about species richness, or species diversity, though the generality of the relationships for the soil fauna remains unknown.

Nine dominance–diversity curves for Collembola were shown by Usher *et al.* (1979): their sites were all of a climax or near climax nature, and the curves corresponded reasonably well with those produced by a log normal distribution (see Fig. 5a for the shapes of such curves under various assumptions about the distribution of individuals amongst species). Data for Collembola from a variety of ecosystems are shown in Fig. 5b. The two steepest curves relate to a moss-turf habitat on Signy Island (Usher & Booth 1984) and a *Deschampsia antarctica* sward on Lynch Island (Usher & Edwards 1984), both Antartica, and they correspond to a steep, geometric series distribution. The shallower line relates to Sparkes' (1982) samples from a successional ecosystem in the Yorkshire Wolds: it also approximates to a geometric series distribution. Finally, two coniferous woodland curves, one from North Wales (Poole 1961) and the other from Caledonian pine forest in Scotland (Usher 1970), are shown. Such curves have been used in the interpretation of plant communities (Greig-Smith 1983): it would seem that similar use can be made for curves plotted using soil arthropod data.

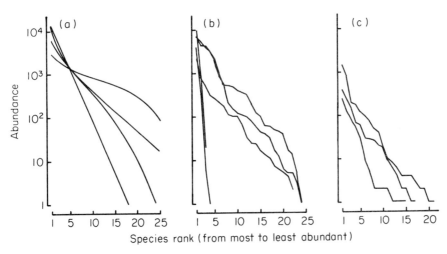

Abundance

Species rank (from most to least abundant)

FIG. 5. Dominance–diversity curves for communities of Collembola. (a) Four theoretical distributions representing, from bottom to top at species rank 15, a geometric series, a log normal distribution, a logarithmic series, and MacArthur's distribution respectively. Drawn from a diagram in Whittaker (1972). (b) Data for five Collembolan communities, representing, from left to right at the 10^2 level, a Signy Island moss-turf, the Lynch Island grass sward, Wharram Quarry Nature Reserve, a Scots pine forest, and a mixed coniferous forest respectively. Sources of data are listed in the text. (c) Data for three successional communities of Moor House National Nature Reserve, representing, from left to right at the 10^2 level, the communities in *Eriophorum angustifolium*, *Eriophorum vaginatum* and *Calluna vulgaris*. Data from Hale (1963).

Relationship of species complement to site type

Gisin's (1943) pioneering work in Switzerland, relating faunal lists of Collembola to site types, led to numerous similar faunal studies, including other groups of the soil fauna, e.g. Rajski (1961) for Cryptostigmata in Poland and Huţu (1982) for Mesostigmata (Uropodidae) in north-eastern Romania. Other studies have attempted to relate arthropod communites directly to various environmental factors. Buxton (1981) related both the structure of a termite community, and its activity, to local variations in rainfall. Cotton & Curry (1982) looked at the distribution of earthworms along a gradient from mineral soil to peat, the peat having 75% of the species (compared with the mineral soil), 15% of the numbers, but only 9% of the biomass. Acidic pH and lack of suitable food apparently restricted the earthworm community at the peat end of the gradient. A principle components analysis to investigate the relationships between the environment and the mite and Collembola communities showed that the mites were related to soil organic matter content, moisture content, base status, but not to the site vegetation (Curry 1978). The Collembola community was related to neither the soil nor the vegetation. In another study (Curry & Ganley 1977) no relationship was shown between the presence of either mite or Collembola species and the roots of higher plants in an Irish pasture. In contrast Hågvar (1982) demonstrated a relationship between the Collembola communities and vegetation in a Norwegian coniferous forest. The relationship became stronger as environmental conditions became more extreme, and this may explain why Curry's studies failed to demonstrate relationships in Irish lowland pastures.

There are three interacting factors of a site: the soil environment, the vegetation, and the community of soil animals. It is clear that little is yet understood about how strong the relationships are, or under what circumstances a relationship might or might not be expected. A study of ecosystems which are actively changing, either naturally or experimentally, may aid the unravelling of these relationships.

Changes in communities

Three categories of changes can occur in communities. First, there is the natural process of succession which, as primary or secondary succession, affects the whole community in which the soil animals live. This form of succession should be distinguished from the sequence of organisms associated with the decomposition of a dead organic product (Usher & Parr 1977) such as arthropods in an acorn (Winston 1956) or nematodes in cow dung (Sudhaus 1981). Second, changes follow an alteration in the management of the site (e.g. sheep grazing or forestry). Third, it is possible to alter communities deliberately as part of an experimental programme to investigate community structure. Such perturbation experiments are, unfortunately, rare.

Succession

There are relatively few descriptions of successon in animal communities. Huhta *et al.* (1979), who documented the fauna of some artificial soils made of crushed pine or spruce bark and sewage sludge, noted that the first colonists were flying insects, phoretic mites and nematodes, arriving within a few days. Collembola arrived after a few weeks, and the Cryptostigmata arrived later still. Succession of Collembola in agricultural soil (Hermosilla 1982) was accompanied by an increase in diversity over a 494-day period, though an increase is virtually inevitable when one starts from nothing or nearly nothing. Two further examples of succession can be investigated in detail. One relates to the cyclical process of peat build-up in heather moorlands and its subsequent erosion, and the other relates to a primary succession on the floor of an abandoned chalk quarry.

Hale (1963, 1966) and Block (1965, 1966) described the Collembola and mites respectively of moorland in northern England. The position of *Eriophorum vaginatum* in the successional sequence (Table 3) is that adopted by Hale (1963). In general, the number of species and the number of individuals of Cryptostigmata, Prostigmata and Collembola increased during the successional sequence, but the density of Mesostigmata reached an upper limit more speedily. Whelan (1978) similarly showed a rapid increase in the number of Mesostigmata colonizing peat, but this is in contrast to the colonization of colliery spoil (Hutson & Luff 1978)

TABLE 3. Arthropods associated with the succession on eroded peat at Moor
Reserve. The successional sequence of the vegetation is shown, together with tl
and density in thousands per m² (N). The data are taken from Hale (196:

Vegetation type	Collembola		Mesostigmata			
	S	N	S	N		
Bare peat*	2	0	0	0		
Eriophorum angustifolium	17	4·7	—	0·6		
Eriophorum vaginatum	14	24·4	—	4·1		
Mixed moor	26	31·0	20	3·2	45	45·5
Hummock top	25	38·8	—	2·5	—	95·0

* No permanent populations of mites or Collembola on bare peat.

where the predatory Mesostigmata were slow to colonize the new habitats.
Dominance-diversity curves for the stages of the moorland succession (Fig. 5c),
however, show little change in shape.

In the chalk quarry (Parr 1980), no distinct stages in the ecological succession
had been determined quantitatively *a priori*. Parr ordinated the arthropod data for
the thirty areas that he sampled: the horizontal axes in Figs 3 and 6 are the first
ordination axis which was identified as representing the successional sequence.
Figure 3 demonstrates that the species complement changes with succession:
density also increases with successional age (Fig. 6). The number of species of mites
increased during the succession, but that of Collembola remained virtually constant. A measure of diversity, the Brillouin Index H, also increased during succession, although in the Collembola there was an indication that H decreased in the
centre of the range due to the dominance of a few species.

Parr (1980) had only a small segment of a full successional sequence, but, like
Hale (1963), he demonstrated that the total community tended to increase in
abundance, species richness and diversity, although there may be periods of relative
constancy in one or more of these measures of the community structure.

Management

The effects of silvicultural practices in Finland on the soil fauna (e.g. Huhta et al.
1967; Huhta 1976), of manuring increasing populations but reducing diversity (Weil
& Kroontje 1979), of ploughing having only short-term effects (Loring, Snider &
Robertson 1981), and of organic farming increasing both species richness and
abundance of carabid beetles (Dritschilo & Erwin 1982), have all been described.
The effects of management can be demonstrated by the Collembola in six pastures
in New South Wales, Australia: pairs of pastures had sheep stocking intensities of
10, 20 and 30 sheep ha^{-1} (King, Hutchinson & Greenslade 1976). The most
noticeable effect of the increased stocking was on the overall abundance of Collembola, which decreased from 117 100 m^{-2} to 63 200 m^{-2} to 5100 m^{-2} respectively. The number of species was also reduced, but by a smaller factor, decreasing
from 21 to 17 to 15 respectively. The changes in diversity were not so clear cut,
although there was some evidence that it increased slightly with increased stocking.

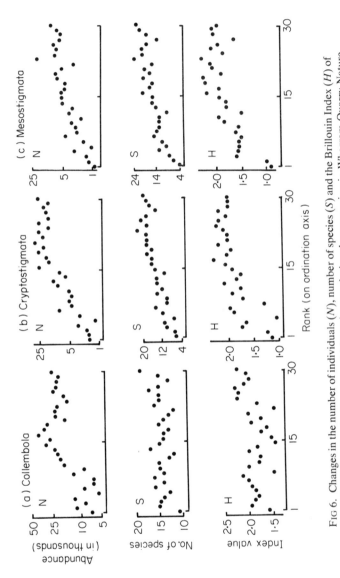

FIG 6. Changes in the number of individuals (*N*), number of species (*S*) and the Brillouin Index (*H*) of the Collembola, Cryptostigmata and Mesostigmata during the succession in Wharram Quarry Nature Reserve. The successional rank (1 = youngest, 30 = oldest) is used in Fig. 3 and described in the text. Data from Parr (1980).

Of the 28 species recorded on the pastures, 15 species were found in pastures of all stocking intensities, and a further 9 were in the lightly and intermediately stocked pastures. Two rare species were not found in the least stocked pastures, and two other rare species were found on the most and least intensively stocked pastures. The evidence suggested that species were lost from the pastures as the stocking intensity increased.

To understand the results of management one should consider both changes in quantities of food for the soil fauna, and structural changes in the soil habitat. Manuring and organic farming resulted in increased arthropod populations, and clear felling of forests (Huhta 1976) led to a temporary (about one decade) increase in Lumbricid and Enchytraeid worm populations: all these increases could be attributed to extra organic matter. On the other hand trampling (Garay & Nataf 1982), compaction or drastic disturbance destroyed the soil habitat, decreasing populations of soil animals.

Perturbation

The dividing line between management and perturbation experiments is small: perturbation experiments are, however, designed specifically to investigate inter-actions. The design of a series of replicated experiments was discussed by Usher, Booth & Sparkes (1982), but results of these experiments are not yet available. An experiment in a Canadian spruce forest with nitrogen (as urea) resulted in very minor changes in the species composition, density, and vertical distribution of soil arthropods (Behan, Hill & Kevan 1978). The comparisons were between untreated and treated (nitrogen at 220 kg ha^{-1}) plots, but whether the small changes are general or due to resilience of the old spruce forest remains unknown. Insecticides have frequently been used; for example, application of DDT to cultivated forest soil in Nigeria resulted in a decrease of Mesostigmata (10 260 to 590 m^{-2}) and an increase in Collembola (9060 to 35 310 m^{-2}; Perfect *et al.* 1981), but whether the latter is coincidental or the result of release from predation is unknown. A study in a Swedish pine forest by Bååth *et al.* (1980) showed that acidification of a soil resulted in a decrease in Mesostigmata (6180 to 2900 m^{-2}) and an increase in Collembola (33 270 to 48 270 m^{-2}). The study investigated many aspects of soil biology, and showed that acidification also decreased populations of Enchytraeids and of some species of Cryptostigmata, and radically changed the community of bacteria.

Such perturbation experiments, designed to investigate the concurrent changes in the populations of many species in the soil, and suitably replicated, provide an experimental tool that soil biologists have hardly started to use. This kind of experimentation is likely to be a profitable field for future research.

DISCUSSION

To conclude this review there is one aspect of the population dynamics of soil organisms that should be discussed. This is concerned with the links between the decomposer food web and the more general theories of population dynamics in ecology. To what extent are the general theories applicable to the decomposer ecosystem, and what might a study of decomposers contribute more widely in ecology?

It is true that the majority of aspects of population dynamics—fecundity, mortality, movement into or out of populations—and of interactions between populations—competition, predation, herbivory—are essentially the same in any food web, be it above- or below-ground, and hence general theoretical developments can be applied to any of the systems. There are, however, three possible differences when the soil arthropod communities of temperate latitudes are considered. One of these differences concerns a diapause, and whether hatching is synchronous or staggered over a long period of time leading to pulses of juveniles entering a population. A second difference concerns the ability of the soil fauna to function throughout the year: the soil environment is buffered from climatic extremes, the temperature at a depth of 25 cm being recorded as not dropping below 2°C during the British winter (Frazer 1983). Third, there has been some discussion as to whether the soil fauna show features of r- or K-selection (Gerson & Chet 1981), but a more interesting concept is that of A-selection (Greenslade & Greenslade 1983). *Adversity-selection* would be shown in species in communities of continuously harsh, and hence predictable, environments, and it is postulated that diversity is low and there are few interactions between species in such communities. The concept of A-selection is thus very similar to that of stress-tolerance in plant ecology (see Grime 1979). Greenslade & Greenslade (1983) suggest that one should look for A-strategists in Australia on mountain tops, in cold wet forests, in the arid zone, and deep in the soil. The first three habitats are places where the majority of the local fauna actually live in the soil, and it is perhaps the soil fauna that have adopted a form of A-strategy more than any other fauna. Certainly the data for the two Antarctic sites (Fig. 5b) represent communities from extremely harsh environments, and would probably represent communities of A-strategists: can one postulate that the individuals in such communities would have log series distributions amongst the species?

The soil ecosystem tends to be one that has been forgotten by ecologists, perhaps because soil unlike water or air is opaque and perhaps because there are few attractive species (colourful, cuddly, or cute) that live within it. However, there are a number of interesting aspects of the population and community dynamics that have been discovered, and probably many more to be discovered, especially where interactions between more than two trophic levels are considered. It has the one great advantage for researchers that it exists, and is experimentable with, for 12 months of each year in all but the most extreme climates.

ACKNOWLEDGMENTS

I should like to thank Drs R. G. Booth, T. W. Parr and K. E. Sparkes for allowing me to use data in their doctoral theses, and Professor W. G. Hale for providing his data from Moor House National Nature Reserve. Helpful suggestions on the manuscript have been made by Professor A. Macfadyen and Drs R. G. Booth and A. H. Fitter, all of whom I should like to thank.

REFERENCES

Allmen, H. von & Zettel, J. (1982). Populationsbiologische Untersuchungen zur Art *Entomobrya nivalis* (Collembola). *Revue Suisse de Zoologie*, **89**, 919–926.

Allmen, H. von & Zettel, J. (1983). Embryonic development and postembryonic growth in a population of *Entomobrya nivalis* (Collembola). *Revue d'Écologie et de Biologie du Sol*, **20**, 77–86.

Anderson, J.M. (1975). The enigma of soil animal species diversity. *Progress in Soil Zoology*, pp. 51–58. Proceedings of the 5th International Colloquium on Soil Zoology, 1973. Czechoslovak Academy of Sciences, Prague.

Anderson, J.M. & Healey, I.N. (1972). Seasonal and inter-specific variation in major components of the gut contents of some woodland Collembola. *Journal of Animal Ecology*, **41**, 359–368.

Athias-Binche, F. (1978/79). Étude quantitative des Uropodides (Acariens: Anactinotriches) d'un arbre mort de la hêtraie de la Massane. 2. Eléments démographiques d'une population d'*Allodinychus flagelliger* (Berlèse, 1910). *Vie Milieu*, 38/9, 35–60.

Bååth, E., Berg, B., Lohm, U., Lundgren, B., Lundkvist, H., Rosswall, T., Söderström, B. & Wiren, A. (1980). Effects of experimental acidification and liming on soil organisms and decomposition in a Scots pine forest. *Pedobiologia*, **20**, 85–100.

Baker, G.H. (1978). The population dynamics of the millipede, *Ommatoiulus moreletii* (Diplopoda: Iulidae). *Journal of Zoology*, **186**, 229–242.

Bauer, T. (1982). Predation by a carabid beetle specialized for catching Collembola. *Pedobiologia*, **24**, 169–179.

Behan, V.M. & Hill, S.B. (1978). Feeding habits and spore dispersal of Oribatid mites in the North American Arctic. *Revue d'Écologie et de Biologie du Sol*, **15**, 497–516.

Behan, V.M., Hill, S.B. & Kevan, D.K. McE. (1978). Effects of nitrogen fertilizer, as urea, on Acarina and other arthropods in Quebec black spruce humans. *Pedobiologia*, **18**, 249–263.

Behan-Pelletier, V.M. & Hill, S.B. (1983). Feeding habits of sixteen species of Oribatei (Acari) from an acid peat bog, Glenamoy, Ireland. *Revue d'Écologie et de Biologie du Sol*, **20**, 221–267.

Bengtsson, G. & Rundgren, S. (1983). Respiration and growth of a fungus, *Mortierella isabellina*, in response to grazing by *Onychiurus armatus* (Collembola). *Soil Biology and Biochemistry*, **15**, 469–473.

Berthet, P.L. (1964). Field study of the mobility of Oribatei (Acari), using radioactive tagging. *Journal of Animal Ecology*, **33**, 443–449.

Blancquaert, J.P., Mertens, J. & Coessens, R. (1982). Annual cycle of populations of *Sphaeridia pumilis* (Collembola). *Revue d'Écologie et de Biologie du Sol*, **19**, 605–611.

Block, W.C. (1965). Distribution of soil mites (Acarina) on the Moor House National Nature Reserve, Westmorland, with notes on their numerical abundance. *Pedobiologia*, **5**, 244–251.

Block, W.C. (1966). The distribution of soil Acarina on eroding blanket bog. *Pedobiologia*, **6**, 27–34.

Booth, R.G. (1979). *The nutritional importance of microorganisms in the diet of soil microarthropods.* Ph.D. thesis, University of Exeter.

Booth, R.G. (1983). Effects of plaster-charcoal substrate variation on the growth and fecundity of *Folsomia candida* (Collembola, Isotomidae). *Pedobiologia*, **25**, 187–195.

Burn, A.J. (1982). Effects of temperature on the feeding activity of *Cryptopygus antarcticus*. *Colloque sur les Ecosystèmes Subantarctiques* (Ed. by P. Jouventin, L. Massé & P. Trehen), pp. 209–217. Comité National Français des Récherches Antarctiques, Paimpont.

Buxton, R.D. (1981). Changes in the composition and activities of termite communities in relation to changing rainfall. *Oecologia*, **51**, 371–378.

Cancela da Fonseca, J.P., Kiffer, E. & Poinsot-Balaguer, N. (1979). Sur les rapports entre microarthropodes et micromycètes d'un sol forestier, II. Quelques observations sur les acariens

Gamasides (non Uropodes) et les Collemboles. *Revue d'Écologie et de Biologie du Sol*, **16**, 181–194.

Connor, E.F. & McCoy, E.D. (1979). The statistics and biology of the species-area relationship. *American Naturalist*, **113**, 791–833.

Cotton, D.C.F. & Curry, J.P. (1982). Earthworm distribution and abundance along a mineral-peat soil transect. *Soil Biology and Biochemistry*, **14**, 211–214.

Curry, J.P. (1978). Relationships between microarthropod communities and soil and vegetation types. *Scientific Proceedings of the Royal Dublin Society, Series A*, **6**, 131–141.

Curry, J.P. & Ganley, J. (1977). The arthropods associated with the roots of some common grass and weed species of pasture. *Ecological Bulletins (Stockholm)*, **25**, 330–339.

Davis, P.R. (1978). *Approaches towards modelling competition and predation in a simple community of soil arthropods*. D.Phil. thesis, University of York.

Dennison, D.F. & Hodkinson, I.D. (1983). Structure of the predatory beetle community in a woodland soil ecosystem, I. Prey selection. *Pedobiologia*, **25**, 109–115.

Dritschilo, W. & Erwin, T.L. (1982). Responses in abundance and diversity of cornfield carabid communities to differences in farm practices. *Ecology*, **63**, 900–904.

Finlay, R.D. (1983). *Interactions between the endomycorrhizal associations of higher plants and soil-dwelling Collembola*. D.Phil. thesis, University of York.

Frazer, D. (1983). *Reptiles and Amphibians in Britain*. Collins, London.

Garay, I. & Nataf, L. (1982). Microarthropods as indicators of human trampling in suburban forests. *Urban Ecology* (Ed. by R. Bornkamm, J.A. Lee & M.R.D. Seaward), pp. 201–207. Blackwell Scientific Publications, Oxford.

Gerson, U. & Chet, I. (1981). Are allochthonous and autochthonous soil microorganisms r- and K-selected? *Revue d'Écologie et de Biologie du Sol*, **18**, 285–289.

Ghilarov, M.S. (1977). Why so many species and so many individuals can coexist in the soil. *Ecological Bulletins (Stockholm)*, **25**, 593–597.

Gisin, H. (1943). Ökologie und Lebensgemeinschaften der Collembolen im Schweizerischen Exkursionsgebiet Basels. *Revue Suisse de Zoologie*, **50**, 131–224.

Greenslade, P. & Greenslade, P.J.M. (1980). Relationships of some Isotomidae (Collembola) with habitat and other soil fauna. *Proceedings of the 7th International Soil Zoology Colloquium*, 491–506.

Greenslade, P.J.M. & Greenslade, P. (1983). Ecology of soil invertebrates. *Soils: an Australian Viewpoint* (Ed. by Division of Soils, CSIRO), pp. 645–669. CSIRO, Melbourne and Academic Press, London.

Grégoire-Wibo, C. (1979). Cycle phénologique de ≪*Folsomia quadrioculata*≫ en forêt (Insecte: Collembole). *Annales de la Société Royale de Zoologie Belgique*, **109**, 43–65.

Greig-Smith, P. (1983). *Quantitative Plant Ecology* (3rd edn). Blackwell Scientific Publications, Oxford.

Grime, J.P. (1979). *Plant Strategies and Vegetation Processes*. Wiley, Chichester.

Grimnes, K.A. & Snider, R.M. (1981). An analysis of egg production in four strains of *Folsomia candida* (Collembola). *Pedobiologia*, **22**, 224–231.

Haeck, J., Hengeveld, R. & Turin, H. (1980). Colonization of road verges in three Dutch polders by plants and ground beetles (Coleoptera: Carabidae). *Entomologia Generalis*, **6**, 201–215.

Hågvar, S. (1982). Collembola in Norwegian coniferous forest soils, I. Relations to plant communities and soil fertility. *Pedobiologia*, **24**, 255–296.

Hale, W.G. (1963). The Collembola of eroding blanket bog. *Soil Organisms* (Ed. by J. Doekson & J. van der Drift), pp. 406–413. North-Holland Publishing Co., Amsterdam.

Hale, W.G. (1966). The Collembola of the Moor House National Nature Reserve, Westmorland: a moorland habitat. *Revue d'Écologie et de Biologie du Sol*, **3**, 97–122.

Harris, J.R.W. (1974). *Aspects of the kinetics of predation with reference to a soil mite*, Pergamasus longicornis Berlese. D.Phil. thesis, University of York.

Harris, J.R.W. & Usher, M.B. (1978). Laboratory studies of predation by the grassland mite *Pergamasus longicornis* Berlese and their possible implications for the dynamics of populations of Collembola. *Scientific Proceedings of the Royal Dublin Society, Series A*, **6**, 143–153.

Healey, I.N. (1965). *Studies on the production biology of soil Collembola with special reference to Onychiurus*. Ph.D. Thesis, University College of Swansea.

Hermosilla, W. (1982). Sukzession und Diversität der Collembolenfauna eines rekultivierten Ackers. I. *Revue d'Écologie et de Biologie du Sol*, **19**, 225–236.

Higgins, R.C. (1982). Predation by *Notiophilus* (Coleoptera: Carabidae) on Collembola as a predator-prey teaching model. *Journal of Biological Education*, **16**, 128–130.

Hislop, R.G. & Prokopy, R.J. (1981). Mite predator responses to prey and predator-emitted stimuli. *Journal of Chemical Ecology*, **7**, 895–904.

Holling, C.S. (1966). The functional response of invertebrate predators to prey density. *Memoirs of the Entomological Society of Canada*, **48**, 1–86.

Huhta, V. (1976). Effects of clear-cutting on numbers, biomass and community respiration of soil invertebrates. *Annales Zoologici Fennici*, **13**, 63–80.

Huhta, V., Karppinen, E., Nurminen, M. & Valpas, A. (1967). Effect of silvicultural practices upon arthropod, annelid and nematode populations in coniferous forest soil. *Annales Zoologici Fennici*, **4**, 87–143.

Huhta, V., Ikonen, E. & Vilkamaa, P. (1979). Succession of invertebrate populations in artificial soil made of sewage sludge and crushed bark. *Annales Zoologici Fennici*, **16**, 223–270.

Huhta, V. & Mikkonen, M. (1982). Population structure of Entomobryidae (Collembola) in a mature spruce stand and in clear-cut reforested areas in Finland. *Pedobiologia*, **24**, 231–240.

Hutson, B.R. (1978). Effects of variations of the plaster-charcoal culture method on a collembolan, *Folsomia candida*. *Pedobiologia*, **18**, 138–144.

Hutson, B.R. (1981). Age distribution and the annual reproductive cycle of some Collembola colonizing reclaimed land in Northumberland, England. *Pedobiologia*, **21**, 410–416.

Hutson, B.R. & Luff, M.L. (1978). Invertebrate colonization and succession on industrial reclamation sites. *Scientific Proceedings of the Royal Dublin Society, Series A*, **6**, 165–174.

Huțu, M. (1982). Strukturelle Eigenschaften von Uropodiden-Zönosen in der Streuschicht verschiedener Waldtypen längs eines Höhengradienten. *Pedobiologia*, **23**, 68–89.

Joosse, E.N.G. (1981). Ecological strategies and population regulation of Collembola in heterogeneous environments. *Pedobiologia*, **21**, 346–356.

Joosse, E.N.G. (1981). New developments in the ecology of Apterygota. *Pedobiologia*, **25**, 217–234.

Kaczmarek, M. (1975). An analysis of Collembola communities in different pine forest environments. *Ekologia Polska*, **23**, 265–293.

Karg, W. (1983). Verbreitung und Bedeutung von Kaubmilben der Cohors Gamasina als Antagonisten von Nematoden. *Pedobiologia*, **25**, 419–432.

King, K.L., Hutchinson, K.J. & Greenslade, P. (1976). The effects of sheep numbers on associations of Collembola in sown pastures. *Journal of Applied Ecology*, **13**, 731–739.

Lacy, R.C. (1984). Predictability, toxicity, and trophic niche breadth in fungus-feeding Drosophilidae (Diptera). *Ecological Entomology*, **9**, 43–54.

Lawton, J.H. & Hassell, M.P. (1981). Asymmetrical competition in insects. *Nature*, **289**, 793–795.

Leinaas, H.P. & Bleken, E. (1983). Egg diapause and demographic strategy in *Lepidocyrtus lignorum* Fabricius (Collembola; Entomobryidae). *Oecologia*, **58**, 194–199.

Longstaff, B.C. (1976). The dynamics of collembolan populations: competitive relationships in an experimental system. *Canadian Journal of Zoology*, **54**, 948–962.

Longstaff, B.C. (1980). The functional response of a predatory mite and the nature of the attack rate. *Australian Journal of Ecology*, **5**, 151–158.

Loring, S.J., Snider, R.J. & Robertson, L.S. (1981). The effect of three tillage practices on Collembola and Acarina populations. *Pedobiologia*, **22**, 172–184.

McMillan, J.H. (1975). Interspecific and seasonal analyses of the gut contents of three Collembola (family Onychiuridae). *Revue d'Écologie et de Biologie du Sol*, **12**, 449–457.

Mitchell, M.J. (1977). Population dynamics of oribatid mites (Acari, Cryptostigmata) in an aspen woodland soil. *Pedobiologia*, **17**, 305–319.

Mitchell, M.J. & Parkinson, D. (1976). Fungal feeding of Oribatid mites (Acari: Cryptostigmata) in an aspen woodland soil. *Ecology*, **57**, 302–312.

Muraleedharan, V. & Prabhoo, N.R. (1978). Observations on the feeding habits of some soil Collembola from an abandoned field in Kerala. *Entomon*, **3**, 207–213.

Parkinson, D. (1983). Functional relationships between soil organisms. *New Trends in Soil Biology* (Ed.

by P. Lebrun, H.M. André, A. de Medts, C. Grégoire-Wibo & G. Wauthy), pp. 153–165. Dieu-Brichart, Ottignies-Louvain-la-Neuve.

Parkinson, D., Visser, S. & Whittaker, J.B. (1979). Effects of collembolan grazing on fungal colonization of leaf litter. *Soil Biology and Biochemistry*, **11**, 529–535.

Parr, T.W. (1980). *The structure of soil microarthropod communities with particular reference to ecological succession.* D.Phil. thesis, University of York.

Perfect, T.J., Cook, A.G., Critchley, B.R. & Russel-Smith, A. (1981). The effect of crop protection with DDT on the microarthropod population of a cultivated forest soil in the sub-humid tropics. *Pedobiologia*, **21**, 7–18.

Petersen, H. (1971). Collembolernes ernaeringsbiologi og dennes økologiske betydning. *Entomologiske Meddelelser*, **39**, 97–118.

Petersen, H. (1978). Sex-ratios and the extent of parthenogenetic reproduction in some Collembolan populations. *First International Seminary on Apterygota* (Ed. by R. Dallai), pp. 19–35. Accademia delle Scienze di Siena detta de' Fisiocritici, Siena.

Ponge, J.-F. & Charpentie, M.-J. (1981). Étude des relations microflore–macrofaune: expériences sur *Pseudosinella alba* (Packard), Collembole mycophage. *Revue d'Écologie et de Biologie du Sol*, **18**, 291–303.

Poole, T.B. (1961). An ecological study of the Collembola in a coniferous forest soil. *Pedobiologia*, **1**, 113–137.

Rajski, A. (1961). Studium ekologiczno-faunistyczne nad mechowcami (Acari, Oribatei), w kilku zespołach roślinnych, I. Ekologia. *Poznańskie Towarzystwo Przyjaciół Nauk, Prace Komisji Biologicznéj*, **25**, 123–283.

Rusek, J. & Weyda, F. (1981). Morphology, ultrastructure and function of pseudocelli in *Onychiurus armatus* (Collembola: Onychiuridae). *Revue d'Écologie et de Biologie du Sol*, **18**, 127–133.

Sager, G. & Stelter, H. (1981). Zur mathematischen Formulierung der Relation zwischen Anfangs- und Endverseuchung mit *Globodera rostochiensis* (Nematoda) beim Anbau von Wirtspflanzen. *Pedobiolgia*, **22**, 304–311.

Snider, R.J. & Butcher, J.W. (1973). Response of *Onychiurus justi* (Denis) (Collembola: Onychiuridae) to constant temperatures and variable relative humidity. *Proceedings of the First Soil Microcommunities Conference* (Ed. by D.L. Dindal), pp. 176–184. US Atomic Energy Commission.

Sparkes, K.E. (1982). *Studies on ecological succession with particular reference to soil microarthropods.* D.Phil. thesis, University of York.

Stamou, G.P., Kattoulas, M., Cancela de Fonseca, J.P. & Margaris, N.S. (1981). Observations on the biology and ecology of *Achipteria holomonensis* (Acarina, Oribatida). *Pedobiologia*, 53–58.

Stanton, N.L. (1979). Patterns of species diversity in temperate and tropical litter mites. *Ecology*, **60**, 295–304.

Stefaniak, O. & Seniczak, S. (1981). The effect of fungal diet on the development of *Oppia nitens* (Acari, Oribatei) and on the microflora of its alimentary tract. *Pedobiologia*, **21**, 202–210.

Straalen, N.M. van (1983a). Demographic analysis of soil arthropod populations. A comparison of methods. *Pedobiologia*, **25**, 19–26.

Straalen, N. van (1983b). *Vergelijkende demografie van springstaarten.* Doctoral Thesis, Free University of Amsterdam.

Sudhaus, W. (1981). Über die Sukzession von Nematoden in Kuhfladen. *Pedobiologia*, **21**, 271–297.

Takeda, H. (1983). A long term study of life cycles and population dynamics of *Tullbergia yosii* and *Onychiurus decemsetosus* (Collembola) in a pine forest soil. *Pedobiologia*, **25**, 175–185.

Takeda, H. & Ichimura, T. (1983). Feeding attributes of four species of Collembola in a pine forest soil. *Pedobiologia*, **25**, 373–381.

Tosi, L. (1977). Alimenti animali nelle diete di alcune specie di collemboli: ricerche preliminari sul cannibalismo in *Sinella coeca* (Schott). *Ateneo Parmense, Acta Naturalia*, **13**, 445–455.

Ulber, B. (1983). Einfluss von *Onychiurus fimatus* Gisin (Collembola, Onychiuridae) und *Folsomia fimetaria* (L.) (Collembola, Isotomidae) auf *Pythium ultimum* Trow., einen Erreger des Wurzelbrandes der Zuckerrübe. *New Trends in Soil Biology* (Ed. by P. Lebrun, H.M. André, A. de Medts. C. Grégoire-Wibo & G. Wauthy), pp. 261–268. Dieu-Brichart, Ottignies-Louvain-la-Neuve.

Usher, M.B. (1970). Seasonal and vertical distribution of a population of soil arthropods: Collembola.

Pedobiologia, **10**, 224–236.

Usher, M.B. (1976). Aggregation responses of soil arthropods in relation to the soil environment. *The Role of Terrestrial and Aquatic Organisms in Decomposition Processes* (Ed. by J.M. Anderson & A. Macfadyen), pp. 61–94. Blackwell Scientific Publications, Oxford.

Usher, M.B. (1984). Spiders in the Falkland Islands. *Newsletter from the Falkland Islands Foundation*, **2**, 4–6.

Usher, M.B. & Balogun, R.A. (1966). A defence mechanism in *Onychiurus* (Collembola, Onychiuridae). *Entomologists' Monthly Magazine*, **102**, 237–238.

Usher, M.B. & Parr, T.W. (1977). Are there successional changes in arthropod decomposer communities? *Journal of Environmental Management*, **5**, 151–160.

Usher, M.B., Davis, P.R., Harris, J.R.W. & Longstaff, B.C. (1979). A profusion of species? Approaches towards understanding the dynamics of the populations of the microarthropods in decomposer communities. *Population Dynamics* (Ed. by R.M. Anderson, B.D. Turner & L.R. Taylor), pp. 359–384. Blackwell Scientific Publications, Oxford.

Usher, M.B., Booth, R.G. & Sparkes, K.E. (1982). A review of progress in understanding the organization of communities of soil arthropods. *Pedobiologia*, **23**, 126–144.

Usher, M.B. & Davis, P.R. (1983). The biology of *Hypoaspis aculeifer* (Canestrini) (Mesostigmata): is there is a tendency towards social behaviour? *Acarologia*, **24**, 243–250.

Usher, M.B. & Booth, R.G. (1984). Arthropod communities in a maritime Antarctic moss-turf habitat: three-dimensional distribution of mites and Collembola. *Journal of Animal Ecology*, **5**, 427–442.

Usher, M.B. & Bowring, M.F.G. (1984). Laboratory studies of predation by the Antarctic mite *Gamasellus racovitzai* (Acarina: Mesostigmata). *Oecologia*, **62**, 245–249.

Usher, M.B. & Edwards, M. (1984). The terrestrial arthropods of the grass sward of Lynch Island, a specially protected area in Antarctica. *Oecologia*, **63**, 143–144.

Valpas, A. (1969). Hot rod technique, a modification of the dry funnel technique for extracting Collembola especially from frozen soil. *Annales Zoologici Fennici*, **6**, 269–274.

Vegter, J.J. (1983). Food and habitat specialization in coexisting springtails (Collembola, Entomobryidae). *Pedobiologia*, **25**, 253–262.

Verhoef, H.A. & Selm, A.J. van (1983). Distribution and population dynamics of Collembola in relation to soil moisture. *Holarctic Ecology*, **6**, 387–394.

Waldorf, E. (1981). The utilization of pollen by a natural population of *Entomobrya socia*. *Revue d'Écologie et de Biologie du Sol*, **18**, 397–402.

Warnock, A.J., Fitter, A.H. & Usher, M.B. (1982). The influence of a springtail *Folsomia candida* (Insecta, Collembola) on the mycorrhizal association of leek *Allium porrum* and the vesicular-arbuscular mycorhizal endophyte *Glomus fasciculatum*. *New Phytologist*, **90**, 285–292.

Webb, N.R. & Elmes, G.W. (1979). Variations between populations of *Steganacarus magnus* (Acari; Cryptostigmata) in Great Britain. *Pedobiologia*, **19**, 390–401.

Weil, R.R. & Kroontje, W. (1979). Effects of manuring on the arthropod community in an arable soil. *Soil Biology and Biochemistry*, **11**, 669–679.

Whelan, J. (1980). Acarine succession in grassland on cutaway raised bog. *Scientific Proceedings of the Royal Dublin Society, Series A*, **6**, 175–183.

Whittaker, R.H. (1972). Evolution and measurement of species diversity. *Taxon*, **21**, 213–251.

Wicklow, D.T. (1979). Hair ornamentation and predator defence in *Chaetomium*. *Transactions of the British Mycological Society*, **72**, 107–110.

Wicklow, D.T. & Yocom, D.H. (1982). Effect of larval grazing by *Lycoriella mali* (Diptera: Sciaridae) on species abundance of coprophilous fungi. *Transactions of the British Mycological Society*, **78**, 29–32.

Winston, P.W. (1956). The acorn microsere, with special reference to arthropods. *Ecology*, **37**, 120–132.

trees. The resulting growth reduction due to nitrogen stress may be unimportant in stable natural forest ecosystems but in commercial plantations is cause for concern.

A number of factors contribute to the development of these organic layers, not least the poor substrate quality (Swift, Heal & Anderson 1979) of coniferous litter and the immobilization of nitrogen in the soil microbial biomass. However, if their development is indeed related to the closing of nutrient cycles then all components of the ecosystem may be expected to contribute. In 1935 Romell suggested that the high densities of mycorrhizal fine roots and associated mycelium found in surface horizons were a contributory factor. Accelerated decomposition rates of litter in the absence of living fine roots and mycorrhizas have since been demonstrated (Gadgil & Gadgil 1975; Babel 1977; Berg & Lindberg 1980) and attributed to competition for nutrients between mycorrhizal fungi and the saprophytic micro-flora or increased moisture stress in the substrate caused by transpiration.

Trenching (i.e. root exclusion) experiments can shed light on the interaction between roots and soil processes. For example, Romell (1938) attributed the increased lushness of ground vegetation within trenched plots to greater nitrogen availability resulting from reduced competition and green-manuring from dead roots. More recently, Vitousek *et al.* (1982) used trenching to examine the control of nitrogen mineralization and nitrate production in the range of North American forests. In this paper we report the effects of trenching on decomposition rates, fungal activity, and nitrogen mineralization and transformations in the organic horizons of a Sitka spruce (*Picea sitchensis* (Bong.) Carr) stand during the first 8 months following treatment.

METHODS

Study site

In November 1982 three 0·04 ha plots were established in 37-year-old Sitka spruce, yield class 16, growing on an acid brown forest soil of the Countesswells series in Durris forest (National Grid reference NJ733944). There was no understorey and brash was cleared from the plots to facilitate sampling. At six selected sites in each plot a zero-tension lysimeter constructed from $300 \times 92 \times 19$ mm PVC guttering was installed at the junction of the organic horizons and the mineral soil. The depth of the lysimeters varied between 5 and 10 cm below the soil surface. In March 1983 trenches were dug to isolate three irregular $2-3$ m^2 areas in each plot, each containing a lysimeter. Soil was excavated to the indurated layer (40–70 cm), all roots severed, and the trenches were backfilled after the inside walls had been lined with a double layer of 800 gauge polythene sheeting.

Sampling

Samples were collected on a monthly basis beginning in April, approximately 2 weeks after trenching. Lysimeter leachate was stored in 1 litre polythene bottles containing 10 mg $HgCl_2$. Throughfall was collected in similar bottles fitted with a 142 mm diameter polythene funnel plugged with glass wool. Three throughfall bottles were randomized on each plot and rerandomized each month. When rainfall was heavy, intermediate collections were made and saved for analysis with the normal monthly sample. A 1 m^2 permanent quadrat was laid out within each trenched area and three similar quadrats were established in the untrenched parts of each plot. Each month four 40 mm diameter cores were taken in each quadrat to the organic/mineral interface using a collapsible corer. One core was wrapped in plastic film food wrapping and replaced *in situ* for estimation of net nitrogen mineralization and a duplicate core was used for exchangeable mineral nitrogen determination. The six remaining cores from the three trenched and three untrenched quadrats in each plot were bulked, passed through a 5·6 mm sieve and weighed prior to determination of moisture content, fungal activity and loss on ignition.

Analyses

Throughfall and lysimeter leachate was filtered (Whatman No. 44 + 0·22 μm membrane filter) prior to analysis for NH_4^+-N and NO_3^--N. An aliquot of the filtrate was concentrated by boiling and digested with H_2SO_4/H_2O_2 (Allen *et al.* 1974). Soluble organic nitrogen was estimated as the increase in NH_4^+-N in the filtrate following digestion. Exchangeable NH_4^+-N and NO_3^--N in whole soil cores were extracted with 200 cm^3 of 1 M KCl. Extracts were filtered (Whatman No. 42) prior to analyses. Nitrogen determinations were carried out by automated flow colorimetric analysis using Technicon procedures.

Fungal activity in a 5 g subsample of the sieved bulked cores was estimated by counting fluorescein diacetate (FDA)-stained mycelium (Söderström 1977, 1979a). Fluorescing hyphae $\geqslant 1$ μm in diameter were counted. Moisture content was determined by oven drying for 24 hours at 103°C and organic matter by loss on ignition at 450°C for 3 hours.

Analysis of variance showed that exchangeable nitrogen content and rates of mineralization did not differ significantly between the three replicate plots, and data for these have been bulked for graphical representation.

RESULTS

Throughfall and soil moisture content

Between April and November 250 mm of throughfall were collected, distributed as shown in Fig. 1. May and September were the wettest months; August was unusually dry. The moisture content of the organic horizons as estimated from the single monthly sample was closely related to throughfall volume and was lowest in August and November. The effect of root exclusion, and the consequent elimination of

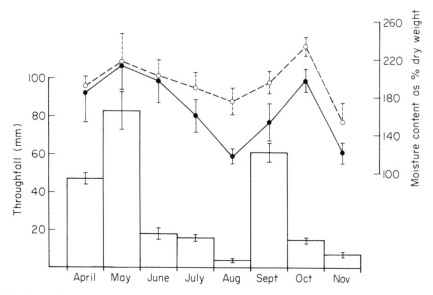

FIG. 1. Throughfall, in mm ± standard errors (open bars) and moisture content as a percentage of dry weight of forest floor (± standard errors) in trenched (○) and untrenched (●) areas between April and November.

evapotranspiration, was apparent between July and October when the moisture content of trenched areas was higher than that of rooted areas.

Nitrogen in throughfall and lysimeter samples

The soluble nitrogen input in throughfall to the forest floor over the study period amounted to 9·18 kg ha^{-1} consisting of 2·35 kg ha^{-1} soluble organic N (26%), 3·12 kg ha^{-1} NH$_4^+$-N (34%) and 3·71 kg ha^{-1} NO$_3^-$-N (40%).The pattern of input was related to throughfall volume (Fig. 2a), although there was some compensatory increase in nitrogen concentration when throughfall volume was low, most obviously during August. The relative proportions of the three forms of soluble nitrogen remained relatively constant throughout the period.

There was considerable variation in leachate output between individual lysimeters. Overall the volume collected from trenched areas (19·9 litres) was greater than that from untrenched areas (14·5 litres). Although low volume flow was associated with higher nitrogen concentrations, total output of nitrogen was directly related to the volume discharged by the lysimeters and ultimately to the input from throughfall (Fig. 2b). No lysimeter leachate was collected in August.

The output of nitrogen from lysimeters in rooted areas was 1·38 kg ha^{-1}, 85% less than the input from throughfall, and consisted of 0·60 kg ha^{-1} soluble organic N (43%) 0·47 kg ha^{-1} NH$_4^+$-N (34%) and 0·31 kg ha^{-1} NO$_3^-$-N (23%). The effect of trenching on lysimeter nitrogen output was immediate. There was a 60% increase with 1 month and over the study period 4·58 kg ha^{-1} was discharged, over three

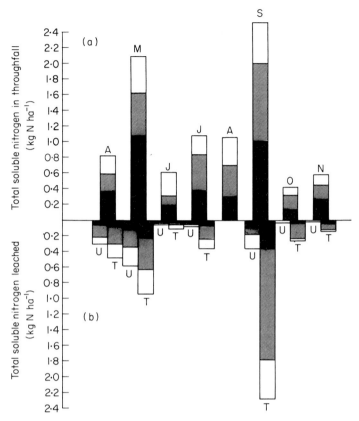

FIG. 2. (a) Total soluble nitrogen input in throughfall (kg N ha^{-1}) between April and November. (b) Total soluble nitrogen (kg N ha^{-1}) leached from untrenched (U) and trenched (T) areas between April and November. Where bars are compartmentalized components of total kg N ha^{-1} are NO$_3^-$-N (solid), NH$_4^+$-N (hatched) and soluble organic nitrogen (open).

times that from untrenched lysimeters, but still only 50% of the throughfall input. The output was made up of 1·20 kg ha^{-1} soluble organic N (26%), 2·25 kg ha^{-1} NH$_4^+$-N (56%) and 0·82 kg ha^{-1} NO$_3^-$-N (18%), representing an increase in the proportion of NH$_4^+$-N after trenching. Almost 50% of the output followed heavy rains during September.

Exchangeable nitrogen and net nitrogen mineralization

The amount of exchangeable NH$_4^+$-N in the organic horizons outside the trenches, c. 2 kg ha^{-1}, did not change significantly throughout the study period (Fig. 3). Exchangeable NO$_3^-$-N was always less than 0·1 kg ha^{-1}. Trenching had an immediate effect on exchangeable NH$_4^+$-N which rose to c. 4 kg ha^{-1} within 2 months and after June rose more rapidly to a maximum of 21·3 kg ha^{-1} by November. This upward trend was interrupted in October, the temporary drop at that time being

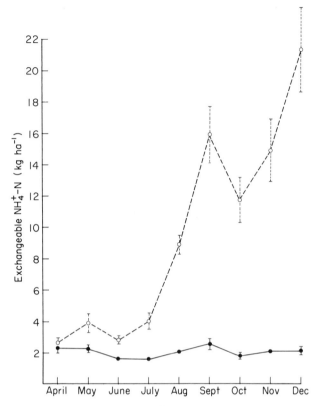

FIG. 3. Quantities of exchangeable NH_4^+-N (kg ha^{-1}) in trenched (○) and untrenched (●) areas between April and December. When not shown standard errors are contained within the data point.

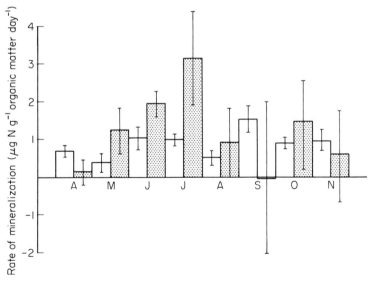

FIG. 4. Rate of mineralization (μg N g^{-1} organic matter day^{-1}, ± standard errors) in trenched (stippled) and untrenched (open) areas between April and November.

reflected in the wash-out of NH_4^+-N in the lysimeter samples during September. Despite this large accumulation of NH_4^+-N, levels of NO_3^--N remained low after trenching and had reached only 0.3 kg ha^{-1} by November.

There was no significant increase in the amount of exchangeable NO_3^--N during incubation of cores either from inside or outside the trenches. In the absence of nitrification net mineralization is therefore equivalent to the increase in the level of NH_4^+-N. Outside the trenches the mean monthly rate ranged from 0.39 to 1.53 μg N g^{-1} organic matter day^{-1}. The rate was lowest in May and August and highest in September (Fig. 4). The rate of mineralization within the trenches rose steadily to a peak of 3.16 μg N g^{-1} during July. Thereafter it fell, there was increased variation between cores, and in a significant number net immobilization of nitrogen occurred. No relationship could be found between mineralization rates and the forest floor moisture content value for the month during which incubation took place.

The total net nitrogen mineralization over the study period amounted to 11.6 kg ha^{-1} outside, and 17.5 kg ha^{-1} inside, the trenches.

FDA-stained hyphae

The mean length of FDA-stained mycelium in the samples ranged from 130 to 230 m g^{-1} organic matter (Fig. 5). Although the variation was considerable samples from outside the trenched areas showed distinct peaks of activity in May and

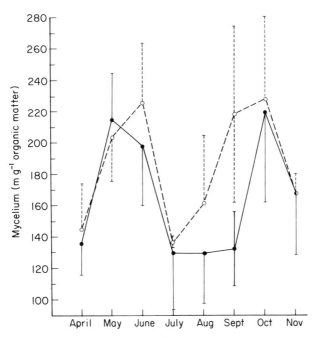

FIG. 5. Mean length of FDA-stained mycelium (m g^{-1} organic matter, $\pm$ standard errors) for trenched (○) and untrenched (●) areas between April and November.

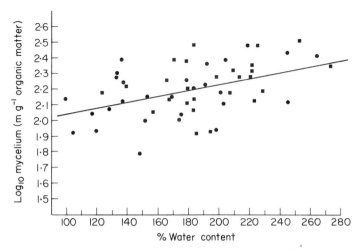

FIG. 6. Relation of total length of FDA-stained mycelium ($\log_{10}$ metres g^{-1} organic matter) to percentage water content of forest floor, (trenched areas, ■; untrenched areas, ●). $Y = 0.001x + 1.85$; $r = 0.45$; $(P = 0.001)$.

October and a trough from July to September. Samples from within the trenches followed the same pattern except during August and September when activity rose sharply. At this time the reduced moisture content outside the trenches was most marked (Fig. 1) and Fig. 6 shows that there was indeed a significant relationship between moisture content and the length of FDA-stained mycelium over the study period.

Organic matter accumulation

The mean weight of organic matter on the forest floor in April was $59\,000 \pm 8800$ kg ha^{-1}. There was no significant change over the course of the experiment or in response to trenching.

DISCUSSION

The current study has shown that the trenching has a substantial effect on both the amount of inorganic nitrogen and rate of mineralization within the forest floor. Whilst this may be due to the exclusion of functioning roots, trenching may have other confounding effects. Reduction of water loss by the elimination of transpiration elevated the humus water content of trenched plots to levels often greater than those of rooted areas (Fig. 1). Although this may not have a significant effect on suppression of litter decomposition (Gadgil & Gadgil 1975) water availability is known to influence rates of nitrogen mineralization in mineral soils (Myers, Campbell & Weier, 1982; Matson & Vitousek 1981; Miller & Johnson 1964) and humus. Laboratory incubations of Scots pine humus showed a 6–7 fold reduction in its mineralization rate when humus water content was reduced to 80% of its maximum

(Clarholm *et al.* 1981). The minimum moisture content of rooted areas occurred during August and was about 20% less than the maximum recorded in October, thus mineralization may have been affected. However, there was no correlation between moisture content and either exchangeable nitrogen or rates of mineralization; moreover effects of trenching were immediate, and occurred when trenched and rooted areas had similar water contents (Figs 1 and 4).

Although roots severed during trenching may act as a green manure and enhance the availability of nitrogen (Romell 1938) this may be of little significance in Sitka humus. The mean standing crop of fine roots and mycorrhizas less than 5 mm in diameter for 35-year old Sitka spruce is about 1000 kg ha^{-1} (Alexander & Fairley 1983). Assuming these to contain 1·5% nitrogen, trenching would provide *c.* 15 kg N ha^{-1}. These amounts are very small when compared to the quantity of *c.* 1000 kg N ha^{-1} contained in a forest floor with a dry weight of *c.* 59 000 kg ha^{-1} organic matter.

Moisture did appear to account for the slight difference in FDA-active hyphal lengths attributable to trenching. In general, lengths followed a similar seasonal pattern to that found elsewhere (Söderström 1979b) irrespective of trenching, although the values (250 m g^{-1}) are low compared to other studies. In August and September it seemed that activity in rooted areas was depressed by low soil moisture content (Figs 1 and 5). However, a number of other unknown factors also influence hyphal activity. It might be expected that trenching would reduce hyphal lengths as mycorrhizal fungi decline but the persistence of excised roots over the study period to date lessens that possibility (Ferrier & Alexander, p. 175).

The forest floor contains a large amount of organic nitrogen, decomposition of which is generally assumed to be an important source for tree growth. Nitrogen mineralization in the trenched areas between April and November was estimated to be *c.* 17·5 kg N ha^{-1} an increase of approximately 35% over rooted areas for the same period. On a yearly basis rooted areas mineralized *c.* 14·3 kg N ha^{-1}, which was very low when compared with the 67 kg N ha^{-1} reported for Sitka spruce litter and humus by Williams (1983). This large discrepancy may result from differences in experimental methodology. Whereas Williams measured accumulation by subsampling sieved litter within enclosed pots in the field, the current study utilized intact cores isolated at monthly intervals.

There was no evidence of significant NO_3^--N in soil extracts, soil incubations or lysimeter samples. Vitousek *et al.* (1982) have discussed the processes which might account for such a situation. In view of the ready availability of NH_4^+-N, particularly in trenched areas, competition for NH_4^+-N cannot be the cause. Rapid denitrification seems equally unlikely. Delayed or inhibited nitrification due to allelochemic suppression of nitrifiers, competition for some other limiting nutrient, or low initial populations of nitrifiers are more probable. In the latter circumstance an increase in NO_3^--N might be expected in trenched plots as the study proceeds.

Input of nitrogen in throughfall was relatively high when compared to other coniferous systems. When estimated on annual basis the recorded quantities of 3·71 kg NO_3^--N ha^{-1} and 3·12 kg NH_4^+-N ha^{-1} were above the median values for the ranges 0·22–20·00 kg NO_3^--N and 0·2–6·4 kg NH_4^+-N given by Parker (1983). The total soluble nitrogen input of 9·18 kg ha^{-1} was similarly high and when compared to the 11·6 kg ha^{-1} N mineralized within the forest floor, represents an important source of nitrogen. Input in throughfall was about six times greater than leaching from rooted areas suggesting utilization or retention of nitrogen within the forest floor. Exclusion of roots increased loss by leaching to 50% of input and was mainly due to the raised concentrations of NH_4^+-N in the leachate. This is frequently observed in trenched areas (Vitousek *et al.* 1982).

The sum of inorganic nitrogen input in throughfall and that estimated to be mineralized in trenched areas amounted to 24·33 kg ha^{-1}. Allowance for leaching loss of 3·37 kg ha^{-1} gives a theoretical increase of 20·96 kg ha^{-1} of exchangeable nitrogen which compares with a measured increase of 18·7 kg ha^{-1} (21·3–2·6, Fig. 3). As quantities of exchangeable inorganic nitrogen within rooted areas changed little between April and November, the figure of 17·65 kg ha^{-1} derived by similar calculation represents the amount of nitrogen available for uptake by roots. Estimated over 1 year the forest floor may provide 26·5 kg N ha^{-1}. This represents about 50% of the average annual uptake requirement of coniferous trees (Cole & Rapp 1981) or 35% of an estimated requirement of 75 kg N ha^{-1} for yield class 16 Sitka spruce (H. G. Miller, pers. comm.). This shortfall may be made good from sources in the mineral soil. Alternatively this large discrepancy between input and uptake requirement may be caused by immobilization of nitrogen by death of fine roots within the incubation (Popović 1980). However, excised fine roots may remain alive for a considerable time, and fine roots can be found several years after clear felling Scots pine (Persson 1982). Mycorrhizal fine roots of Sitka spruce may survive 8 months in litter bags and FDA-stained mycorrhizal sheath tissue was found in root excluded areas several months after trenching (Ferrier & Alexander, p. 175). There is a strong possibility that during short-term incubations, particularly when using undisturbed cores, mineralized nitrogen is absorbed by living mycorrhizal roots. The greatest rate of mineralization measured for rooted areas occurred during September (Fig. 4), when the mean number of live root tips within the incubating cores was at a minimum (R. Harmer & I. J. Alexander, unpubl.).

On our study site over an 8 month period, trenching did not appear to increase rates of organic matter decomposition as suggested by Gadgil & Gadgil (1975) or Berg & Lindberg (1980). Moreover, as readily extractable NH_4^+-N accumulated in trenched areas it is unlikely that in our system microbial growth or breakdown processes are limited by nitrogen availability and therefore not surprising that removal of root competition had no effect. This would also explain the similarity in FDA-stained hyphal lengths inside and outside trenched areas. It seems much more likely that microbial activity and organic matter decomposition are governed by the intractable nature of Sitka spruce litter as a carbon source. The low FDA-stained hyphal lengths recorded here support this conclusion.

ACKNOWLEDGMENTS

We thank NERC for financial support, the Forestry Commission for permission to work in Durris Forest, and Dorothy McKinnon for technical assistance.

REFERENCES

Alexander, I.J. & Fairley, R.I. (1983). Effects of N fertilisation on populations of fine roots and mycorrhizas in spruce humus. *Plant and Soil*, **71**, 49–53.

Allen, S.E., Grimshaw, H.M., Parkinson, J.A. & Quarmby, J. (1974). *Chemical Analysis of Ecological Material*. Blackwell Scientific Publications, Oxford.

Babel, U. (1977). Influence of high densities of fine roots of Norway spruce on processes in humus covers. *Soil Organisms as Components of Ecosystems* (Ed. by U. Lohm and T. Persson), *Ecological Bulletins (Stockholm)*, **25**, 584–586.

Berg, B. & Lindberg, T. (1980). Is litter decomposition retarded in the presence of mycorrhizal in forest soil? *Swedish Coniferous Forest Project Internal Report*, **95**, 10 pp.

Clarholm, M., Popović, B., Rosswall, T., Söderström, B., Sohlenius, B., Staff, H. & Wiren, A. (1981). Biological aspects of nitrogen mineralisation in humus from a pine forest podsol incubated under different moisture and temperature conditions. *Oikos*, **37**, 137–145.

Cole, D.W. & Rapp, M. (1981). Elemental cycling in forest ecosystems. *Dynamic Properties of Forest Ecosystems* (Ed. by D.E. Reichle), pp. 341–409. Cambridge University Press, Cambridge.

Gadgil, R.L. & Gadgil, P.D. (1975). Suppression of litter decomposition by mycorrhizal roots of *Pinus radiata*. *New Zealand Journal of Forestry Science*, **5**, 33–41.

Matson, P.A. & Vitousek, P.M. (1981). Nitrogen mineralisation and nitrification potentials following clearcutting in the Hoosier National Forest, Indiana. *Forest Science*, **27**, 781–791.

Miller, R.D. & Johnson, D.D. (1964). The effect of soil moisture tension on carbon dioxide evolution, nitrification and nitrogen mineralisation. *Soil Science Society of America Proceedings*, **28**, 644–647.

Myers, R.J.K., Campbell, C.A. & Weier, K.L. (1982). Quantitative relationship between net nitrogen mineralisation and moisture content of soils. *Canadian Journal of Soil Science*, **62**, 111–124.

Parker, G.G. (1983). Throughfall and stemflow in the forest nutrient cycle. *Advances in Ecological Research*, **13**, 1–134.

Persson, H. (1982). Changes in the tree and dwarf shrub fine-roots after clear cutting in a mature Scots pine stand. *Swedish Coniferous Forest Project Technical Report*, **31**, 19 pp.

Popović, B. (1980). Mineralisation of nitrogen in incubated soil samples from an old Scots pine forest. *Structure and Function of Northern Coniferous Forests—An Ecosystem Study* (Ed. by T. Persson). *Ecological Bulletins (Stockholm)*, **32**, 411–418.

Romell, L.G. (1935). Ecological problems of the humus layer in the forest. *Cornell University Agricultural Experimental Station Memoir* No. 170.

Romell, L.G. (1938). A trenching experiment in spruce forest and its bearing on problems of mycotrophy. *Svensk Botanisker Tidskrift*, **32**, 89–99.

Söderström, B.E. (1977). Vital staining of fungi in pure cultures and in soil with fluorescein diacetate. *Soil Biology and Biochemistry*, **9**, 59–63.

Söderström, B.E. (1979a). Some problems in assessing the fluorescein diacetate active fungal biomass in the soil. *Soil Biology and Biochemistry*, **11**, 147–148.

Söderström, B.E. (1979b). Seasonal fluctuations of active fungal biomass in horizons of a podzolised pine-forest soil in central Sweden. *Soil Biology and Biochemistry*, **11**, 149–154.

Swift, M.J., Heal, O.W. & Anderson, J.M. (1979). *Decomposition in Terrestrial Ecosystems*. Blackwell Scientific Publications, Oxford.

Vitousek, P.M., Gosz, J.R., Grier, C.C., Melillo, J.M. & Reiners, W.A. (1982). A comparative analysis of potential nitrification and nitrate mobility in forest ecosystems. *Ecological Monographs*, **52**, 155–177.

Williams, B.L. (1983). Nitrogen transformations and decomposition in litter and humus from beneath closed-canopy Sitka spruce. *Forestry*, **56**, 17–32.

Relationships between Collembola and their environment in a maritime Antarctic moss-turf habitat

R. G. BOOTH AND M. B. USHER

Department of Biology, University of York, York YO1 5DD

INTRODUCTION

Compared with temperate or boreal acidic peat habitats, the arthropod fauna of Antarctic moss-turf habitats is species poor. Geographical isolation and the lack of a continental pool of potential colonizers are at least as important as the harshness of the Antarctic environment in maintaining this low species diversity.

Low species diversity is also a feature of the plant populations. Moss-turves in the maritime Antarctic develop on gently sloping, stable sites, and are usually formed by two mosses, *Chorisodontium aciphyllum* (Hook, f. & Wils.) Broth. and *Polytrichum alpestre* Hoppe. A semi-ombrogenous peat accumulates below the green surface zone of production (Allen & Northover 1967). The relative simplicity of this naturally occurring system lends itself to observation, sampling and experimental manipulation, which are designed to investigate the inter-relationships between the soil fauna, the plant species and their environment.

SITE AND METHODS

The study site, on Signy Island, South Orkney Islands (60°43'S, 45°36'W), was described by Tilbrook (1973), and Davis (1981) summarized the biological studies. Only two species of Collembola occur regularly in this habitat: the more numerous, *Cryptopygus antarcticus* Willem, which is a widespread and abundant Antarctic species, and *Friesea woyciechowskii* Weiner, which is currently known only from the South Orkney and South Shetland Islands. *C. antarcticus* is a typical isotomid Collembola, with mandibles that possess a well developed molar plate, and it feeds on algae, fungi and other plant material (Broady 1979). *F. woyciechowskii* has the biting type of mouthparts typical of the genus, and although its diet is unknown, it is probably carnivorous. Thus, the two species are unlikely to be competing for the same food resource. Two other species of Collembola, *Parisotoma octooculata* (Willem) and *Archisotoma brucei* (Carpenter), occur rarely at the site, although both are numerous in other habitats on Signy Island.

Contiguous samples, rather than random cores, were collected to study the vertical and horizontal distributions of the Collembola and to investigate the effects of the environmental variables on these observed distributions. A single block of moss and peat, 20 cm square and at least 9 cm deep, was collected on each of four sampling occasions during the 1980–81 austral summer. Each block was cut verti-

279

cally into sixteen square cores, each 5 cm square and arranged in a contiguous pattern of four by four. Each core was sectioned horizontally into six samples, yielding ninety-six samples for each block. The surface sample comprised the green moss and upper dying leaves and was approximately 1·5 cm thick, and the five subsequent samples were each 1·5 cm thick. The moss cover of the surface layer of each core was estimated, the arthropods were extracted from the samples by high gradient heat extraction, and the dried plant material was returned to the UK for chemical analysis (analytical methods follow Allen *et al.* 1974).

RESULTS

Figure 1 shows the vertical distribution of *Cryptopygus* and *Friesea* in the moss-turf. *Cryptopygus* occurs predominantly at or near the surface, with 55% of its population in the surface layer (0–1·5 cm). Only very few individuals are likely to occur below 9 cm deep. In contrast, *Friesea* occurs commonly over a broader depth range, although it was absent from the surface layer. An unknown proportion of the population occurs below 9 cm.

Calculation of the variance to mean ratio showed that *Cryptopygus* was strongly aggregated, especially in the upper three layers, and that *Friesea* was strongly aggregated in the deeper layers of the moss-turf. However, when only small numbers of individuals were present in a layer, it was not possible to examine dispersion using this ratio. An estimate of b, the slope of the regression of the logarithm of variance on the logarithm of mean, was used as an index of aggregation (Taylor 1961). Values for b of 1·55 and 1·50 were obtained for *Cryptopygus* and *Friesea* respectively. Both values were significantly greater than unity ($P < 0·001$), the value expected for a Poisson distribution (which is the null hypothesis, i.e. randomness). Thus, overall, both species were strongly aggregated. The physical and chemical characteristics of the moss-turf were examined to investigate some possible

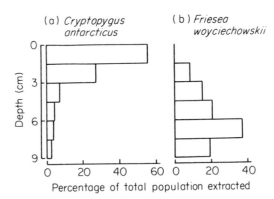

FIG. 1. The vertical distribution of the two most abundant species of Collembola. The numbers are expressed as percentages in each of the six layers for the four blocks combined: the total numbers extracted were 2119 *C. antarcticus* and 121 *F. woyciechowskii*.

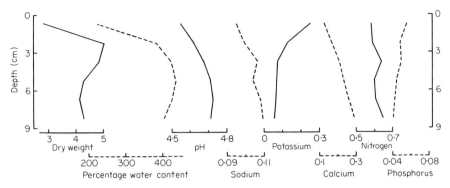

FIG. 2. The vertical distributions of eight measures of the physical and chemical environment of a block of moss-turf collected from Signy Island on 23 January 1981. For each variable, the six layer means (*n* = 16) have been plotted. Solid lines represent the upper set of horizontal axes, dashed lines the lower set of axes. Chemical concentrations are expressed as a percentage of sample dry weight.

causes for both the aggregations and the differences in vertical distribution of the two species.

Eight physical and chemical characteristics of the environment were determined for all samples. Figure 2 shows the vertical distributions of these environmental data for 23 January 1981, i.e. during the middle of the summer. Vertical trends are evident in many of the variables in this block, but, in particular, dry weight, percentage water content, potassium and calcium showed consistent trends in all four blocks.

The effect of the physical and chemical environmental factors on the distributions of *Cryptopygus* and *Friesea* was examined in two zones. A principal components analysis had shown that the arthropod populatios could be subdivided (Usher & Booth 1984): the green moss zone comprised the top layer data, and the dead moss zone comprised the moss-turf below 3 cm in depth. The transition between the two zones occurred within the second layer (1·5–3 cm), so these data were omitted from the analyses because they were intermediate.

In the December 1980 block, there was a positive correlation between percentage water content and counts of *Cryptopygus* in the surface layer, but a negative correlation in the dead moss zone. There was a significant positive correlation in the green moss zone on 20 February 1981, but all other correlation coefficients were not significant. Figure 3 suggests an optimum range of percentage water content (correlations not significant), and extremes where an influence was exerted on the distribution of *Cryptopygus*. Percentage water content appeared to have no influence on the distribution of *Friesea*.

In the green moss zone, multiple regression equations relating the numbers of *Cryptopygus* to the environmental factors were statistically significant in three of the four blocks. Percentage water content and sodium consistently contributed positive and negative regression coefficients respectively, whilst *Polytrichum* cover contributed one positive and one negative coefficient. All the remaining variables (dry weight, pH, potassium, calcium, phosphorus and nitrogen) were incorporated

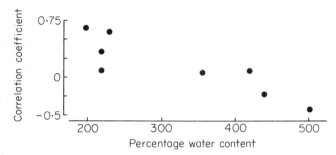

FIG. 3. Scatter plot of the correlation coefficients (between percentage water content and the numbers of *Cryptopygus*) plotted against the mean percentage water contents of two zones for four blocks of samples. The four points to the left relate to the green moss zone, and the four points to the right to the dead moss zone.

only once each (Booth & Usher 1984). In the dead moss zone, again three of the four multiple regression equations were significant. The coefficients for calcium were consistently negative in all three blocks. Potassium was incorporated into two of the equations and both coefficients were negative. Three other variables (*Polytrichum* cover, sodium and phosphorus) were incorporated once each. The percentage of the variance accounted for by the regression equations varied between 14 and 25%

The influence of the environment on the distribution of *Friesea* in the dead moss zone appeared to be slight. Although the regression equations relating *Friesea* to the environmental variables were all statistically significant (the December 1980 block was excluded as it contained only a single *Friesea*), they accounted for only between 8 and 12% of the variance.

DISCUSSION

The strongly aggregated distributions of the two Collembola species are typical of many soil arthropods. Can the causes of these aggregations be identified? Usher (1976) reviewed a variety of field studies and concluded that both the physical and chemical environment could influence the location of arthropod aggregations. Food location was an important factor in laboratory studies, but its influence in the field was uncertain.

Cryptopygus and *Friesea* aggregated to a similar extent, but they differed markedly in their response to the environmental factors. The influence of *Polytrichum* cover was not consistent, although it was incorporated into five of the nine regression equations. Plotting the values of the correlation coefficients between *Polytrichum* cover and the Collembola counts against the mean *Polytrichum* cover for each block (as for percentage water content in Fig. 3) showed no recognizable trend for *Cryptopygus*, but suggested that *Friesea* may favour areas of high or low *Polytrichum* cover (the latter implying high *Chorisodontium* cover), and avoid areas of intermediate cover. Percentage water content only appeared to influence the distribution of *Cryptopygus*. The chemical characteristics of the environment appeared to have a greater effect on the distribution of *Cryptopygus* than did the physical factors. Sodium was important in the green moss zone, and calcium and potassium were important in the dead moss zone. The observed influence of the chemical variables on *Friesea* was small compared to their influence on *Cryptopygus*.

Which other environmental features might be structuring the arthropod community? Competition for food supply and predation pressure can be considered. Competition between these two species of Collembola is unlikely, and thus cannot explain the differences in their vertical distributions. The small prostigmatid mite, *Nanorchestes berryi* Strandtmann, is a surface dwelling, algal grazer, whose food requirements and distribution overlap with those of *Cryptopygus*. Another prostigmatid mite, *Ereynetes macquariensis* Fain, occurs in the dead moss zone and is a potential competitor with *Friesea*, although little is known of the food requirements of either species.

The moss-turf supports a single species of arthropod predator, the mesostigmatid mite *Gamasellus racovitzai* (Trouessart). *Cryptopygus* contributed about 80% of all food traces found in a field population of *Gamasellus* (Lister, in press) and Usher & Bowring (1984) estimated, from laboratory studies, that *Gamasellus* could remove annually some 6–8% of the population of *Cryptopygus* from a moss-turf habitat. Since the predator captures prey mainly from the larger size classes of *Cryptopygus*, it may have a disproportionate effect on the potential breeding population. In the moss-turf *Gamasellus* had a similar vertical distribution to *Cryptopygus*, its preferred prey (Usher & Booth 1984).

Does *Friesea* avoid predation by occurring deeper in the moss-turf profile or does it rely on a means of chemical defence? Unfortunately, there are no laboratory or field data available on interactions between *Friesea* and *Gamasellus* to answer these questions. The value of flocking behaviour as an anti-predator strategy by vertebrates has been widely discussed (e.g. Krebs & Davis 1981): is it possible that predation pressure on invertebrate populations (e.g. *Cryptopygus* in the moss-turf) is a contributory influence to microarthropod aggregation?

ACKNOWLEDGMENTS

We are grateful to the British Antarctic Survey and its staff for logistical support and practical help. The chemical analyses were undertaken by the Institute of Terrestrial Ecology, and thanks are due to Mr S. E. Allen and his staff. This work was funded by the Natural Environment Research Council.

REFERENCES

Allen, S.E., Grimshaw, H.M., Parkinson, J.A. & Quarmby, C. (1974). *Chemical Analysis of Ecological Materials*. Blackwell Scientific Publications, Oxford.
Allen, S.E. & Northover, M.J. (1976). Soil types and nutrients on Signy Island. *Philosophical Transactions of the Royal Society of London, Series B*, 252, 179–185.

Effects on the plants

Lethal effects

Figure 1 shows the survival of sugar-beet, not protected by a pesticide, grown in a field heavily infested with the soil-pest complex. Here, plants died at a decreasing rate over the first 30 days after emergence as they grew through the cotyledon stage; then there was a drop in the number of plants surviving. As the surviving plants developed beyond this stage and grew two, four and then six true leaves, they became tolerant of the damage and the numbers of plants reached an asymptote at between 60 and 65% of the seeds sown. All the dead plants examined had extensive damage to their root systems. This pattern of plant survival was typical of eleven other sites studied during spring 1981.

The survival of sugar-beet plants in plots protected from the soil-pest complex by carbofuran is shown in Fig. 2. Like the untreated beet, the treated ones die at a

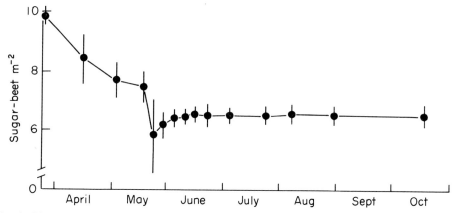

FIG. 1. The temporal pattern of sugar-beet failures in plots unprotected by pesticide at Ramsey Forty Foot, 1981. Mean population densities ±SE are shown.

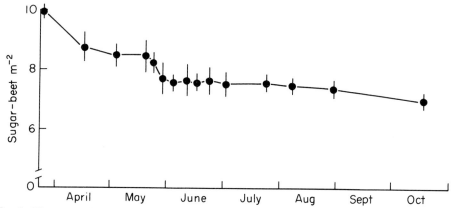

FIG. 2. The temporal pattern of sugar-beet failures in plots protected from the soil-pest complex by the pesticide carbofuran at Ramsey Forty Foot, 1981. Mean population densities ±SE are shown.

decreasing rate over the 30 days after emergence, followed by a small drop in numbers during the two true leaf stage. The population density of sugar-beet plants appears to stablilize at about 50–60 days after emergence, there then seems to be a slight decrease again just before harvest.

There is a spatial pattern of plant survival, as well as the temporal one described above, that typifies a sugar-beet crop damaged by the soil pest complex (Fig. 3). Jaggard (1979) found, through surveys and through experiments with crops undamaged by the soil-pest complex, that plant failures occur at random. Brown (1981) found this to be the case at two sites infested with the soil-pest complex, where an effective pesticide was used. Here the number of consecutive seed sta-

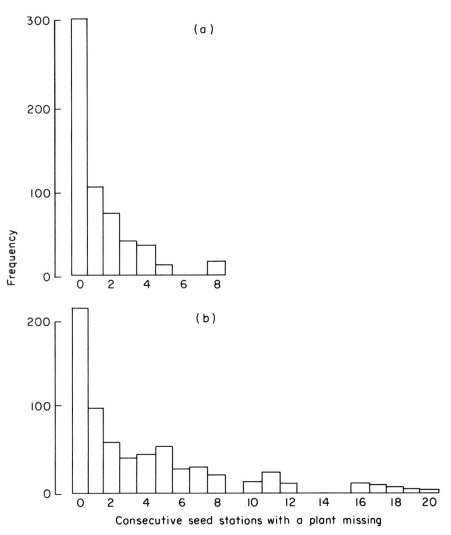

FIG. 3. The spatial pattern of sugar-beet seedling failures at North Newbald, 1981; (a) treated with carbofuran, (b) untreated.

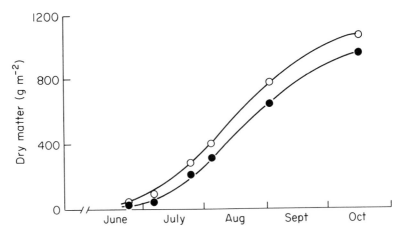

FIG. 5. The accumulation of dry matter in the roots of sugar-beet protected from the soil-pest complex by the pesticide carbofuran (○) and unprotected (●). The dry matter per unit area is shown.

It appears that individual sugar-beet plants can compensate for damage to their roots through an increased growth rate in the latter half of the growing season; this is perhaps due to reduced competition between plants because of lower plant populations where there has been damage by the soil-pest complex. This compensation, however, is less than would be expected. Individual roots from the untreated plots were of a similar dry weight to the treated ones at final harvest, we would expect them to be larger as they were growing in a less dense plant population (Figs 1 and 2). This discrepancy is consistent with the hypothesis that the productivity of damaged plants is lowered, the final yield of individual pest-damaged plants being less than would be expected; it is not possible to say whether this is due to excessive gappiness or to reduced root efficiency. Though individual plants compensated, to an extent, for damage done at an early stage by the soil-pest complex, this was not sufficient to replace the yield of the plants that died as a result of pest damage; this resulted in the yield per square metre remaining significantly less in the untreated plots throughout the season.

CONSEQUENCE OF THIS DAMAGE

For plant competition

Intraspecific competition

The plants at Ramsey Forty Foot suffered much higher mortality from soil-inhabiting herbivores (and possibly pathogens) at the seedling stage on the untreated rather than on the treated plots. No further mortality occurred on the untreated plots after the plants had reached the four true-leaf stage. A small but significant ($t_4 = 3.45$, $P < 0.05$) mortality was recorded on the treated plots towards the end of the season.

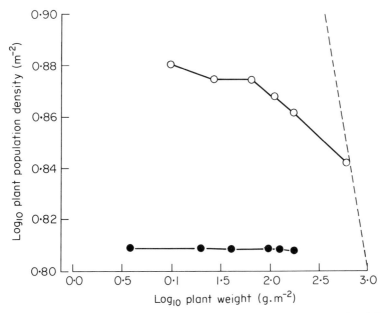

FIG. 6. Self-thinning of sugar-beet at Ramsey Forty Foot. The trajectory in treated plots (○) and in untreated ones (●) is shown. The dotted line is the self thinning line of plant population density against plant weight per unit area, with the predicted slope of −2.

It is likely that these deaths are due to self-thinning in the crop: the deaths of apparently established plants, most often in the smallest size classes (Harper & White 1974), due to intraspecific competitions. This phenomenon was first quantified by Yoda *et al.* (1963), and their prediction that the ratio of the logarithm of mean plant weight to the logarithm of surviving plant density is −3/2 became known as Yoda's rule. Figure 6 shows the self-thinning effect in the treated plots (open circles) between June and October. Here, the logarithm of surviving plant density is plotted against the logarithm of mean plant weight per unit area, so the (dotted) self-thinning line should have a slope of −2 rather than −3/2 (Crawley 1983). What little self-thinning was noticed occurred in the treated, rather than the untreated plots; in this instance root grazing by the herbivores decreased the number of plant deaths by relaxing intraspecific competition. Before these deaths occurred, the sugar-beet in the treated plots were not at a population density where this would be expected (7·2, SE 0·3 m^{-2}). It is possible that the low numbers of contagiously distributed herbivores in the treated plots could make the plants differentially competitive, without severely lowering the surviving plant population density, therefore accentuating self-thinning.

Interspecific competition

Table 2 shows that though weed diversity was not significantly affected by the different amounts of grazing on the treated and untreated plots, weeds were signifi-

TABLE 2. The effects of grazing by the soil-pest complex on the number of weed species, weed abundance and sugar-beet 'dominance' at Ramsey Forty Foot, Cambridgeshire, August 1981. Weed abundance is expressed as weed plants m^{-2} and sugar-beet 'dominance' as the ratio of sugar-beet population density to total plant population density. Means $\pm$ SE are given

	Treated	Untreated	n	F ratio	Significance
Number of species	$6 \cdot 75 \pm 0 \cdot 48$	$7 \cdot 00 \pm 0 \cdot 41$	8	$0 \cdot 16$	NS
Weed abundance	$10 \cdot 00 \pm 1 \cdot 47$	$23 \cdot 25 \pm 4 \cdot 82$	8	$6 \cdot 91$	$P < 0 \cdot 05$
Sugar-beet 'dominance'	$0 \cdot 42 \pm 0 \cdot 08$	$0 \cdot 28 \pm 0 \cdot 09$	8	$46 \cdot 81$	$P < 0 \cdot 001$

cantly more abundant on the untreated plots. It appeared that the larger gaps in the untreated plots, caused by the soil-pest complex killing seedlings, allowed the establishment of a greater number of weeds later in the season. When the soil-pest complex were allowed to graze the roots of the sugar-beet, greater numbers of weeds were able to establish both in and between the rows than where root-grazing was limited by the use of a pesticide. Sugar-beet is often more attractive to the soil-pest complex than many common weeds (Brown 1982). It is possible that there was no effect on weed species diversity because the soil-pest complex do not distinguish between the different weed species at this site or that all the weeds found were equally well represented in the seed bank in this particular field. The dominance of sugar-beet at this site, defined as the ratio of the sugar-beet population density to the total plant population density, is decreased by the feeding of the soil-pest complex.

ACKNOWLEDGMENTS

I would like to thank M. J. Crawley, R. A. Dunning, M. J. Durrant, R. E. Green, K. W. Jaggard and G. H. Winder for discussion; J. M. Cooper and J. M. Wardman for assistance, and the farmers, Mr Oswald Rose and Mr Walter Turton. This work was financed in part by the Ministry of Agriculture's Sugar-Beet Research and Education Committee, and by grants from FMC Corporation (UK) Ltd, Union Carbide UK Ltd, Cyanamid of Great Britain Ltd., Rhom and Haas (UK) Ltd, Bayer UK Ltd, Ciba Geigy Agrochemicals, Sandoz Products Ltd, Eli Lilly International Corporation, Hoechst UK Ltd, Pan Britanica Industries Ltd, Murphy Chemicals Ltd and BASF UK Ltd.

REFERENCES

Bleasdale, J.K.A. (1966). The effect of plant spacing on the yield of bulb onions (*Allium cepa* L.) grown from seed. *Journal of Horticultural Science*, **41**, 145–153.

Brown, R.A. (1981). Gappiness, sugar-beet yield loss and soil-inhabiting pests. *Proceedings of the 11th British Insecticide and Fungicide Conference*, **3**, 803–810.

Brown, R.A. (1982). *The ecology of soil-inhabiting pests of sugar-beet, with special reference to Onychiurus armatus*. Ph.D. Thesis, University of Newcastle-upon-Tyne.

Crawley, M.J. (1983). *Herbivory: the Dynamics of Animal–Plant Interactions*. Blackwell Scientific Publications, Oxford.

Edwards, C.A. (1959). The ecology of the Symphyla, II. Seasonal soil migrations. *Entomologia Experimentalis et Applicata*, **2**, 257–267.

Green, R.E. (1978). *The ecology of bird and mammal pests of sugar beet.* Ph.D. Thesis, University of Cambridge.

Harley, J.L. (1969). *The Biology of Mycorrhiza.* Leonard Hill, London.

Harper, J.L. & White, J. (1974). The demography of plants. *Annual Review of Ecology and Systematics*, **5**, 419–463.

Jaggard, K.W. (1979). *The effect of plant distribution on the yield of sugar beet.* Ph.D. Thesis, University of Nottingham.

Jepson, P.C. & Green, R.E. (1983). Prospects for improving the control of sugar-beet pests in England. *Advances in Applied Biology* **7**, 175–250.

Jones, F.G.W. & Dunning, R.A. (1972). *Sugar Beet Pests.* MAFF Bulletin 162, HMSO, London.

Sokal, R.R. & Rolhf, F.J. (1969). *Biometry. The Principles and Practice of Statistics in Biological Research.* Freeman & Co. San Francisco.

Ulber, B. (1983). Enfluss von *Onychiurus fimatus* Gisin Collembola, Onychiuridae) und *Folsomia fimetaria* (L). auf *Pythium ultimum* Trow., einen Erreger des Würzelbrandes der Zückerrube. *New Trends in Soil Biology* (Ed. by Ph. Lebrun, H.M. Andre, A. de Mets, C. Gregoire-Wibo & G. Wauthy), pp. 261–268. Proceedings VIII International Colloquium of Soil Zoology, 1982. Dieu-Brichart, Louvain-la-Neuve, Belgium.

Warnock, A.J., Fitter, A.H. & Usher, M.B. (1982). The influence of a springtail *Folsomia candida* (Insecta, Collembola) on the mycorrhizal associations of leek *Allium porrum* on the vesicular-arbuscular endophyte *Glomus fasciculatum*. *New Phytologist*, **90**, 285–292.

Yoda, K., Kira, T., Ogawa, H. & Hozumi, K. (1963). Self thinning in overcrowded pure stands under cultivated and natural conditions. *Journal of Biology of Osaka City University*, **14**, 107–129.

Role of the soil invertebrates in determining the composition of soil microbial communities

SUZANNE VISSER

Kananaskis Centre for Environmental Research and Department of Biology, University of Calgary, Calgary, Alberta, Canada T2N 1N4

SUMMARY

1 The various mechanisms by which soil fauna can influence the composition of microbial communities are separated into three main categories: (i) comminution, channelling and mixing, (ii) grazing, (iii) dispersal, and discussed.

2 Comminution, channelling and mixing may reduce fungal species numbers and divert fungal successional patterns on decaying plant residues. This may be due to the sensitivity of various fungi to destruction of their thallus and microhabitat, thereby giving a competitive advantage to fast-growing, short life-cycle micro-organisms such as bacteria and *Mortierella* spp.

3 Grazing by the fauna on selected fungi may significantly alter fungal distribution and succession on decomposing leaf litter and may deleteriously affect plant growth by reducing the effectiveness of the mycorrhizal symbiosis. The impact of the fauna on their microbial food sources appears to be dependent on the grazing pressure of the invertebrate and possibly on the growth rate of the organism being grazed.

4 Many soil micro-organisms, particularly those commonly associated with the invertebrate body, may rely on the soil fauna for dispersal of their propagules. However, the success of the transmitted propagules in establishing themselves in new microhabitats requires further investigation. The interaction of the dispersal, grazing and comminution/mixing activities of the soil fauna in determining dynamic microbial communities is speculated upon.

INTRODUCTION

Despite the low estimated contribution of soil fauna to total soil respiratory metabolism in comparison with the soil microflora (Reichle 1977; Persson *et al.* 1980), the density of various faunal groups can be very high (Petersen & Luxton 1982). Therefore, it has been speculated that the roles of the soil fauna may be more subtle than is evident from their contributions to total soil metabolic activity. These roles include the regulation of decomposition and mineral cycling processes and direct and indirect effects on soil microbial activities (Seastedt 1984).

The efficiency of the soil microflora in decomposing organic debris is determined by resource quality and quantity, and by both the inoculum potential, and the competitive ability of the decomposer organisms (Garrett 1970). By altering resource quality and influencing inoculum potential and competitive saprophytic ability of decomposers and root-infecting fungi, soil/litter invertebrates can affect microbial community structure as well as decomposition/mineralization processes.

The type of influence of soil fauna on the microflora will depend largely on the animal's size and its method of feeding. Generally, a direct relationship exists between animal size and the size of ingested food particles. Therefore, it seems likely that smaller animals have a high chance of feeding on specific components of the microflora. Based on their size, numbers, mobility, and wide diversity of feeding strategies (Burges & Raw 1967; Wallwork 1970; Petersen & Luxton 1982), the activities of the soil fauna in relation to their potential influence on soil micro-organisms can be separated into three main categories:

    (i) comminution, mixing and channelling of litter and soil;
    (ii) grazing on the microflora;
    (iii) dispersal of microbial propagules.

Information regarding the specific roles of the macro-, meso-, and microfauna in structuring microbial communities is sparse and has been generated from studies conducted in a wide variety of systems including litter/soil microcosms, cattle and rabbit dung, and woodland streams. An attempt is made here to review these studies within the context of the three categories of faunal activities mentioned previously. Since much of the data has been obtained from laboratory studies, extrapolation to field conditions entails much speculation

COMMINUTION, CHANNELLING AND MIXING

In the present context comminution is defined as the fragmentation and mastication of plant debris including the microflora growing within the plant residues. In contrast, grazing implies that the invertebrate has a feeding mechanism which allows separation of the microflora from its food base, with little destruction to the microbial food base.

It has been suggested that fragmentation of organic matter, channelling, and mixing of soil components are key roles by which soil fauna stimulate microbial activity, thereby enhancing the rate of organic matter decomposition (Kevan 1962; Wallwork 1970; Crossley 1977; Swift, Heal & Anderson 1979). Both uningested litter fragments and ingested (later defaecated) materials result from faunal comminution activities. The size of the uningested fragments will presumably be larger than the size of the ingested fragments where the degree of reduction will depend on the size of the invertebrate. For example, plant debris is reduced to particles 200–300 μm diameter by earthworms and millipedes, 100–200 μm by medium-sizes arthropods and 20–50 μm by micro-arthropods (Swift *et al.* 1979). This consumed material is expelled as faeces, with the constituent particle size varying with different faunal groups; a factor which may also determine the persistence of the faeces (Webb 1977). The pelletization of organic debris, and resultant effects through changed pore volume, on moisture retention and aeration, will alter the balance between bacteria and fungi. Maximal bacterial development has been recorded in fine textured soils (Bhaumik & Clark 1947), and Parr, Parkinson &

Norman (1967) showed bacterial respiration to be greatest in artificial systems with small pore size, whereas fungal vegetative activity (*Trichoderma viride*) was greatest in systems of large pore size. The importance of pore size in allowing fungal sporulation has been shown for *Cunninghamella*, *Botrytis*, *Rhizopus* and *Mucor* by Kubiena (1938) and for *Curvularis* and *Pythium* by Griffin (1963a, b). The movement of fungal spores under changes in moisture tension is also limited by the size of the pores in soil and by spore size (Dickinson & Parkinson 1970) and limitations of space may be of considerable importance for the development of sequential microbial populations on organic substrata (Stotzky & Norman 1961).

Extrapolating these data to sites of faecal accumulation (e.g. F_2, and H layers of forest soils) it might be expected that in macro-arthropod faeces the constituent particles would be large enough to allow fungal growth and sporulation, whereas the compaction of particles in micro-arthropod pellets would only allow bacterial growth. Support for this extrapolation has been given by Hanlon (1981b).

In comparison with undigested resources where no alteration of chemical quality would occur, increased NH_4-N levels (Parle 1963b; Bocock 1963; Anderson, Ineson & Huish 1983), pH (Hanlon 1981c), and carbohydrate levels (Parle 1963b) have been observed in millipede faeces and earthworm casts. A higher bacterial biomass has been recorded from invertebrate faeces than from uningested food materials (Parle 1963a; Reyes & Tiedje 1975; Brown, Swift & Mitchell 1978; Anderson & Bignell 1980, Hanlon 1981c) and this may be attributed to altered chemical quality of the faeces or to bacterial multiplication within the invertebrate gut (Parle 1963a; Anderson & Bignell 1980; Hanlon 1981c).

Fungi appear to be susceptible to damage as a result of the comminution and digestion process (Hanlon 1981b), this possibly being related to the size of the animal involved. Thus, fungal viability may be less affected after passage through millipede and earthworm guts (as indicated by the fungal isolation data of Nicholson, Bocock & Heal 1966; Parle 1963b) than after passage through guts of Collembola (Ponge & Charpentie 1981). It appears that the main effect of macroinvertebrate feeding (and resource comminution) is to stimulate bacterial activity and reduce fungal biomass, an effect exemplified in the microcosm study of Hanlon & Anderson (1980) and the dung study of Lussenhop *et al.* (1980).

Moisture conditions and their variations at faunal feeding times may influence the subsequent microbial colonization of casts and faeces. Wet conditions may give an advantage to bacteria, which may decrease as the food base dries (Nicholson, Bocock & Heal 1966). Alternating wet and dry conditions may favour fungal activity.

There have been very few studies regarding the effects of resource comminution on the specific nature of the microflora. In a laboratory study, Wicklow & Yocum (1982) recorded a reduction in the number of coprophilous fungal species in rabbit dung with increased density of dipteran larvae (Table 1). It was suggested that the decrease was due to damage to the mycelial thallus, with some fungi (*Podospora tetraspora*, *Lasiobolus intermedius*) being more sensitive than others. The feeding behaviour of the larvae consisted of both channelling within the dung pellets and

TABLE 1. Dipteran larvae density effects on the number of species within coprophilous fungal communities (from Wicklow & Yocum 1982)

	Initial larval density per dung sample							
	0	5	10	25	50	70	100	125
Total fungal species	14	14	12	8	9	6	9	7
Mean species per 2 g sample	6	6	5·2	5·2	5·2	3·0	3·2	3·2

grazing on the pellet surface which, in contrast to selective grazing on a particular component of the microflora, was thought to have caused the reduction in fungal species. This study emphasizes the importance of distinguishing between comminution/channelling activities and selective grazing since these two feeding behaviours may influence microbial communities in completely different ways.

In another study, Lussenhop *et al.* (1980) observed that arthropods added to cattle dung in the field had no significant effect on numbers of fungal species which fruited although hyphal density was reduced. Unlike conditions in the laboratory, where moisture levels of experimental material can be controlled, dung in the field eventually dries out, resulting in emigration and pupation of insects. Thus, climatic properties were believed to exert a major influence on invertebrate–microfloral interactions. The inconsistency of results obtained in the laboratory and the field may be largely due to the interactions of abiotic variables with insect and microbial life cycles. These interactions should be taken into account when designing laboratory microcosm experiments intended to elucidate field observations.

In addition to potentially altering microbial community structure by affecting fungal species abundance and richness, the comminutive activities of the soil fauna may also affect fungal successional patterns on decaying plant debris. The mechanisms of replacement involved in fungal successions have been thoroughly reviewed by Frankland (1981) and summed up by her as follows 'a decomposer can replace another species when changes in the substrate (or site) have interacted with changes in its relative competitive saprophytic ability and inoculum potential giving it a decisive advantage.' By altering substrate quality and a colonizer's inoculum potential, soil invertebrates may have a controlling influence on successional patterns. For example, Swift (1982) observed that when decomposing branchwood was invaded by dipteran larvae, the Basidiomycete community colonizing the wood was largely replaced by fungi typical of the soil microflora such as moniliaceous and mucoraceous forms. The sensitivity of the Basidiomycete biomass to comminution and mixing thereby reducing its inoculum potential was believed to be one of the factors causing the shift in fungi—a factor also considered by Wicklow & Yocum (1982) for coprophilous fungi in rabbit dung. Other factors included a change in the physical structure of the resource as a consequence of invertebrate activity and the transport of microbial propagules from the litter and soil into the wood via Collembola and mites.

Fungal successional sequences on leaf litter may also be diverted as a result of passage of microbially colonized litter through the invertebrate gut. Nicholson, Bocock & Heal (1966) found that when faeces of millipedes previously fed on 6–12

TABLE 2. Frequently-isolated fungi from ingested, defaecated hazel leaves and uningested hazel leaves in the field (from Nicholson, Bocock & Heal 1966; Hering 1965)

Fungi isolated	Age of ingested matter in faeces (mo)			Age of uningested organic matter (mo)		
	12	14	18	12	14	18
Cladosporium herbarum	+	+	−	+	+	+
Geomyces cretaceous	+	+	+	−	−	−
Mucor hiemalis	+	+	−	+	+	−
M. ramannianus	+	−	−	+	+	+
Mortierella isabellina	+	+	−	−	−	−
Trichoderma sporulosum	−	+	+	−	−	−
Phoma sp.	−	+	+	−	−	−
Paecilomyces farinosus	−	−	−	+	+	+
Aureobasidium pullulans	−	−	−	+	+	+
Penicillium frequentans	−	−	−	+	+	−
P. thomii	−	−	−	+	+	+
Trichoderma viride	−	−	−	−	+	+

month old hazel litter were placed in the field, they were initially colonized by Phycomycetes such as *Mortierella* and *Mucor* (fungi capable of rapid exploitation of easily assimilable materials), followed by Ascomycetes and Fungi Imperfecti such as *Phoma* and *Trichoderma sporulosm* (Table 2). This successional pattern did not correspond with the pattern of fungal colonization observed on decaying, uningested hazel leaves (Hering 1965) (Table 2), but was thought to resemble more closely that found on dung.

As is the case for ingested plant debris, the size of the uningested particles will depend to a large extent on the size of the animal concerned. The exposure of uncolonized tissues in uningested leaf fragments will provide new surfaces for microbial colonization and may also result in the development of more continuous water films thereby promoting bacterial colonization (Griffin 1969). The decay of uningested fragments may be stimulated if they are withdrawn into earthworm burrows where environmental conditions are more conducive to microbial growth and organic matter decomposition. Burrowing in soil, mining in plant tissues, and channelling of wood may result in greater aeration which stimulates fungal growth and sporulation (Kubiena 1938; Warcup 1965; Griffin 1969), although Swift (1981) suggested that comminution of wood could also lead to higher moisture contents causing a drop in oxygen levels and a subsequent inhibition of Basidiomycete growth.

In aquatic systems, where leaf-eating invertebrates selectively feed on parts of leaves heavily colonized by micro-organisms, Bärlocher (1980) observed that invertebrate feeding on oak and larch leaves reduced fungal species number. He hypothesized that this was due to the removal of portions of leaf which, if they had not been consumed, would be available for colonization by later successional species, i.e. invertebrate consumption of both the resource and the microbes colonizing this resource, deleted the late successional phase and prolonged the persistence of early successional species. No equivalent studies have been done in terrestrial ecosystems, but study is required to determine if the fungal successional pattern on rejected plant debris (e.g. lignified materials) is altered as a result of selective consumption of micro-organisms associated with the more palatable plant material

GRAZING

Grazing, as defined earlier, is the selective removal of microbial tissue from its resource by members of the microfauna (protozoa, nematodes) and mesofauna (enchytraeids, Collembola, mites, dipteran larvae) without extensive destruction of the resource. There is much documentation of exploitation of microbial biomass by these members of the soil fauna as a result of gut content analyses and feeding preference studies under laboratory conditions. The present discussion will deal mainly with data collected for organisms in temperate woodlands.

Microbial food sources of microfauna

Bacteria and yeasts are major food sources for protozoa in soil (Stout & Heal 1967; Stout 1974), and laboratory studies have shown preferential feeding by these organisms (Stout & Heal 1967) for members of the Pseudomonadaceae and Enterobacteriaceae, and yeasts such as *Rhodotorula*, *Saccharomyes* and *Kloeckera*. Some bacteria, actinomycetes (*Streptomyces griseus*) and fungi (*Trichoderma viride*) produce extracellular substances toxic to protozoa. Because protozoa generally prefer fast-growing, zymogenous micro-organisms, Stout (1974) suggested that they would flourish in habitats such as the rhizosphere where such micro-organisms would be stimulated by root exudates. In the soil, where a mainly autochthonous microflora was believed to predominate, micropredation could influence the composition of the bacterial flora (Stout 1974). Research is necessary to support these statements.

Nematodes use 'protoplasm' as their major food (Overgaard Nielsen 1967) and obtain this from a variety of sources including bacteria and fungi by applying a range of feeding mechanisms. There is little evidence of preferential feeding by bacterivorous nematodes but their growth rates may vary with different bacteria (Twinn 1974). Mycophagous nematodes feed on many root-inhabiting fungi (Riffle 1971) including vesicular-arbuscular mycorrhizal fungi (Salawu & Estey 1979), endomycorrhizal fungi of Ericaceae (Shafer, Rhodes & Riedel 1981) and ectomycorrhizal fungi (Sutherland & Fortin 1968; Riffle 1967, 1971).

Microbial food sources of mesofauna

Gut content analyses of field-collected mites and Collembola have shown the majority of animals to be non-specific feeders (Harding & Stuttard 1974; Anderson & Healey 1972; Anderson 1975; Behan-Pelletier & Hill 1983; Takeda & Ichimura

1983) with some mite species appearing to be either predominantly or exclusively mycophagous (Anderson & Healey 1972; Behan-Pelletier & Hill 1983). However, many of the micro-arthropods classified as panphytophages (Luxton 1972) have been observed to be mainly mycophagous (Poole 1959; Hägvar & Kjondal 1981; Mitchell & Parkinson 1976). Factors such as animal size (Anderson 1975) and availability of food items in particular microhabitats (Wallwork 1970; Petersen 1971) may influence the amount of fungus material consumed.

In contrast to the data from gut content analyses, laboratory experiments have generally shown not only preferential feeding by mites and Collembola on fungi but also their capability for selecting specific fungal species (e.g. Hartenstein 1962; Mills & Sinha 1971; McMillan 1976; Mitchell & Parkinson 1976; Visser & Whittaker 1977; Addison & Parkinson 1978). While the relevance of these studies to field situations may be questioned, they do not allow clarification of possible microflora–fauna relationship in the field. The relevance of this type of study is enhanced if the test organisms have the potential of interacting in the field. Newell (1984a) showed that *Onychiurus latus*, a common mycophagous Collembola at her field site, preferred mycelium of *Marasmius androsaceus* to that of *Mycena galopus*, not only in laboratory experiments but also in field tests.

Enchytraeids and dipteran larvae have been shown to accelerate organic matter decay rates (Standen 1978) and this may be related to their selective grazing on fungi (Dash & Cragg 1972). Many dipteran species are associated with fungal fruit bodies (Buxton 1960) but studies of their influence on fungal growth and development have concentrated on their effects on cultivated mushrooms (Binns 1980).

Boddy, Coates & Rayner (1983) reported that a volatile material, produced in zones of intraspecific antagonism between certain wood-decaying Basidiomycotina and Ascomycotina, was attractive to Mycetophilid fungus gnats. They suggested that this could significantly increase the consequences of grazing if it occurred in the field.

Effects of grazing in structuring microbial communities

If selective animal grazing on specific groups of the microflora occurs it could have a significant influence on microbial community structure. This influence, both in soil/litter systems and on roots will depend on: (i) grazing pressure, which is dependent on the density of grazers, the growth and turnover rates of the grazer and the time spent by the grazer on its food sources; (ii) growth rate, nutritional quality and physiological condition of the grazed organism which will determine its competitive ability; (iii) the ability of the food source to escape grazing, for example, by the production of toxic secondary metabolites.

Selective grazing by protozoa may alter the population density of specific bacteria (Habte & Alexander 1975). Given the potential high growth and turnover rates of protozoa and nematodes, their impact on the soil microflora, particularly the impact of nematodes on slow growing fungi, could be considerable.

Laboratory studies have shown that the effect of collembolan grazing on senescent fungi can be dependent on the nutritional quality of the fungal food base, with fungi grown under low nutrient regimes exhibiting a reduction in activity in the presence of Collembola and fungi grown on nutrient rich substrates being stimulated by faunal grazing (Hanlon 1981a). Under conditions of high grazing pressure it appears that bacteria have a competitive advantage over fungi (Hanlon & Anderson 1979); however, the environmental conditions used (e.g. moisture), the time course of the study and the potential role of Collembola as dispersal agents of bacteria could give the bacteria a competitive advantage regardless of direct grazing by Collembola on the fungus.

Destruction of fungal thalli through grazing may alter not only the balance between bacteria and fungi but, if selective destruction occurs, also the dynamics of fungal colonization of both decaying leaf litter and roots. Thus, in a microcosm study, selective fungal feeding by Collembola in *Populus tremuloides* leaf litter effectively reduced the competitive colonizing ability of a preferred fungus (sterile dark form), giving a less preferred fungus (Basidiomycete) the competitive advantage (Parkinson, Visser & Whittaker 1979). In this way fungal succession appeared to be advanced and the impact of this in increasing litter decay rate was considered likely since the basidiomycete had an active cellulolytic potential not exhibited by the sterile dark form.

Laboratory and field studies performed by Newell (1984b) have demonstrated the effects of selective grazing by Collembola on competitive saprophytic colonization of Sitka spruce needles by two common litter-decaying Basidiomycetes, *Marasmius androsaceus* and *Mycena galopus*. Selective grazing on *M. androsaceus* (the fungus with higher colonizing ability) gave *M. galopus* a competitive advantage. Exclusion experiments in the field indicated an inverse relationship between collembolan density and the activity of the preferred fungus. Depending on the fungal species, collembolan grazing either stimulated litter decomposition (*M. androsaceus*) or reduced litter decay (*M. galopus*). Possibly, *M. androsaceus* is less sensitive than *M. galopus* to a reduction in biomass resulting from grazing because the former has a faster growth rate.

Invertebrate grazing on root region microfloras may have significant consequences on both the micro-organisms and plant growth. Selective grazing by the mycophagous nematode, *Aphelenchus avenae*, prevented the establishment of the mycorrhizal relationship between *Pinus resinosa* and *Suillus granulatus*; however, plant growth was not significantly reduced (Sutherland & Fortin 1968). In a study by Warnock, Fitter & Usher (1982) it was observed that the rate of growth of the vesicular-arbuscular mycorrhizal fungus, *Glomus fasciculatum*, within pre-infected leek roots was not substantially reduced by the presence of the Collembola, *Folsomia candida*. However, plant growth of both mycorrhizal and nonmycorrhizal leeks was adversely affected by the activities of *F. candida* with the effect being more pronounced for mycorrhizal plants. It was believed that the reduction in growth of the mycorrhizal plants was the result of *F. candida* feeding on the extramatrical hyphae of the fungus, thereby negating the beneficial effects usually conferred upon the host by the fungus. The observations made in this study suggest that the influence of the soil fauna on the microflora may be very subtle and illustrate how little we understand the potential impact of grazers on microbial processes in the below-ground system.

Collembolan grazing may reduce the effectiveness of root pathogens or alter the species composition of the root region microflora thereby affecting interactions with symbionts (Wiggins & Curl 1979). Root exudates are a source of easily assimilable carbon compounds which if not immobilized by mycorrhizal fungi, will presumably stimulate the non-symbiotic root region microflora. This in turn could cause an increase in invertebrates feeding on the microflora (Clarholm 1981). More attention is required on the impact of grazers on root region microfloras in the presence and absence of mycorrhizal fungi.

The litter–soil–plant root system comprises a complex mosaic of microsites for microbial development, superimposed on which are aggregations of soil invertebrates. Hence, consideration of the mechanism by which soil fauna may alter fungal community structure is very difficult, particularly when substantive data are very scarce. In view of this, it may be instructive to consider the possible implications of other plant–herbivore systems to trophic interactions in soil.

In a study of protected intertidal communities, Lubchenco (1983) showed that preferential grazing by periwinkle snails on early successional ephemeral algae allowed an accelerated establishment of later successional algal species, by removing the inhibition caused by the rapid growth and reproductive rate of the ephemerals. A similar situation may occur during fungal colonization of leaf debris or plant roots, although here the successional sequences are complicated by microenvironmental and resource quality factors which cause much instability during early successional phases. Nevertheless, preferential faunal grazing on early successional fungi, e.g. *Cladosporium* spp. on leaf litter or *Thelephora terrestris* on pine roots, could allow accelerated fungal succession on these resources.

It was also observed in intertidal communities that the grazers not only preferred early ephemeral species but avoided consuming the later successional algal species, possibly because the algae contained higher levels of polyphenolic compounds. The production of secondary metabolites by some fungi, particularly Basidiomycetes, may serve the same purpose. Sutherland & Fortin (1968) found that *Rhizopogon luteolus*, an ectomycorrhizal fungus, was toxic to nematodes, an effect which may be due to the production of calcium oxalate by this fungus (Malajczuk & Cromack 1982). Binns (1980) considered that calcium oxalate limited feeding on mycelium of *Agaricus bisporus* by sciarid larvae and caused subsequent migration of the adults. In the same way, Richter (1980) observed that slugs, fed on immature *Amanita muscaria* died and suggested that toxic alkaloids, similar to those known to deter invertebrate herbivory of higher plants, could be responsible. Many slow growing Basidiomycetes which generally predominate late in the successional sequence, and for which destruction of the thallus could significantly reduce their competitive abilities and opportunities to reproduce, may deter grazing by producing toxic substances.

In addition to the production of toxic metabolites by older plants, Lubchenco

(1983) observed that, in the case of young plants which were susceptible to grazing, the heterogeneity of the substratum, plant density and patchy grazing behaviour of the herbivore were important factors in the successful establishment of later successional algae. In the below-ground system an increase in heterogeneity of micro-habitats due to a reduction in litter particle size (with attendant decrease in pore size) and the penetration of roots or plant debris by fungal hyphae could provide a mechanism by which young colonies and actively growing hyphal tips of later successional fungi are protected from faunal grazing. Patchy grazing may not only allow the escape of some fungal species, but could significantly stimulate their activity (presumably by release of nutrients) as demonstrated by Bengtsson & Rundgren (1983). Consequently, preferential invertebrate grazing on early successional fungi, because later successional species have evolved strategies to deter grazing, could result in accelerated fungal succession.

The role of faunal grazers in regulating fungal species diversity in litter–soil systems or on roots has not been investigated. Lubchenco (1978) concluded that the impact of an unspecialized grazer on the species diversity of its food source depended on grazing intensities and on the food preferences of the grazer, coupled with the competitive relationships between the food plants. The effect of faunal grazing on fungal species diversity may also be controlled by these factors.

There are various special cases of invertebrates causing local changes in fungal community structure in soil. A number of arthropods secrete volatile fungistatic and fungicidal materials (Roth 1961). *Scaptocoris divergens*, a soil-burrowing hemipteran, produces substances which are fungistatic or fungicidal to some soil-borne pathogenic fungi and several soil saprophytes (Timonin 1961).

The mutualistic association between some ant species and fungi sometimes results in extensive fungus gardens where a dominant single fungus is selectively maintained even though it is exposed to potentially high contamination. The supplying of suitable food resources, 'weeding', and the production of substances inhibitory to saprophytic bacteria and fungi (Weber 1972; Schildknecht & Koob 1971) by the ants are the methods by which the dominance of one species of fungus is maintained.

DISPERSAL

In addition to their roles in comminuting plant debris and grazing, soil invertebrates have also been considered important as dispersal agents of microbial inoculum.

Although the Phallales have been considered to be the only group of fungi which have evolved total dependence on insects for spore dispersal (Ingold 1971), there are a number of cases where fungi have developed specialized relationships to ensure insect dispersal of their spores (e.g. various rusts and smuts, *Ceratocystis ulmi*, *Stereum sanguinolentum*). However, all these cases involve the dispersal of spores between resources above-ground, where wind and water are considered the major dispersal mechanisms. Below-ground, members of the soil fauna may well be primary agents of spore transport. This would apply particularly in dry climates where propagule transport via water percolation through the profile would be expected to be minimal.

The spread of microbial inoculum may occur by means of propagules carried on external parts of the fauna and by inoculum passed through the gut and expelled as faeces (provided viability of this inoculum is maintained). Jacot (1930) observed fungal spores attached to the exterior of various Phthiracarid mite species, and similar observations have been made for a range of mite and Collembola species by Witkamp (1960), Warcup (1965) and others.

MacNamara in 1924 observed fungal spores in the guts of mandibulate Collembola, and since then numerous investigators have commented on the presence of fungal spores and hyphal fragments in the guts and faeces of the soil mites, Collembola and enchytraeids. The probable importance of fungal spores as a major component of the diet of these soil animals has prompted researchers to quantify the spore component of gut contents (Poole 1959; Gilmore & Raffensperger 1970; Anderson 1975, 1978; Behan-Pelletier & Hill 1983). Laboratory experiments on food preferences have suggested that some Collembola species prefer feeding on fungal spores, depending on spore type, rather than on hyphae (Knight & Angel 1967; McMillan 1976).

Much of the work on the role of soil animals as dispersal agents of fungi in litter–soil systems has emphasized the qualitative nature of the fungi associated with the animals or in their faecal pellets. Talbot (1952) recorded many species of fungi associated with wood (including Basidiomycetes) in the guts of woodlice, mites, springtails and slugs. Viability tests of spores of the wood decay fungus, *Merulius lacrymans*, after passage through woodlice guts suggested that there was a reduction in spore viability after digestion, and that the spores which retained viability required a longer time to germinate than uningested spores. It was postulated that insect activity was a major spore dispersal mechanism for some resupinate Hymenomycetes.

The majority of studies done to date indicate that mites and Collembola extracted from a range of habitats have associated with them approximately twenty species of fungi (Christen 1975; Behan & Hill 1978; Pherson & Beattie 1979). These species are comprised mainly of heavily sporulating forms including soil saprophytes (*Mortierella*, *Cladosporium*, *Penicillium*, *Chrysosporium*, *Aspergillus*), potential invertebrate parasites (*Beauveria*, *Paecilomyces*), and potential plant pathogens (*Fusarium*). They have the ability to metabolize a wide variety of substrates (Behan & Hill 1978) and generally reflect the most common genera isolated from the material from which the animals were extracted (Pherson & Beattie, 1979). The number of fungal genera associated with each animal appears to be related to body surface area, with the Collembola and Acari having fewer generally per individual (0·5) than the Coleoptera and Oligochaeta (1–2) (Pherson & Beattie 1979).

S. Visser, M. Hassall & D. Parkinson (unpubl.) examined the fungal taxa associated with the Collembolan, *Onychiurus subtenuis* extracted from the L, F_1, F_2, and H layers of an aspen woodland. Individual animals (105 per layer) were placed on

TABLE 3. Numbers of fungal propagules and taxa associated with *O. subtenuis* extracted from the L, F_1, F_2, and H layers of an aspen poplar woodland. Values are ranges of arithmetic means for thirty-five animals from each of three cores

		Litter layer		
	L	F_1	F_2	H
Number of propagules	3·1–5·5	2·4–5·0	2·5–5·0	2·5–3·1
Number of taxa	2·8–3·8	1·8–2·9	1·8–2·4	1·9–2·1

2% malt extract agar (with and without addition of bactericidal antibiotics), allowed to wander over the plates for 24 hours and then squashed into the medium to ensure that microbial propagules from exterior and interior body parts could be recorded. Bacteria and fungi were isolated from the tracks and bodies of all individuals, with the greatest number of fungal propagules and taxa being obtained from animals from the L layer and lowest from those from the H layer (Table 3). A total of 120 taxa of fungi were isolated (Table 4), with the most frequently occurring species (*Beauveria bassiana*, *Cladosporium cladosporioides*, *C. herbarum*, *Mortierella alpina*, *M. elongata* and *Penicillium raistrickii*) being similar to those isolated by previous investigators. This group of fungi was isolated from animals extracted from all layers; however, some species showed increasing or decreasing

TABLE 4. The percentage of occurrence (± SD) of the most common species isolated from the tracks and body of *O. subtenuis*

		Litter layer		
Fungal species	L	F_1	F_2	H
Beauveria bassiana (Bals.) Vuill.	11[a] (13)	62[a] (64)	51[a] (58)	5[a] (6)
Cladosporium cladosporioides (Fres.) de Vries	124[d] (53)	38[c] (21)	15[b] (3)	4[a] (2)
C. herbarum (Pers.) Link ex Gray	42[a] (4)	19[a] (10)	13[a] (2)	34[a] (27)
Mortierella alpina Peyronel	34[a] (10)	79[a] (19)	142[b] (17)	144[b] (28)
M. elongata Linnem.	5[a] (3)	19[a] (16)	11[a] (13)	25[a] (4)
M. exigua Linnem.	4 (3)	3 (0)	1 (2)	2 (3)
Penicillium raistrickii Smith	27[b] (12)	24[b] (8)	68[b] (70)	1[a] (2)
Penicillium spp. (14 forms)	12 (4)	12 (7)	14 (7)	11 (13)
Verticillium leptobactrum W. Gams	8[a] (3)	9[a] (6)	9[a] (7)	4[a] (4)
V. psalliotae Treschow	6[a] (3)	1[a] (2)	8[a] (7)	5[a] (4)
Sterile dark (9 forms)	21 (29)	9 (3)	2 (3)	11 (3)
Sterile hyaline (24 forms)	28 (9)	20 (13)	8 (2)	6 (3)
Acremonium strictum W. Gams	4 (3)	2 (3)	1 (2)	0
Phoma eupyrena Sacc.	11 (13)	3 (5)	1 (2)	0
Calcarisporium arbuscula Preuss	9 (4)	19 (18)	0	0
Coleophoma cylindrospora (Desm.) Hohn.	12 (12)	3 (5)	0	0
Phialophora sp. KS 88	1 (2)	11 (11)	0	0
Eleutheromyces subulatus (Fr.) Fckl.	17 (30)	0	0	0
Total species and forms isolated/layer	45	35	24	27

Values are means of the number of occurrences for each species/35 animals × 100. Where possible data analysed by one-way ANOVA with blocking. Values in each row followed by the same letter do not differ significantly ($P < 0·05$).

frequency of occurrence with depth, e.g. the occurrence of *C. cladosporioides* on Collembola from the L layer was significantly greater than on animals from the H layer, while the reverse was true for the occurrence of *M. alpina*. Many fungal taxa were restricted to animals extracted from one or two litter layers, and included not only sporulating species but also slow-growing non-sporulating dark and hyaline forms. No clamped Basidiomycete forms were isolated. Generally the pattern of fungal occurrences on *O. subtenuis* was similar to that previously recorded on L, F_1, F_2 and H layer organic material (Visser & Parkinson 1975). Two exceptions were *Beauveria bassiana* and *Trichoderma* spp. The former, an insect parasite, was common on the animals extracted from the F_1 layer (62% frequency of occurrence) but much less fequent in the F_1 layer itself (7% frequency of occurrence), while the latter were predominant in the H layer material (71% frequency of occurrence), but never isolated from the animals extracted from the H layer. Some species of *Trichoderma* have been shown to be antagonistic to such soil invertebrates as nematodes (Miller & Anagnostakis (1977) and hence may also be toxic to *O. subtenuis*. From this study it was concluded that *O. subtenuis* was capable of dispersing a wide variety of fungi throughout the LFH layers of the organic horizon.

Since faecal pellets are often assumed to be major sources of fungal inoculum, pellets from *O. subtenuis* extracted from each litter layer were plated on nutrient agar. Approximately 50% of these pellets yielded fungi. Van der Drift (1965) reported that, after feeding *Onychiurus quadrocellatus* on a sporulating colony of *Cladosporium* sp., hyphae in the pellets were non-viable and only 13% of the visually undamaged spores germinated compared with 83% germination for uningested spores. Ponge & Charpentie (1981) reported significant reductions in spore viability of a range of soil and litter fungi as a result of gut passage through Collembola. Digestion resulted in the disappearance of spore protoplasm and caused a considerable degree of physical damage to large-sized spores such as those of *Ulocladium consortiale*. Thus, it appears that faecal pellets, particularly those of micro-arthropods, may not be as important fungal dispersal agents as is transport via exterior body parts. Macro-invertebrates may cause less damage to fungal propagules, and therefore their faeces may exhibit greater fungal viability. For example, viable spores of the Endogonaceae have been reported from earthworm casts, leading to the suggestion that earthworm activity could have a significant impact on the distribution of vesicular-arbuscular mycorrhizal fungi within the soil profile (McIlveen & Cole 1976).

Although a wide variety of micro-organisms have been found associated with soil-dwelling invertebrates, evidence that the associated organisms become established in habitats different from those where the propagules were gathered is only available from sterile systems. Earthworms were observed to increase significantly the rate of spread of *Cephalosporium*, *Trichoderma viride*, *Rhizopus nigricans* and *Penicillium spinulosum* in autoclaved garden soil (Hutchinson & Kamel 1956), and Witkamp (1960) found that growth of fungal mycelium from unsterilized forest soil into sterilized forest soil was stimulated by the activities of oribatid mites. Bacteria are particularly successful after introduction of their faunal carriers into sterilized and inoculated litter systems (Parkinson, Visser & Whittaker 1979).

The dissemination and establishment of fungi and bacteria by Collembola into the rhizosphere of cotton plants growing in sterilized soil has also been shown (Wiggins & Curl 1979). *Aspergillus flavus*, *Trichoderma harzianum*, *Fusarium oxysporum f. sp. vasinfectum* readily colonized rhizosphere soil and root tissue after the introduction of Collembola reared on these particular organisms. Bacteria were readily transported into aseptic systems to which field-collected Collembola were introduced.

Soil-dwelling invertebrates are often found associated with the fruit-bodies of ectomycorrhizal fungi; however, their importance in the dissemination of mycorrhizal propagules and the establishment of the transported fungi on susceptible host roots has not been determined.

DISPERSAL AND MICROBIAL COMMUNITY STRUCTURE

The establishment of the transported propagules in non-sterile conditions requires that the propagules must germinate, colonize and exploit a resource in the presence of competition from the indigenous microflora. Therefore, before an organism can successfully colonize a substrate which it has the metabolic potential to exploit, its inoculum potential and competitive saprophytic ability must be high enough to overcome this competition unless these same features of the indigenous microorganisms are reduced by faunal activity or by adverse microclimatic conditions. Through their comminution and grazing activities, fauna may not only reduce competition from micro-organisms they graze upon, but may at the same time provide uncolonized microhabitats for colonization by propagules carried on their body.

Many of the organisms frequently associated with soil invertebrates have the capacity to grow rapidly and sporulate profusely. This group of organisms, consisting of the bacteria and such fungi as *Penicillium* spp., *Cladosporium* spp., *Mortierella* spp. and *Trichoderma* spp., exhibit characteristics which resemble those of an opportunist or *r*-selected group of organisms (Pianka 1970). It is possible that these micro-organisms are dependent to a large degree on the soil fauna for their persistence in the litter–soil–root system. The fauna not only provide them with transport, but also with new areas and substrates for rapid colonization (e.g. grazed patches, faeces, surfaces opened by comminution). Some of the fungal species regularly found associated with the body parts of the fauna, are also highly preferred faunal food sources (e.g. *Cladosporium* spp.), consequently there may be a mutualistic relationship between animal and fungus in this case. Swift (1976) felt that fungi such as mucoraceous species could effectively avoid competition by having a life cycle based on a repeated pattern of short time spans between spore and spore. He postulated that such organisms would benefit from the constant opening of 'pioneer' microhabitats by the soil animals.

Organisms capable of colonizing freshly opened resource areas would presumably be those with the highest inoculum potential on or in the vicinity of the soil animal, e.g. the bacteria and fungi mentioned previously. Presumably both groups of micro-organisms would rapidly exploit new resources, with the bacteria dominating if extensive water films are present. If microbial propagules associated with soil fauna are predominantly those which exhibit rapid germination and growth, this would provide a mechanism by which nutrients made available by the fauna could be quickly immobilized rather than be lost through leaching. Caution should be exercised when assuming that it is only the heavy sporulating fungi which are most frequently transported by soil fauna because much of the available data is based on isolations made on synthetic media which select for fast growing, heavily sporulating fungi (i.e. *r*-selected organisms). It should be kept in mind that many of the mycorrhizal fungi, due to their specific growth requirements, cannot be isolated, hence the role of the fauna in their dispersal may be considerably underestimated. This applies particularly to the vesicular-arbuscular mycorrhizal fungi.

Soil invertebrates may alter fungal community structure by increasing the heterogeneity of microhabitats for colonization by the exposure of previously uncolonized microhabitats and the production of faecal pellets. It is frequently overlooked that the soil fauna donate substantial amounts of chitin and other substrates to litter–soil systems through exuviae and dead tissues, thus increasing resource heterogeneity. Cornaby (1973, in Seastedt & Tate 1981) estimated that there were approximately 984 mg m^{-2} of arthropod exoskeletons in a southern Appalachian pine ecosystem, but since this was based on hand sorting of litter it was believed to be an underestimate. Most of the data on fungi associated with animal debris have come from incidental observations, e.g. Warcup (1965) observed that the fungi included animal parasites (*Conidiobolus*, *Entomophthora*, and *Beauveria*) and many saprophytes (*Absidia*, *Cunninghamella*, *Mucor*, *Penicillium*, *Aspergillus*). Among fungi commonly transmitted by micro-arthropods are *Mortierella* spp., many of which are highly chitinolytic, and their frequent presence on animals may be one mechanism by which they can successfully compete for chitin substrates when the animal moults or dies. A similar suggestion has been made by Pherson & Beattie (1979). *Mortierella* spp. may also be instrumental in the turnover of chitin in dead fungal mycelium and may aid in the decay of hyphal fragments in faecal pellets. Insect parasites such as *Beauveria bassiana* are also frequently isolated from the exterior parts of soil fauna and if their inoculum potential is high enough they may successfully infect the animal. Thus, the distribution of fungi such as these within the soil profile is determined by the distribution of their hosts.

Through their dispersal activities, soil animals may affect fungal successional sequences on both roots and decaying litter. Mixing of microbial inoculum may ensure that species found in later stages of succession are redistributed to microhabitats in earlier stages of decay.

Members of the soil fauna move up and down the litter profile in response to changes in moisture conditions (Springett, Brittain & Springett 1970; Metz 1971; Whitford *et al.* 1981; Hassall, Visser & Parkinson 1983). Thus, they may introduce late successional organisms into areas where competition by indigenous mycoflora has been reduced by microclimatic effects (e.g. prolonged drought in the L layer) or where such a mycoflora has not yet become established. In this way the microbial successional sequence can be advanced and decay rates altered. Similarly the colonization of new short conifer roots may be affected by fungal inoculum introduced by micro-arthropods which previously had been in contact with the fruit-bodies of potential fungal symbionts.

The number of species within fungal communities may be increased as a result of faunal grazing and dispersal activities depending on the grazing behaviour of the animal and the colonizing ability of the organism being grazed. Patchy grazing of a preferred fungus by soil animals could reduce the inoculum potential and hence colonizing ability of the fungus long enough to allow species dispersed by the grazer to colonize the grazed patch. This could result in a short-term increase in species number. If the growth rate of a preferred fungus allows it to tolerate grazing without detrimental effects to its colonizing ability (and, perhaps, even being stimulated due to relief of biostasis), then the dispersed fungi may have little chance of establishing themselves. Therefore, grazing pressure and the competitive ability and growth rate of a preferred food source will determine if species number is altered as a result of establishment of disseminated propagules. Species number may also be increased as a result of comminution and dispersal if the transported propagules are able to colonize newly opened microhabitats. Consequently, it is possible that the comminution, grazing and dispersal activities of the soil-dwelling fauna results in more dynamic microbial communities whose composition is constantly varied in both space and time.

CONCLUSIONS

It is clear from the body of data assembled to date that there are a variety of mechanisms by which soil fauna can influence microbial communities. Comminution and channelling may reduce fungal species richness and divert fungal successional patterns on decaying plant residues. This appears to be a result of the sensitivity of specific fungi to destruction of their thallus and microhabitat, thereby giving a competitive advantage to fast-growing, short life-cycle species commonly associated with the invertebrate body (i.e. bacteria, *Mortierella* spp., *Penicillium* spp.). Grazing by the mesofauna on selected fungi (which may occur during periods of high fungal activity or in systems where hyphal density is high) may have a significant effect on fungal distribution patterns as exemplified by the field studies of Newell (1984a,b), or may deleteriously affect plant growth by reducing the effectiveness of the mycorrhizal symbiosis (Warnock *et al.* 1982). The degree of impact of the fauna on their microbial food sources is probably dependent not only on the grazing pressure of the invertebrate (Hanlon & Anderson 1979), but also on the growth rate of the organism being grazed. More studies are needed to determine if the sensitivity of various micro-organisms (particularly fungi) to a range of grazing pressures is related to the growth rate of the micro-organisms.

Perhaps the most important function of the soil invertebrates is that of dispersal of microbial propagules. It is now well established that soil fauna have associated with them a wide variety of organisms; however, the success of these organisms in establishing themselves in microhabitats differing from those they originally occupied remains undetermined.

Much of the research elucidating the relationship between fauna and the micro-flora has been restricted to laboratory experiments where extended confinement of experimental animals under optimum temperature and moisture conditions can result in observations which bear little relevance to the field situation. Microcosm studies should be substantiated by field research if we are to gain a more thorough understanding of how fauna affect microbial communities in the presence of vary-ing abiotic factors.

Numerous studies have been performed dealing with the distribution of micro-organisms such as fungi on decaying litter, soil and plant roots, and many inferences have been made regarding the impact of the soil fauna on these distributions. It is only in the last decade that attempts have been made to unravel some of the complex interactions occurring between the wide variety of soil fauna and micro-flora in the below-ground system. A more precise view of the faunal impact on micro-organisms will depend on separating the effects of the soil fauna from the effects of other factors. This formidable task will demand the combined expertise of soil zoologists and microbiologists.

ACKNOWLEDGMENTS

I am sincerely grateful to Dennis Parkinson for his support and advice during the preparation of this manuscript. This work was funded by a Natural Sciences and Engineering Research Council of Canada research operating grant (No. A2257).

REFERENCES

Addison, J.A. & Parkinson, D. (1978). Influence of collembolan feeding activities on soil metabolism at a high arctic site. *Oikos*, 30, 529–538.

Anderson, J.M. (1975). Succession, diversity and trophic relationships of some soil animals in decomposing leaf litter. *Journal of Animal Ecology*, 44, 475–495.

Anderson, J.M. (1978). Competition between two unrelated species of soil Cryptostigmata (Acari) in experimental microcosms. *Journal of Animal Ecology*, 47, 787–803.

Anderson, J.M. & Bignell, D.E. (1980). Bacteria in the food, gut contents and faeces of the litter-feeding millipede *Glomeris marginata* (Villers). *Soil Biology and Biochemistry*, 12, 251–254.

Anderson, J.M. & Healey, I.N. (1972). Seasonal and inter-specific variation in major components of the gut contents of some woodland Collembola. *Journal of Animal Ecology*, 41, 359–368.

Anderson, J.M., Ineson, P. & Huish, S.A. (1983). Nitrogen and cation mobilization by soil fauna feeding on leaf litter and soil organic matter from deciduous woodlands. *Soil Biology and Biochemistry*, 15, 463–467.

Bärlocher, F. (1980). Leaf-eating invertebrates as competitors of aquatic hyphomycetes. *Oecologia*, 47, 303–306.

Behan, V.M. & Hill, S.B. (1978). Feeding habits and spore dispersal of Oribatid mites in the North American arctic. *Révue d'Écologie et de Biologie du Sol*, 15, 497–516.

Behan-Pelletier, V.M. & Hill, S.B. (1983). Feeding habits of sixteen species of Oribatei (Acari) from an acid peat bog, Glenamoy, Ireland. *Révue d'Écologie et de Biologie du Sol*, 20, 221–267.

Bengtson, G. & Rundgren, S. (1983). Respiration and growth of a fungus, *Mortierella isabellina*, in response to grazing by *Onychiurus armatus* (Collembola). *Soil Biology and Biochemistry*, 15, 469–473.

Bhaumik, H.D. & Clark, F.E. (1947). Soil moisture tension and microbial activity. *Soil Science Society of America Proceedings*, 12, 234–238.

Binns, E.S. (1980). Field and laboratory observations on the substrates of the mushroom fungus gnat *Lycoriella auripila* (Diptera: Sciaridae). *Annals of Applied Biology*, 96, 143–152.

Boddy, L., Coates, D. & Rayner, A.D.M. (1983). Attraction of fungus gnats to zones of intraspecific antagonism on agar plates. *Transactions of the British Mycological Society*, 81, 149–151.

Bocock, K.L. (1963). The digestion and assimilation of food by *Glomeris*. *Soil Organisms* (Ed. by J. Doekson & J. van der Drift), pp. 85–91. North Holland Publishing, Amsterdam.

Brown, B.A., Swift, B.L. & Mitchell, M.J. (1978). Effect of *Oniscus asellus* feeding on bacterial and nematode populations of sewage sludge. *Oikos*, 30, 90–94.

Burges, A. & Raw, F. (1967). *Soil Biology*. Academic Press, London.

Buxton, P.A. (1960). British Diptera associated with fungi, III. Flies of all families reared from 150 species of fungi. *Entomologists' Monthly Magazine*, 96, 61–94.

Christen, A.A. (1975). Some fungi associated with Collembola. *Révue d'Écologie et de Biologie du Sol*, 12, 723–728.

Clarholm, M. (1981). Protozoan grazing of bacteria in soil. *Microbial Ecology*, 7, 343–350.

Cornaby, B.W. (1973). *Population parameters and systems models of litter fauna in a White pine ecosystem*. Dissertation. University of Georgia, Athens, Georgia, USA.

Crossley, D.A. (1977). The roles of terrestrial saprophagous arthropods in forest soils: Current status of concepts. *The Role of Arthropods in Forest Ecosystems* (Ed. by W.J. Mattson), pp. 49–56. Springer-Verlag, New York.

Dash, M.C. & Cragg, J.B. (1972). Selection of microfungi by Enchytraeidae (Oligochaeta) and other members of the soil fauna. *Pedobiologia*, 12, 282–286.

Dickinson, C.H. & Parkinson, D. (1970). Effects of mechanical shaking and water tension on survival and distribution of fungal inoculum in glass microbead media. *Canadian Journal of Microbiology*, 16, 549–552.

Frankland, J.C. (1981). Mechanisms in fungal successions. *The Fungal Community. Its Organization and Role in the Ecosystem* (Ed. by D.T. Wicklow & G.C. Carroll), pp. 403–426. Marcel Dekker, Inc. New York.

Garrett, S.D. (1970). *Pathogenic Root-Infecting Fungi*. Cambridge University Press, Cambridge.

Gilmore, S.K. & Raffensperger, E.M. (1970). Foods ingested by *Tomocerus* spp. (Collembola, Entomobryidae), in relation to habitat. *Pedobiologia*, 10, 135–140.

Griffin, D.M. (1963a). Soil physical factors and the ecology of fungi, I. Behaviour of *Curvularia ramosa* at small soil water suctions. *Transactions of the British Mycological Society*, 46, 273–280.

Griffin, D.M. (1963b). Soil physical factors and the ecology of fungi, II. Behaviour of *Pythium ultimum* at small soil water suctions. *Transactions of the British Mycological Society*, 46, 368–372.

Griffin, D.M. (1969). Soil water in the ecology of fungi. *Annual Review of Phytopathology*, 7, 289–310.

Habte, M. & Alexander, M. (1975). Protozoa as agents responsible for the decline of *Xanthomomas campestris* in soil. *Applied Microbiology*, 29, 159–164.

Hanlon, R.D.G. (1981a). Influence of grazing by Collembola on the activity of senescent fungal colonies grown on media of different nutrient concentration. *Oikos*, 36, 362–367.

Hanlon, R.D.G. (1981b). Some factors influencing microbial growth on soil animal faeces, I. Bacterial and fungal growth on particulate oak leaf litter. *Pedobiologia*, 21, 257–263.

Hanlon, R.D.G. (1981c). Some factors influencing microbial growth on soil animal faeces, II. Bacterial and fungal growth on soil animal faeces. *Pedobiologia*, 21, 264–270.

Hanlon, R.D.G. & Anderson, J.M. (1979). The effects of Collembola grazing on microbial activity in decomposing leaf litter. *Oecologia*, 38, 93–99.

Hanlon, R.D.G. & Anderson, J.M. (1980). Influence of macroarthropod feeding activities on microflora in decomposing oak leaves. *Soil Biology and Biochemistry*, 12, 255–261.

Harding, D.J.L. & Stuttard, R.A. (1974). Microarthropods. *Biology of Plant Litter Decomposition.* Vol. 2 (Ed. by C.H. Dickinson & G.J.G. Pugh), pp. 489–532. Academic Press, London.

Hartenstein, R. (1962). Soil Oribatei, I. Feeding specificity among forest soil Oribatei (Acarina). *Annals of the Entomological Society of America*, **55**, 202–206.

Hassall, M., Visser, S. & Parkinson D. (1983). Vertical migration of *Onychiurus subtenuis* in relation to rainfall and microbial activity. Abstract. *New Trends in Soil Biology* (Ed. by Ph. Lebrun, H.M. Andre, A. De Medts, C. Gregoire-Wilbo & G. Wauthy), p. 612. Proceedings of the VIII. International Colloquium of Soil Zoology, Dieu-Brichart, Louvain-la-Neuve, Belgium.

Hägvar, S. & Kjondal, B.R. (1981). Succession, diversity and feeding habits of microarthropods in decomposing birch leaves. *Pedobiologia*, **22**, 385–408.

Hering, T.F. (1965). Succession of fungi in the litter of a Lake District oakwood. *Transactions of the British Mycological Society*, **48**, 391–408.

Hutchinson, S.A. & Kamel, M. (1956). The effect of earthworms on the dispersal of soil fungi. *Journal of Soil Science*, **7**, 213–218.

Ingold, C.T. (1971). *Fungal Spores: their liberation and dispersal.* Clarendon Press, Oxford.

Jacot, A.P. (1930). Moss-mites as spore-bearers. *Mycologia*, **22**, 94–95.

Kevan, D.K. McE. (1962). *Soil Animals.* Witherby Ltd, London.

Knight, C.B. & Angel, R.A. (1967). A preliminary study of the dietary requirements of *Tomocerus* (Collembola). *The America Midland Naturalist*, **77**, 511–517.

Kubiena, W.L. (1938). *Micropedology.* College Press Inc., Ames, Iowa.

Lubchenco, J. (1978). Plant species diversity in a marine intertidal community: importance of herbivore food preference and algal competitive abilities. *The American Naturalist*, **112**, 23–39.

Lubchenco, J. (1983). *Littorina* and *Fucus*: effects of herbivores, substratum heterogeneity, and plant escapes during succession. *Ecology*, **64**, 1116–1123.

Lussenhop, J., Kumar, R., Wicklow, D.T. & Lloyd, J.E. (1980). Insect effects on bacteria and fungi in cattle dung. *Oikos*, **34**, 54–58.

Luxton, M. (1972). Studies on the oribatid mites of a Danish beech-wood soil, I. Nutritional biology. *Pedobiologia*, **12**, 434–463.

Malajczuk, N. & Cromack, Jr. K. (1982). Accumulation of calcium oxalate in the mantle of ectomycorrhizal roots of *Pinus radiata* and *Eucalyptus marginata*. *New Phytologist*, **92**, 527–531.

MacNamara, C. (1924). The food of Collembola. *The Canadian Entomologist*, **56**, 99–105.

McIlveen, W.D. & Cole, H. Jr. (1976). Spore dispersal of Endogonaceae by worms, ants, wasps and birds. *Canadian Journal of Botany*, **54**, 1486–1489.

McMillan, J.H. (1976). Laboratory observations on the food preference of *Onychiurus armatus* (Tullb.) Gisin (Collembola, Family Onychiuridae). *Révue d'Écologie et de Biologie du Sol*, **13**, 353–364.

Metz, L.J. (1971). Vertical movement of Acarina under moisture gradients. *Pedobiologia*, **11**, 262–268.

Miller, P.M. & Anagnostakis, S. (1977). Suppression of *Pratylenchus penetrans* and *Tylenchorhynchus dubius* by *Trichoderma viride*. *Journal of Nematology*, **9**, 182–183.

Mills, J.T. & Sinha, R.N. (1971). Interactions between a springtail, *Hypogastrura tullbergi*, and soil-borne fungi. *Journal of Economic Entomology*, **64**, 398–401.

Mitchell, M.J. & Parkinson, D. (1976). Fungal feeding of oribatid mites (Acari: Cryptostigmata) in an aspen woodland soil. *Ecology*, **57**, 302–312.

Newell, K. (1984a). Interaction between two decomposer Basidiomycetes and a collembolan under Sitka spruce: distribution, abundance and selective grazing. *Soil Biology and Biochemistry*, **16**, 227–233.

Newell, K. (1984b). Interaction between two decomposer Basidiomycetes and a collembolan under Sitka spruce: grazing and its potential effects on fungal distribution and litter decomposition. *Soil Biology and Biochemistry*, **16**, 235–239.

Nicholson, P.B., Bocock, K.L. & Heal, O.W. (1966). Studies on the decomposition of the faecal pellets of a millipede (*Glomeris marginata* (Villers)). *Journal of Ecology*, **54**, 755–766.

Overgaard Nielsen, C. (1967). Nematoda. *Soil Biology* (Ed. by A. Burges & F. Raw), pp. 197–211. Academic Press, London.

Parkinson, D., Visser, S. & Whittaker, J.B. (1979). Effects of Collembolan grazing on fungal colonization of leaf litter. *Soil Biology and Biochemistry*, **11**, 529–535.

Parle, J.N. (1963a). Micro-organisms in the intestines of earthworms. *Journal of General Microbiology*, **31**, 1–11.

Parle, J.N. (1963b). A microbiological study of earthworm casts. *Journal of General Microbiology*, **31**, 13–22.

Parr, J.F., Parkinson, D. & Norman, A.G. (1967). Growth and activity of soil microorganisms in glass microbeads, II. Oxygen uptake and direct observations. *Soil Science*, **103**, 303–310.

Persson, T., Booth, E., Clarholm, M., Lundkvist, H., Soderstrom, B.E. & Sohlenius, B. (1980). Trophic structure, biomass dynamics and carbon metabolism of soil organisms in a Scots Pine forest. *Structure and Function of Northern Coniferous Forests—An Ecosystem Study* (Ed by T. Persson). *Ecological Bulletins (Stockholm*, **32**, 419–459.

Petersen, H. (1971). The nutritional biology of Collembola and its ecological significance. *Entomologiske Meddelelser* **39**, 97–118.

Petersen, H. & Luxton, M. (1982). A comparative analysis of soil fauna populations and their role in decomposition processes. *Quantitative Ecology of Microfungi and Animals in Soil and Litter* (Ed. by H. Petersen), *Oikos*, **39**, 287–388.

Pherson, D.A. & Beattie, A.J. (1979). Fungal loads of invertebrates in beech leaf litter. *Révue d'Écologie et de Biologie du Sol*, **16**, 325–335.

Pianka, E.R. (1970). On *r*- and *K*-selection. *The American Naturalist*, **104**, 592–597.

Ponge, J.F. & Charpentie, M.-J. (1981). Etude des relations microflore–microfaune: experiences sur *Pseudosinella alba* (Packard), Collembole mycophage. *Révue d'Écologie et de Biologie du Sol*, **18**, 291–303.

Poole, T.B. (1959). Studies on the food of Collembola in a Douglas fir plantation. *Proceedings of the Zoological Society of London*, **132**, 71–82.

Reichle, D.E. (1977). The role of soil invertebrates in nutrient cycling. *Soil Organisms as Components of Ecosystems*. Proceedings 6th International Colloquium of Soil Zoology. *Ecological Bulletin (Stockholm)*, **25**, 145–156.

Reyes, V.G. & Tiedje, J.M. (1976). Metabolism of C^{14}-labelled plant materials by woodlice (*Tracheoniscus rathkei* Brandt) and soil micro-organisms. *Soil Biology and Biochemistry*, **8**, 103–108.

Richter, K.O. (1980). Evolutionary aspects of mycophagy in *Ariolimax columbianus* and other slugs. *Soil Biology as Related to Land Use Practices* (Ed. by D.L. Dindal), pp. 616–636. Proceedings of the VII International Colloquium of Soil Zoology, Office of Pesticide and Toxic Substances, EPA, Washington, D.C.

Riffle, J.W. (1967). Effect of an *Aphelenchoides* species on the growth of a mycorrhizal and a pseudomycorrhizal fungus. *Phytopathology*, **57**, 541–544.

Riffle, J.W. (1971). Effect of nematodes on root-inhabiting fungi. *Mycorrhizae. Proceedings of the first North American Conference on Mycorrhizae* (Ed. by E. Hacskaylo), pp. 97–113. Misc. Publication 1189, USDA, Forest Service, Washington, D.C.

Roth, L.M. (1961). A study of the odoriferous glands of *Scaptocoris divergens* (Hemiptera, Cydnidae). *Annals of the Entomological Society of America*, **54**, 900–911.

Salawu, E.O. & Estey, R.H. (1979). Observations on the relationships between a vesicular-arbuscular fungus, a fungivorous nematode, and the growth of soybeans. *Phytoprotection*, **60**, 99–102.

Schildknecht, H. & Koob, K. (1971). Myrmicacin, the first insect herbicide. *Angewandte Chemie (International English Edition)* **10**, 124–125.

Seastedt, T.R. (1984). The role of microarthropods in decomposition and mineralization processes. *Annual Review of Entomology*, **29**, 25–46.

Seastedt, T.R. & Tate, C.M. (1981). Decomposition rates and nutrient contents of arthropod remains in forest litter. *Ecology*, **62**, 13–19.

Shafer, S.R., Rhodes, L.H. & Riedel, R.M. (1981). In-vitro parasitism of endomycorhizal fungi of ericaceous plants by the mycophagous nematode *Aphelenchoides bicaudatus*. *Mycologia*, **73**, 141–149.

Springett, J.A., Brittain, J.E. & Springett, B.P. (1970). Vertical movement of Enchytraeidae (Oligochaeta) in woodland soils. *Oikos*, **21**, 16–21.

Standen, V. (1978). The influence of soil fauna on decomposition by microorganisms in blanket bog litter. *Journal of Animal Ecology*, **47**, 25–38.

Stotzky, G. & Norman, A.G. (1961). Factors limiting microbial activities in soil. *Arkiv für Mikrobiologie*, **40**, 341–369.

Stout, J.D. (1974). Protozoa. *Biology of Plant Litter Decomposition*. Vol. 2 (Ed. by C.H. Dickinson & G.J.F. Pugh), pp. 385–420. Academic Press, London.

Stout, J.D. & Heal, O.W. (1967). Protozoa. *Soil Biology* (Ed. by A. Burges & F. Raw), pp. 149–195. Academic Press, London.

Sutherland, J.R. & Fortin, J.A. (1968). Effect of the nematode *Aphelenchus avenae* on some ectotrophic, mycorrhizal fungi and on a red pine mycorrhizal relationship. *Phytopathology*, **58**, 519–523.

Swift, M.J. (1976). Species diversity and the structure of microbial communities in terrestrial habitats. *The Role of Terrestrial and Aquatic Organisms in Decomposition Processes* (Ed. by J.M. Anderson & A. Macfadyen), pp. 185–222. Proceedings of the 17th Symposium of The British Ecological Society, Blackwell Scientific Publications, London.

Swift, M.J., Heal, O.W. & Anderson, J.M. (1979). *Decomposition in Terrestrial Ecosystems*. Studies in Ecology, Vol. 5, Blackwell Scientific Publications, London.

Swift, M.J. (1982). Basidiomycetes as components of forest ecosystems. *Decomposer Basidiomycetes, Their Biology and Ecology* (Ed. by J.C. Frankland, J.N. Hedger & M.J. Swift), pp. 307–333. Cambridge University Press, Cambridge.

Takeda, H. & Ichimura, T. (1983). Feeding attributes of four species of Collembola in a pine forest soil. *Pedobiologia* **25**, 373–381.

Talbot, P.H.B. (1952). Dispersal of fungus spores by small animals inhabiting wood and bark. *Transactions of the British Mycological Society*, **35**, 123–128.

Timonin, M.I. (1961). The interaction of plant, pathogen and *Scaptocoris talpa* Champ. *Canadian Journal of Botany*, **39**, 695–703.

Twinn, D.C. (1974). Nematodes. *Biology of Plant Litter Decomposition*. Vol. 2 (Ed. by C.H. Dickinson & G.J.F. Pugh), pp. 421–465. Academic Press, London.

van der Drift, J. (1965). The effects of animal activity in the litter layer. *Experimental Pedology* (Ed. by E.G. Hallsworth & D.V. Crawford), pp. 227–235. Butterworths, London.

Visser, S. & Parkinson, D. (1975). Fungal succession on aspen poplar leaf litter. *Canadian Journal of Botany*, **53**, 1640–1651.

Visser, S. & Whittaker, J.W. (1977). Feeding preferences for certain litter fungi by *Onychiurus subtenuis* Folsom. *Oikos*, **29**, 320–325.

Wallwork, J.A. (1970). *Ecology of Soil Animals*. McGraw-Hill Publishing Company Limited, London.

Warcup, J.H. (1965). Growth and reproduction of soil micro-organisms in relation to substrate. *Ecology of Soil-borne Plant Pathogens* (Ed. by K.F. Baker & W.C. Snyder), pp. 52–68. University of California Press, Berkeley, Los Angeles.

Warnock, A.J., Fitter, A.H. & Usher, M.B. (1982). The influence of a springtail *Folsomia candida* (Insecta, Collembola) on the mycorrhizal association of leek *Allium porrum* and the vesicular-arbuscular mycorrhizal endophyte *Glomus fasciculatum*. *New Phytologist*, **90**, 285–292.

Webb, D.P. (1977). Regulation of deciduous forest litter decomposition by soil arthropod faeces. *The Role of Arthropods in Forest Ecosystems* (Ed. by W.J. Mattson), pp. 57–69, Springer-Verlag, N.Y.

Weber, N.A. (1972). The Attines: the fungus-culturing ants. *American Scientist* **60**, 448–456.

Whitford, W.G., Freckman, D.W., Elkins, N.Z., Parker, L.W., Parmalee, R., Phillips, J. & Tucker, S. (1981). Diurnal migration and responses to simulated rainfall in desert soil microarthropods and nematodes. *Soil Biology and Biochemistry*, **13**, 417–425.

Wicklow, D.T. & Yocom, D.H. (1982). Effect of larval grazing by *Lycoriella mali* (Diptera: Sciaridae) on species abundance of coprophilous fungi. *Transactions of the British Mycological Society*, **78**, 29–32.

Wiggins, E.A. & Curl, E.A. (1979). Interactions of Collembola and microflora of cotton rhizosphere. *Phytopathology*, **69**, 244–249.

Witkamp, M. (1960). *Seasonal fluctuations of the fungus flora in mull and mor of an oak forest*. Instituut voor Toegepast Biologisch Onderzoek in de Nature. Arnhem, Netherlands. Medd. Nr. 46.

Interactions between soil micro-arthropods and endomycorrhizal associations of higher plants

R. D. FINLAY

Biology Department, University of York, Heslington, York, YO1 5DD

SUMMARY

1 Symbioses between plant roots and mycorrhizal fungi are almost universal in terrestrial plants and have been widely studied in recent years. The effects of these associations on plant growth and nutrient uptake are extensively documented but much less attention has been paid to the way such interactions are influenced by the soil fauna.

2 Existing studies of mycorrhiza–fauna interactions, which relate principally to nematodes and ectomycorrhiza, are briefly reviewed and new data presented concerning interactions between Collembola and vesicular-arbuscular mycorrhiza.

3 Evidence from both field and laboratory experiments now suggests that the yield and phosphorus uptake of infected plants may be influenced by populations of Collembola, and that Collembola population growth, in turn, is influenced by the presence of mycorrhizal hyphae.

4 It is suggested that reductions in the potential yield of mycorrhizal plants are related to grazing of the external mycelium and that such reductions may be induced by naturally occurring population densities of soil animals.

INTRODUCTION

One of the predominant concerns of many mycorrhizal studies has been the stimulatory effect of mycorrhizal fungi on the yield of infected plants, and it is now widely accepted that this phenomenon results from the provision of an increased, and better dispersed, absorbing surface for plant nutrients (Sanders & Sheikh 1983). In nutrient deficient soils roots are typically surrounded by depletion zones (Bieleski 1973; Bhat & Nye 1974) which limit the rate of supply of immobile nutrients, such as phosphate, to the plant root. The external mycelium of mycorrhizal plants extends beyond these depletion zones and is capable of translocating nutrients to the plant over ecologically significant distances (Gray & Gerdemann 1969; Rhodes & Gerdemann 1975). Quantitative aspects of this relationship have been investigated by Bieleski (1973) and Sanders & Tinker (1973) who have shown that the number of hyphal entry points and length of external mycelium are sufficient to account for observed increases in phosphate inflow to infected plants. The enhanced nutrient status of infected plants results in increased yield, improved

*Present address: Department of Botany, University of Sheffield, Sheffield S10 2TN.

319

resistance to a range of ecological stresses including drought (Safir, Boyer & Gerdemann 1972; Allen 1982; Nelsen & Safir 1982) and disease (Marx 1972), and
may alter the balance of interspecific plant competition (Fitter 1977; Hall 1978).

Much of the evidence for the beneficial effects of mycorrhizas on plant growth
has been based on laboratory experiments in which plants are grown in pots of
sterilized soil inoculated with mycorrhizal fungi. These experiments have been
conducted mainly by botanists and mycologists and, whilst possible interactions
with other microflora have often been acknowledged (Marx 1972), the absence of
soil fauna in these studies has been largely ignored.

The influence of soil animals on rates of decomposition and nutrient turnover
has been considered primarily in terms of substrate modification, dispersal of microbial inoculum and altered microbial competition as a consequence of selective
microbivory: selective or differential grazing of soil fungi has been demonstrated in
a number of studies involving nematodes (Sutherland & Fortin 1968), mites
(Mitchell & Parkinson 1976) and Collembola (McMillan 1976; Visser & Wittaker
1977; Addison & Parkinson 1978; Newell 1980). In many soils mycorrhizal
mycelium may constitute a large proportion of the microbial biomass (Hayman
1978) and thus represents a significant potential food source for soil fungivores.
Grazing activity which severs or removes mycorrhizal hyphae will be of particular
significance to plants in that it diminishes the effective absorbing surface of mycorrhizal root systems and should reduce the beneficial effects of the association.

SOIL FAUNA–MYCORRHIZA INTERACTIONS: A REVIEW

Whilst few studies have explicitly considered the mechanisms of interaction between soil fauna and mycorrhiza, a growing body of evidence suggests that such
interactions exist. Early studies document relationships between nematodes and
ectomycorrhizal fungi (ECM); more recent work has, in addition, considered
associations involving vesicular-arbuscular fungi (VAM) and Collembola.

Zak (1965, 1967) noted the ocurrence of root-feeding nematodes and aphids
on mycorrhizal roots of *Pseudotsuga menzesii* and Sutherland & Adams (1964,
1966), who noted the migratory, ectoparasitic existence of *Tylenchorhynchus
claytoni* (Steiner) and other nematodes on roots of *Pinus resinosa* (Ait.), suggested
that the animals may feed on mycorrhizal fungi. Clark (1964) presented circum-
stantial evidence that a *Deleadenus* species of nematode was responsible for non-
establishment of mycorrhizal *Rhododendron* cuttings and postulated direct feeding
on the mycelium.

Feeding by an *Aphelenchoides* species on laboratory cultures of *Suillus
granulatus* (L. ex Fries) Kuntze has been demonstrated by Riffle (1967). The
nematodes caused large reductions in colony diameter and destroyed up to 87% of
the cultures. *Aphelenchoides* species have also been shown to suppress mycorrhizal
associations between *Suillus granulatus* and *Pinus ponderosa* Laws and have been
implicated in the premature death of trees, following prolonged periods of drought
(Riffle 1967, 1975). The nematode *Aphelenchus avenae* Bastian has been shown to

feed on seven different species of ectomycorrhizal fungi grown *in vitro* (Sutherland & Fortin 1968). Significant reductions in colony diameter occurred in all cases and the nematode suppressed associations between *S. granulatus* and *Pinus resinosa*, causing a significant reduction in both the number of mycorrhizal root tips and the length of mycorrhizal root, although there was no significant effect on plant weight in the 5-week time span of the experiment.

More recently a number of studies have investigated the relationships between phytopathogenic nematodes and vesicular-arbuscular mycorrhiza (VAM). Contrasting types of interaction have been reported, including both stimulation (Schenck, Kinloch & Dickson 1975; Atilano, Menge & Van Gundy 1981), and depression (Fox & Spasoff 1972; Hussey & Roncadori 1978) of nematode populations by different mycorrhizal endophytes. Schenck, Kinloch & Dickson found an inverse relationship between numbers of mycorrhizal spores and nematode populations in the field, and better growth of nematode populations in pots with mycorrhizal plants than with uninfected plants, but the nature of this interaction was unclear. Reduced sporulation of VAM in response to population of *A. avenae* has also been reported by Salawu & Estey (1979) but the effects of nematode populations on the mycorrhizal stimulation of plant growth, in a range of studies, are not consistent. Suppression of mycorrhizal plant growth has been reported in experiments by Salawu & Estey (1979) and Atilano *et al.* (1981), but in other experiments nematodes only reduced mycorrhizal plant yields at artificially high densities (Hussey & Roncadori 1981) or had no effect at all (Hussey & Roncadori 1978). Clearly it is not possible to generalize about nematode—VAM interactions from observations of individual species combinations and further studies are required.

Despite the growing body of literature relating to selective fungal feeding in mites and Collembola (Usher, Booth & Sparkes 1982) there have been relatively few studies of interactions between these numerous animals and mycorrhizal fungi. Shaw (1983 and p. 333) describes some preliminary feeding choice experiments in which two species of Collembola were offered a range of ectomycorrhizal and saprophytic fungi. *Paxillus involutus* (Fr.) Fr. was significantly preferred to *Marasmius androsaceus* (L. ex Fr.) and both Collembola species significantly avoided *Heboloma crustiliniforme* (Bull. ex St. Amens) Quel., but the effects of grazing on plant growth and nutrient uptake have not yet been assessed. Warnock, Fitter & Usher (1982) examined the influence of the Collembolan *Folsomia candida* Willem var. *distincta* Bagnall on *Allium porrum* L. plants infected with the mycorrhizal endophyte *Glomus fasciculatum* (Thaxter) Gerdemann & Trappe. Plant growth was stimulated by mycorrhizal infection but in the presence of *F. candida* infected plants grew little better than uninfected plants. Phosphate inflow rates to mycorrhizal plants were significantly reduced in the presence of *F. candida* and the authors conclude that these effects are due to grazing of the external hyphae, rendering the mycorrhizal association ineffective. Interactions between VAM fungi and Collembola have been further investigated by Finlay (1983), and some aspects of these relationships are outlined in the following sections.

MATERIALS AND METHODS

Plant growth and mycorrhizal inoculations

Trifolium pratense L. and *Allium porrum* plants were grown in a (1:1 v/v) mixture of clay loam soil and sand which was steam-sterilized at $10\cdot34$ N cm^{-2} for 1 hour. The resulting mixture had an Olsen extractable P level of 13 μg g^{-1} and a pH $6\cdot2$. Seeds were surface sterilized in a 7% (w/v) sodium hypochlorite solution and germinated aseptically before being transplanted to 9 cm pots. Mycorrhizal inocula were applied as mixtures of sand, spores and infected root fragments and seedlings were infected by placing 10 g of inoculum below the newly emerging root. Plants were grown in growth chambers with a day temperature of 20°C and a night temperature of 15°C; artificial light was supplied for 16 h day^{-1} and at an irradiance of 40 W m^{-2}. Bacterial leachings from non-sterile soil were added to all pots and N was supplied at a rate equivalent to 50 kg ha^{-1}.

Collembola culture, extraction and gut content analysis

Two Collembola species, *Onychiurus ambulans* and *Folsomia candida*, were cultured in 5 cm Petri dishes containing a moist charcoal-plaster of Paris (1:9 v/v) substrate. Animals introduced to experimental pots all exceeded 1 mm in length. Collembola were extracted for counting by flotation in water and transferred to 70% ethanol prior to gut content analysis. Animals were cleared in 30% lactic acid at 90°C, the guts were dissected out using fine needles and placed in $0\cdot05$% lactophenol blue to stain fungal material. The gut wall was ruptured to allow penetration of the stain and squashed under a cover slip prior to microscopic examination.

Laboratory experiments

In four laboratory experiments different mycorrhizal and non-mycorrhizal treatments were combined factorially with the presence and absence of Collembola at different densities. The mean initial densities (dm^{-3}) for the experiments were as follows; Expt A, 33; Expt B, 50; Expt C, 66 and Expt D, 100. Details of the species combinations are summarized in Fig. 1 and full details are given in Finlay (1983). Plants and Collembola were extracted by gentle washing, root length was determined and the dry weights of roots and shoots were recorded. Where soil phosphate was supplemented it was supplied as NaH$_2$PO$_4$ at a rate equivalent to 50 kg ha^{-1}. Phosphate content of plant material was estimated by wet-ashing dried material in a triple acid digest (sulphuric/nitric/perchloric: Allen 1974) and developing the phospho-molybdate blue colour in the diluted extract using ascorbic acid.

Field experiment

The field site was a clay loam soil with a mean level of Olsen extractable P of 13 μg g^{-1} and had been fumigated with methyl bromide 10 months prior to the start of the experiment to remove indigenous mycorrhizal endophytes. A randomized block design with four treatments was chosen. All plots received 75 kg ha^{-1} N, 100 kg ha^{-1} K and 2 kg m^{-2} of *Glomus occultus* mycorrhizal inoculum. *Trifolium pratense* seed was drilled to a depth of 4 cm with a 15 cm spacing between rows and the site was divided into sixteen plots each measuring 4×4 m. Control plots received no further treatments. Insecticide-treated plots received 'chlorfenvinphos' (diethyl 1-(2′,d′-dichlorophenyl)-2-chlorovinyl phosphate) granules at two rates, the recommended normal application rate of 4·5 kg ha^{-1}, and a double rate of 9·0 kg ha^{-1}. Rainfall on the second day following application assisted penetration of the insecticide into the soil volume. Fungicide-treated plots received 'Benomyl' (1-butyl-carbomoyl)-2-(benzimidazol) carbamic acid, methyl ester) as a suspension of 50% wettable powder in 4 litres of water for each plot. The fungicide was applied as a soil drench at 1·1 kg ha^{-1} and plots not receiving benomyl received water at an equivalent rate (2·5×10^3 l ha^{-1}). Harvests were taken at intervals of approximately 20 days, shoot material being sampled from randomly placed 0·25 m^2 quadrats, and root material and soil animals being sampled from soil cores 8 cm in diameter and 15 cm in depth.

RESULTS

Collembola population responses to mycorrhizal endophytes

Final Collembola population densities and the species used in each of the laboratory experiments are displayed in Fig. 1. In three of the experiments (A, B and D)

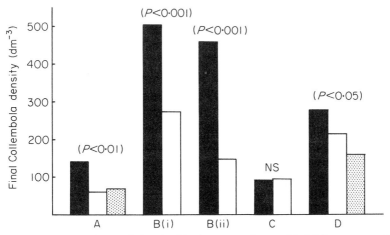

F IG . 1. Final mean Collembola population densities in a range of mycorrhizal (■), non-mycorrhizal (□) and non-mycorrhizal, phosphate supplemented (▨) treatments in four laboratory experiments. Significance levels refer to the effect of mycorrhiza in analyses of variance of ln transformed population data. The plant species was *Allium porrum* in all experiments except D where *Trifolium pratense* was used. The combinations of Collembola and mycorrhizal endophyte were as follows: (A) (*Onychiurus ambulans, Glomus tenue*), B (i) (*Folsomia candida, Glomus fasciculatum*), B (ii) (*O. ambulans, G. fasciculatum*), C (*O. ambulans, G. fasciculatum*), D (*O. ambulans, G. caledonium, G. clarum, G. monosporum, G. occultum*).

final population densities were significantly higher in mycorrhizal treatment than in non-mycorrhizal ones. In a fourth experiment (C) overall population growth was low and there was no significant difference between mycorrhizal and non-mycorrhizal treatments. A phosphate-supplemented, non-mycorrhizal treatment (P+) was included in experiments A and D to determine whether Collembola population growth was influenced by the larger root systems and enhanced P status of mycorrhizal plants. Phosphate fertilized plants had high shoot P levels and large root systems whose length and weight did not differ significantly from those of mycorrhizal plants, but final Collembola densities in P+ treatments did not differ significantly from non-mycorrhizal controls, indicating that the population growth response was not due to changes in the amount or P-status of root material. The root weight ratios of plants were not reduced by Collembola and microscopic examination of the root systems revealed no evidence of root grazing. In experiment A the endophyte *Glomus tenue* did not significantly increase root weight or length and the increased Collembola population growth in this and other experiments suggests that the animals were responding to the availability of mycorrhizal hyphae as food.

TABLE 1. Gut contents of *Onychiurus ambulans* from Experiment A

			Number containing:		
Treatment	Number sampled	fungal hyphae	fungal spores	organic matter	soil
M+ P−	15	10	2	12	15
M− P−	15	1	0	15	15

Collembola gut content analysis

Fifteen Collembola (with visible gut contents) from each of the mycorrhizal and non-mycorrhizal treatments in Experiment A were selected for gut content analysis. The presence or absence of fungal material, organic matter and soil was recorded and these results are displayed in Table 1. Both groups of animals contained organic matter but no fragments were clearly identifiable as root material. All animals with visible gut contents contained soil. Two-thirds of the animals from mycorrhizal pots appeared to have been feeding on fungal hyphae, whereas only one of the animals from the non-mycorrhizal treatments had hyphae in its gut. Two of the animals from mycorrhizal treatments had guts containing fungal spores and the large diameter of these (80–90 μm) suggests they may have been mycorrhizal in origin. Other, non-mycorrhizal, fungi may have been introduced with the soil leachings or the mycorrhizal inoculum but the low incidence of fungal material in the guts of animals from non-mycorrhizal treatments suggests that populations of these fungi were not large or that mycorrhizal hyphae were preferred as food.

Effects of Collembola on mycorrhizal associations

Laboratory experiments

Mycorrhizal infection significantly increased plant growth in all laboratory experiments, but, where final Collembola densities were in excess of 150 dm^{-3} (Expts A, B and D), mycorrhizal stimulation of plant shoot yield was consistently lowered by the presence of Collembola. This phenomenon is demonstrated in Fig. 2, where the significance levels refer to t tests between the means of mycorrhizal and non-mycorrhizal treatments at each Collembola density. Collembola had no significant effect on the shoot yield of non-mycorrhizal plants and no effect on the final levels of internal infection in mycorrhizal plants, but the root weight of plants in experiment B was significantly increased by both *O. ambulans* and *F. candida*.

At low animal densities, below 150 individuals dm^{-3} (Expt C), Collembola density had a markedly non-linear effect on the shoot weight, root length and total shoot P content of mycorrhizal plants. The variance due to Collembola density was partitioned into orthogonal polynomial components and significant ($P<0.01$) quadratic relationships were found in all cases, confirming the non-linear nature of the density effect. The effect is demonstrated for total shoot P in Fig. 3. The beneficial effects of infection were increased with the addition of low to intermediate densities of Collembola but at densities in excess of 100 individuals dm^{-3} these effects appear to decline again. The same trends were observed for shoot weight and root length, with maximum values of these parameters again occurring at an approximate density of 100 individuals dm^{-3}. In all cases non-mycorrhizal plants showed no response to Collembola density.

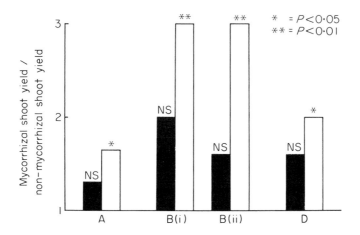

FIG. 2. Mycorrhizal stimulation of shoot yield in a range of laboratory experiments with (■) and without (□) Collembola. Significance levels refers to t tests between the means of ln transformed mycorrhizal and non-mycorrhizal treatments at each Collembola density. Species are given in Fig. 1. D refers to *G. caledonium* alone.

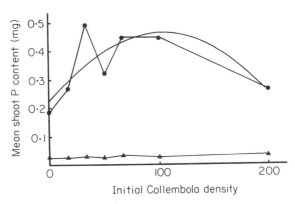

FIG. 3. The relationship between mean shoot P content and initial Collembola density for mycorrhizal (●) and non-mycorrhizal (▲) *Allium porrum* plants grown with seven densities of *Onychiurus ambulans*. The fitted quadratic relationship is shown for mycorrhizal plants.

Field experiment

Applications of the insecticide Chlorfenvinphos resulted in a significant ($P<0·01$), 80% reduction in the numbers of Collembola extracted compared with control plots. Final populations in treated plots did not exceed 10 individuals 100 cm^{-2} and there was no significant difference between the two rates of application. Numbers of animals in benomyl-treated plots were lower than those in control plots but the difference was not significant, suggesting that the fungicide is of negligible toxicity to Collembola.

Plants treated with benomyl had significantly ($P<0·05$) lower shoot yields and lower shoot phosphorus levels than control plants, supporting the suggestion by Bailey & Safir (1978) that benomyl adversely affects growth and nutrient uptake by mycorrhizal plants. Soil density was not measured and it is possible that soil compaction may have resulted from the toxicity of the fungicide to earthworms; however, there was no significant reduction in the root length of fungicide-treated plants and soil compaction does not seem a likely explanation of the reduced yield.

Although a non-mycorrhizal control was not available for comparison, the poor growth of fungicide-treated plants suggested that growth was limited by the low soil phosphate levels (13 μg g^{-1}) and that the *Glomus occultum* inoculum had had a stimulatory effect on plant growth.

Plants grown in chlorfenvinphos-treated plots, with low Collembola densities, showed significant increases in shoot weight ($P<0·001$) (Fig. 4), shoot P concentration ($P<0·01$) and shoot P ($P<0·001$), suggesting that the higher densities of Collembola in control plots may have had a deleterious effect on the growth of mycorrhizal plants. Collembola density had no apparent effect on root growth and no obvious plant pathogens were observed upon microscopic examination of the roots, which appeared intact and healthy. Soil P (Olsen's extraction) was measured at the second and fifth harvests using samples from the extracted soil cores. Whilst variation between blocks was significant, the pesticide treatments had no significant

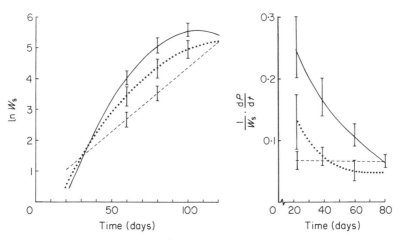

FIG. 4. Changes in (a) mean shoot weight m^{-2} (ln W_s) and (b) rate of shoot phosphate accumulation per unit shoot weight ($1/W_s \cdot dP/dt$) in *Trifolium pratense* plants grown in chlorfenvinphos treated (——), control ($\cdots$) and benomyl-treated (– – –) field plots. Bars indicate 95% confidence intervals.

effect on extractable soil P, and inclusion of the soil P level as a co-variate in analyses of co-variance did not influence the treatment effects demonstrated above in the analysis of variance. Since estimates of total root length were not available it was not possible to calculate specific uptake rates of phosphorus per unit root length; instead a crude measure of the efficiency of P uptake was calculated as the rate of shoot P accumulation per unit shoot weight. This measure of P accumulation was significantly higher in chlorfenvinphos-treated plants, suggesting a more efficient mycorrhizal association where Collembola densities were low (Fig. 4). Fragments of fungal hyphae were found in three-quarters of the animals whose gut contents were analysed and it is likely that a proportion of this fungal material was mycorrhizal in origin.

DISCUSSION

Soil micro-arthropods and mycorrhizal fungi constitute numerically important components of a wide range of soil ecosystems, and in cases where the mycorrhizal mycelium makes up a large proportion of the microbial biomass, there are likely to be animals that graze these hyphae. Mycorrhizal grazing has been demonstrated in Collembola (Warnock *et al.* 1982) and postulated in the Cryptostigmata (St John & Coleman 1983), but studies of these animals have concentrated on their general role in decomposition processes rather than on any direct influence on plant growth and nutrition.

The results of this study suggest that Collembola demonstrate increased population growth in response to the presence of mycorrhizal hyphae, and that the yield and phosphate uptake of mycorrhizal plants growing in phosphate deficient soil are influenced by different densities of Collembola.

In the laboratory experiments, Collembola populations did not respond to increases in the root system size or P-status of phosphate fertilized plants, but increased in response to the presence of mycorrhizal hyphae even when there was no associated change in root yield. Work by Booth (1979) and Booth & Anderson (1979) has shown that changes in the nitrogen levels of *Coriolus versicolor* and *Hypholoma fasciculare* can influence the fecundity of *Folsomia candida*, but in the experiments reported here nitrate was supplied to all treatments and it seems unlikely that nitrogen supply was limiting population growth.

The experimental results are in broad agreement with those of Warnock *et al.* (1982) in suggesting that grazing of the external mycelium may render the mycorrhizal association less efficient, but additionally suggest that the effects on plant yield and nutrient uptake depend upon the density of Collembola. At low grazing intensities mycorrhizal plant growth is stimulated; the reasons for this are not clear but possibilities include the mobilization of nutrients by animals, selective effects on other microbial populations or dispersal of microbial inoculum. Non-linear responses to grazing intensity have been reported in decomposer systems by Hanlon (1981) and it is possible that low grazing intensities may optimize growth of the external mycelium, whilst at higher Collembola densities the effects of fungal feeding outweigh this or any other beneficial effect.

In the field experiment, the increased yield and phosphate accumulation of plants grown in insecticide-treated plots with reduced Collembola densities were consistent with the effects demonstrated in the laboratory experiments. Interpretation of the results, however, is complicated by consideration of the possible side-effects of the pesticides and the absence of a non-mycorrhizal control in the experimental design, which shoud ideally have included a factorial combination of mycorrhizal and insecticide treatments.

Benomyl is known to reduce the yield and phosphate uptake of mycorrhizal plants (Bailey & Safir 1978; Boatman *et al.* 1978), whilst it may have a less pronounced effect on other rhizosphere fungi than non-systemic fungicides such as 'captan' (De Bertoldi *et al.* 1977). The reduced yield and phosphate accumulation of fungicide-treated plants suggest that soil phosphate was a major factor limiting plant growth and that growth in other plots had been stimulated by mycorrhizal infection. Whilst benomyl had little or no effect on Collembola density, Chlorfenvinphos caused large reductions in population size consistent with its reported toxicity to Collembola (Tomlin 1975). Chlorfenvinphos is toxic to a wide range of soil animals, including earthworms (Edwards, Thompson & Benyon 1968) and it is possible that large inputs of organic matter may have occurred following the death of soil animals. No change in the soil phosphate level was detected following chlorfenvinphos application however, and available evidence (Hayman, Macdonald & Spokes 1977) suggests that, in soils where P is not limiting, applications of chlorfenvinphos at the normal rate do not stimulate mycorrhizal plant growth whilst applications at higher rates actually decrease plant yield. Applications of the insecticide at normal and ten times normal rates produced an initial inhibition of mycorrhizal infection but infection levels were similar to those in control plants after 19 weeks.

Whilst pesticide-induced changes in microbial populations must not be ruled out as a possible source of influence on plant growth in the field experiment described here, the increased efficiency of phosphate uptake by insecticide treated plants during the initial stages of the experiment (Fig. 4) and the subsequent yield increases in plots with low Collembola densities, together with the apparent absence of root pathogens in control plots, suggest that natural populations of Collembola may limit the efficiency of mycorrhizal associations in the field. These observations are consistent with the beneficial effects of nematicides on mycorrhizal fungi, reported by Ocampo & Hayman (1978) and Bird, Rich & Glover (1974), suggesting that grazing may be of significance in the field, but the extent to which soil fauna–mycorrhiza interactions are purely localized phenomena, influenced by the overdispersion of animal populations and selective feeding on particular mycorrhizal fungi remains to be determined by further experiments.

The beneficial effects of mycorrhiza on plant growth, demonstrated in the laboratory, are frequently less easy to reproduce in the field and the effects of the soil fauna have been largely overlooked as a possible source of variation in these experiments. Whilst frequent reference has been made to the unreality of ignoring mycorrhizal endophytes when considering plant growth, this study emphasizes the need to consider the possible influence of soil animals associated with these endophytes.

ACKNOWLEDGMENTS

I thank Dr A.H. Fitter and Dr M.B. Usher for many helpful discussions. Dr F.E. Sanders and Dr D.S. Hayman kindly supplied the mycorrhizal inoculum and field-work facilities were made available by the University of Leeds Field Station. This work was supported by a Natural Environment Research Council studentship.

REFERENCES

Addison, J.A. & Parkinson, D. (1978). Influence of Collembolan feeding activities on soil metabolism at a high arctic site. *Oikos*, **30**, 529–538.

Allen, M.F. (1982). Influence of vesicular-arbuscular mycorrhizae on water movement through *Bouteloua gracilis* (H.B.K.) Lag Ex Steud. *New Phytologist*, **91**, 191–196.

Allen, S.E. (Ed.) (1974). *Chemical Analysis of Ecological Materials*. Blackwell Scientific Publications, Oxford.

Atilano, R.A., Menge, J.A. & Van Gundy, S.D. (1981). Interaction between *Meloidogyne arenaria* and *Glomus fasciculatus* in Grape. *Journal of Nematology*, **13**, 52–57.

Bailey, J.E. & Safir, G.R. (1978). The effect of Benomyl on Soybean endomycorrhizae. *Phytopathology*, **68**, 1810–1812.

Bhat, K.K.S. & Nye, P.H. (1978). Diffusion of phosphate to plant roots in soil, III. Depletion around onion roots without root hairs. *Plant and Soil*, **41**, 383.

Bieleski, R.L. (1973). Phosphate pools, phosphate transport and phosphate avavailability. *Annual Review of Plant Physiology*, **24**, 225–252.

Bird, G.W., Rich, J.R. & Glover, S.U. (1974). Increased endomycorrhizae of cotton roots in soil treated with nematicides. *Phytopathology*, **64**, 48–51.

Boatman, N., Paget, D., Hayman, D.S. & Mosse, B. (1978). Effects of systemic fungicides on vesicular-arbuscular mycorrhizal infection and plant phosphate uptake. *Transactions of the British Mycological Society*, **70**, 443–450.

Booth, R.G. (1979). *The nutritional importance of micro-organisms in the diet of soil micro-arthopods.* Ph.D. Thesis, University of Exeter.

Booth, R.G. & Anderson, J.M. (1979). The influence of fungal food quality on the growth and fecundity of *Folsomia candida*. *Oecologia*, **38**, 317–323.

Clark, W.C. (1964). Fungal-feeding nematodes as possible plant pathogens. *New Zealand Journal of Agricultural Research*, 7, 441–443.

De Bertoldi, M., Giovannetti, M., Griselli, M. & Rambelli, A. (1977). Effects of soil applications of benomyl and captan on the growth of onions and the occurrence of endophytic mycorrhizas and rhizosphere microbes. *Annals of Applied Biology*, 86, 111–115.

Edwards, C.A., Thompson, A.R. & Benyon, K.I. (1968). Some effects of chlorfenvinphos, an organophosphorus insecticide, on populations of soil animals. *Revue d'Ecologie et de Biologie du Sol*, 353–364.

Finlay, R.D. (1983). *Interactions between the endomycorrhizal associations of higher plants and soil-dwelling Collembola.* D.Phil. Thesis, University of York.

Fitter, A.H. (1977). Influence of mycorrhizal infection on competition for phosphorus and potassium by two grasses. *New Phytologist*, **79**, 119–125.

Fox, J.A. & Spasoff, L. (1972). Interaction of *Heterodera solanacearum* and *Endogone gigantea* on tobacco. *Journal of Nematology*, 4, 224–225.

Gray, L.E. & Gerdemann, J.W. (1969). Uptake of phosphorus 32 by vesicular-arbuscular mycorrhizae. *Plant and Soil*, 30, 415–422.

Hall, I.R. (1978). Effects of endomycorrhizas on the competitive ability of white clover. *New Zealand Journal of Agricultural Research*, 21, 509–515.

Hanlon, R.D.G. (1981). Influence of grazing by Collembola on the activity of senescent fungal colonies grown on media of different nutrient concentration. *Oikos*, **36**, 362–367.

Hayman, D.S. (1978). Endomycorrhizae. *Interactions Between Non-pathogenic Soil-organisms and Plants* (Ed. by Y.R. Dommergues & S.V. Krupa), pp. 401–442. Elsevier Scientific Publishing Company, Amsterdam.

Hayman, D.S., Macdonald, R.M. & Spokes, J.R. (1977). The effects of pesticides on V.A. mycorrhiza. *Annual Report of Rothamsted Experimental Station for 1977, Part I*, p. 238.

Hussey, R.S. & Roncadori, R.W. (1978). Interaction of *Pratylenchus brachyurus* and *Gigaspora margarita* endomycorrhizal fungus on cotton. *Journal of Nematology*, **10**, 16–20.

Hussey, R.S. & Roncadori, R.W. (1981). Influence of *Aphelenchus avenae* on vesicular-arbuscular endomycorrhizal growth response in cotton. *Journal of Nematology*, **13**, 48–52.

Marx, D.H. (1972). Ectomycorrhizae as biological deterrents to pathogenic root infections. *Annual Review of Phytopathology*, **11**, 171–195.

McMillan, J.H. (1976). Laboratory observations on the food preference of *Onychiurus armatus* (Tulb.) Gisin (Collembola, Family Onychiuridae). *Revue d'Ecologie et de Biologie du Sol*, **13**, 353–364.

Mitchell, M.J. & Parkinson, D. (1976). Fungal feeding of oribatid mites (Acari: Cryptostigmata) in an aspen woodland soil. *Ecology*, **57**, 303–312.

Nelsen, C.E. & Safir, G.R. (1982). Increased drought tolerance of mycorrhizal onion plants caused by improved phosphorus nutrition. *Planta*, **154**, 407–413.

Newell, K. (1980). *The effect of grazing by litter arthropods on the fungal colonization of leaf litter.* Ph.D. Thesis, University of Lancaster.

Ocampo, J.A. & Hayman, D.S. (1978). Effects of pesticides on VA mycorrhiza. Report of Rothamsted Experimental Station (1978) p. 236.

Rhodes, L.H. & Gerdemann, J.W. (1975). Phosphate uptake zones of mycorrhizal and non-mycorrhizal onions. *New Phytologist*, **75**, 555–561.

Riffle, J.W. (1967). Effect of an *Aphelenchoides* species on the growth of a mycorrhizal and a pseudomycorrhizal fungus. *Phytopathology*, **57**, 541–544.

Riffle, J.W. (1957). Two *Aphelenchoides* species suppress formation of *Suillus granulatus* ectomycorrhizae with *Pinus ponderosa* seedlings. *Plant Disease Reporter*, **59**, 951–955.

Safir, G.R., Boyer, J.S. & Gerdemann, J.W. (1972). Nutrient status and mycorrhizal enhancement of water transport on Soybean. *Plant Physiology*, **49**, 700–703.

Salawu, E.O. & Estey, R.H. (1979). Observations on the relationships between a vesicular-arbuscular fungus, a fungivorous nematode, and the growth of soybeans. *Phytoprotection*, **60**, 99–102.

Sanders, F.E. & Sheikh, N.A. (1983). The development of vesicular-arbuscular mycorrhizal infection in plant root systems. *Plant and Soil*, **71**, 233–246.

Sanders, F.E. & Tinker, P.B. (1973). Phosphate flow into mycorrhizal roots. *Pesticide Science*, **4**, 384–395.

Schenk, N.C., Kinloch, R.A. & Dickson, D.W. (1975). Interaction of endomycorrhizal fungi and root-knot nematode on soybean. *Endomycorrhizas* (Ed. by F.E. Sanders, B. Mosse & P.B. Tinker), pp. 607–617. Academic Press, London.

Shaw, P.J.A. (1983). Do Collembola graze mycorrhizas? Abstracts of comunications presented at the mycorrhiza group meeting, Lancaster University, 28–30 March, 1983 (Ed. by J. Dighton) *Merlewood Research & Development Paper No. 95*, Institute of Terrestrial Ecology, Grange-over-Sands, Cumbria.

St. John, T.V. & Coleman, D.C. (1983). The role of mycorrhizae in plant ecology. *Canadian Journal of Botany*, **61**, 1005–1014.

Sutherland, J.R. & Adams, R.E. (1964). The parasitism of red pine and other forest nursery crops by *Tylenchorhynchus claytoni* Steiner. *Nematologica*, **10**, 637–643.

Sutherland, J.R. & Adams, R.E. (1966). Population fluctuations of nematodes associated with red pine seedlings following chemical treatment of the soil. *Nematologica*, **12**, 122–128.

Sutherland, J.R. & Fortin, J.A. (1968). Effect of the nematode *Aphelenchus avenae* on some ectotrophic mycorrhizal fungi and on a red pine mycorrhizal relationship. *Phytopathology*, **58**, 519–523.

Tomlin, A.D. (1975). Toxicity of soil applications of insecticides to three species of springtails (Collembola) under laboratory conditions. *Canadian Entomologist*, **107**, 769–774.

Usher, M.B., Booth, R.G. & Sparkes, K.E. (1982). A review of progress in understanding the organization of communities of soil arthropods. *Pedobiologia*, **23**, 126–144.

Visser, S. & Whittaker, J.B. (1977). Feeding preferences for certain litter fungi by *Onychiurus subtenuis* (Collembola). *Oikos*, **29**, 320–325.

Warnock, A.J., Fitter, A.H. & Usher,M.B.)1982). The influence of a springtail *Folsomia candida* (Insecta, Collembola) on the mycorrhizal association of leek *Allium porrum* and the vesicular-arbuscular mycorrhizal endophyte *Glomus fasciculatum*. *New Phytologist*, **90**, 285–292.

Zak, B. (1965). Aphids feeding on mycorrhizae of Douglas Fir. *Forest Science*, **11**, 410–411.

Zak, B. (1967). A nematode (*Meloidodera* sp.) on Douglas Fir mycorrhizae. *Plant Disease Reporter*, **51**, 264.

Grazing preferences of *Onychiurus armatus* (Insecta: Collembola) for mycorrhizal and saprophytic fungi of pine plantations

P. J. A. SHAW

Department of Biology, University of York, YO1 5DD

INTRODUCTION

It is well established that mycorrhizal infection of plant roots is beneficial to the host plant (Harley 1969), generally due to an increased supply of phosphate (Mosse 1973). The fungal hyphae increase the volume of the depletion zone around the roots, thus increasing the labile phosphorus pool available to the plant (Tinker 1975). Clearly, it is of interest to investigate the factors affecting the size of the depletion zone, and one such factor could be arthropod grazing of the hyphae.

It has been shown that the presence of Collembola in a microcosm experiment reduced the benefits of endomycorrhizal infection to leeks (Warnock, Fitter & Usher 1982), apparently by grazing of the hyphae. Although endomycorrhizal fungi are the most important mycorrhizal symbionts of most plant taxa, the principal mycorrhizal associations of trees are the sheathing or ectomycorrhizas. No work has yet been published on the effects of arthropod grazing on ectomycorrhizal fungi, despite observations that aphids (Zak 1965) and protura (Nosek 1977) suck on ectomycorrhizal sheaths. This paper presents two preliminary lines of investigation relevant to this problem: (i) surveying of the Collembola and fungal communities in a series of pine stands, and (ii) choice experiments using field-collected Collembola and a range of isolates of fungi characteristic of pine forests. The aim of the field survey was to identify the dominant species of Collembola and fungi in the pine plantations, and that of the choice experiments was to determine whether a consistent hierarchy can be found in feeding preferences between fungi.

SITE DESCRIPTION AND METHODS

Fieldwork, based at Spadeadam Forest, Northumberland (National Grid reference NY620740), lasted from 13 September to 16 November 1983. The forest is at an altitude of approximately 300 m, and is mainly lodgepole pine (*Pinus contorta* Dougl.) and Sitka spruce (*Picea sitchensis* (Bong.)). The soil is an acid peat, pH 3.8 ± 0.3, with acetic acid extractable phosphorus (after Allen 1974) 12.6 ± 16.9 mg 100 g^{-1} (mean $\pm$SD, pooled results for seventy-five soil cores). Analyses were performed with the facilities and help of the Chemistry Section at ITE Merlewood.

All fieldwork was done in ten stands of *P. contorta*, representing two replicates of five ages (12, 15, 19, 23 and 26 years). In each stand, sampling was confined to a staked area 10 m square, at least 2 m in from the stand edge (except for one stand where this was not possible). The site was visited weekly, when three stands were sampled, so that each of the ten stands was visited three times during the sampling programme. During sampling, six soil cores (6 cm diameter × 9 cm deep) were taken midway between randomly chosen trees. A 50 cm quadrat was centred on the core location, and two sets of observations were made: the number of sporophores of each saprotrophic fungus in the quadrat, and the number of sporophores of each mycorrhizal fungus within 3 m along the planting ridge. Soil cores were returned to ITE Merlewood upright and intact, then divided into three layers, each 3 cm thick, prior to extraction of the Collembola in a portable high gradient extractor (Usher & Booth 1984).

The choice experiments were carried out using *Onychiurus armatus* (Tullberg) *s. lat.*, being the only dominant Collembola amenable to laboratory culture. The fungi came from stock cultures from ITE Merlewood except for *Marasmius androsaceus*, *Mycena galopus* and *Lactarius rufus* which were isolated from material from the field site during autumn 1983.

All experiments were carried out in 5 cm diameter pill boxes floored with 3 mm of 10:1 plaster-of-Paris:charcoal (Goto 1961). Three holes of 8 mm diameter were sunk in the floor, into which were fitted 8 mm cores of agar so as to be flush with the surface. Two of the agar cores were from cultures of the fungi to be tested, while the third was a sterile control. Twenty Collembola were introduced into each chamber, and the number apparently feeding on each fungus was recorded three times per day. The experiments were kept at 15°C and lasted up to 5 days. At the end of each experiment the number of Collembola alive and dead, the number of faecal pellets present on each agar disc, and the estimated area of each disc remaining ungrazed were counted or estimated.

RESULTS

The site held two dominant and three less common species of mycorrhizal fungi, of which *Lactarius rufus* was the commonest and most widely distributed, although it was recorded most frequently in the younger stands (see Table 1). *L. rufus* is also the only mycorrhizal fungus which has been isolated from Spadeadam material. This is not unexpected since mycorrhizal fungi are typically difficult to isolate (Chu-Chou & Grace 1982; J. Dighton, pers. comm.).

Four species of saprotrophic fungi were recorded, of which two were recorded only once. The commonest species was *Marasmius androsaceus*, while *Mycena galopus* was recorded occasionally. Preliminary results of the Collembola survey indicate that *Onychiurus armatus* (Tullberg) *s. lat.*, *Friesea mirabilis* (Tullberg), *Folsomia brevicauda* Agrell, and *Pseudisotoma sensibilis* (Tullberg) are the most frequent species.

TABLE 1. A list of the species of fungi recorded from a survey of *Pinus contorta* stands in Spadeadam Forest during Autumn 1983. The list contains the number of quadrats where each species was recorded (total number of quadrats is 180), the total number of sporophores counted, and the percentage found in each stand age

	Number of records	Number of sporophores	% of sporophores found in stand age (years)				
			12	15	19	23	26
Mycorrhizal species							
Lactarius rufus (Scop. ex Fr.)	56	222	67	15	10	8	< 1
Inocybe longicystis Fr.	33	88	8	36	2	43	10
Russula emetica (Sceaff. ex Fr.)	9	27	0	69	0	26	15
Laccaria laccata (Scop. ex Fr.)	6	15	7	87	7	0	0
Saprotrophic species							
Marasmius androsaceus (L. ex Fr.)	72	420	22	29	41	2	6
Mycena galopus (Pers. ex Fr.)	12	23	22	7	11	11	48

To date the choice experiments have been run using twelve species of fungi. Table 2 lists the preferences between the eight fungi most intensively studied. From the present results, *L. rufus* would appear to be the most highly preferred species.

DISCUSSION

The fungal community of the field site is depauperate compared to that found in a deciduous wood, but is typical of upland pine plantations on acidic, nutrient poor soils (J. Dighton, pers. comm.). Although the Collembola species composition is

TABLE 2. The relative preferences* shown by *Onychiurus armatus* for eight cultured fungi: three of these (*M. androsaceus, M. galopus* and *M. epipterygia*) are saprotrophic, the others are mycorrhizal

	Fungal species (first letters of generic and specific names)							
	Lr	Ma	Sl	Mg	Me	Pi	Rl	Hc
Lactarius rufus (Scop. ex Fr.)	0	+	+	+	+	+	+	+
Marasmius androsaceus (L. ex Fr.)	<	0	=	+	+	+	+	+
Suillus luteus (Fr.)	<	=	0	+	+	=	+	+
Mycena galopus (Pers. ex Fr.)	<	<	<	0	=	+	+	+
Mycena epipterygia (Scop. ex Fr.)	<	<	<	=	0	=	=	+
Paxillus involutus (Fr.) Fr.	<	<	<	<	=	0	+	=
Rhizopogon luteolus Fr.	<	<	=	<	=	<	0	=
Hebeloma crustiliniforme (Bull. ex St. Amans)	<	<	<	<	<	=	=	0

* In the table the vertical axis will be referred to as the *Y* axis, and the horizontal axis as the *X* axis. The following symbols are used:
0   not a comparison;
=   no significant preference between fungi;
+   *Y* axis fungus is preferred to *X* axis fungus;
<   *Y* axis fungus is avoided relative to *X* axis fungus.

only known approximately, it is comparable with other studies of upland Collembola communities (Usher 1967; Hale 1966).

It is well established that Collembola may exhibit significant preference betwen fungi (McMillan 1976; Visser & Whittaker 1977; Newell 1980), but no work published previously has shown that these preferences may be organized into a consistent hierarchy. The two most highly preferred fungi (*L. rufus* and *M. androsaceus*) are the two likely to have the highest hyphal densities in the field, assuming that hyphal density may be inferred from sporophore frequency.

The results presented here agree with Newell (1980), using *Onychiurus latus* Gisin, who found that *M. androsaceus* was significantly preferred to *M. galopus*, both in laboratory choice experiments and in the field. She also found that selective grazing by *O. latus* altered the outcome of competition between these fungi both in the laboratory and in the field. Parkinson, Visser & Whittaker (1979) also showed that selective grazing by Collembola of the genus *Onychiurus* could affect the outcome of competition between saprotrophic litter fungi. This suggests that a similar effect may be noted with mycorrhizal fungi. Certainly *L. rufus*, a mycorrhizal species highly preferred by *O. armatus*, was the dominant symbiont in stands up to 12 years old and declined in importance thereafter (Table 1). Succession is a well documented phenomenon in ectomycorrhizas (Marks & Foster 1967; Dighton & Mason, in press) but no work published so far has suggested the possibility that arthropod grazing may be involved.

The pronounced preference shown by *O. armatus* and *L. rufus* hyphae combined with the wide distribution of both species in the field suggests that mycorrhizal grazing is most likely to occur to some extent. Any such grazing will reduce the mean length of the mycorrhizal hyphae, thus the volume of soil available to the root system. The results of Warnock *et al.* (1982) indicate that even a moderate level of hyphal grazing might have a deleterious effect on the plant, although the two systems are so different that accurate analogies may not be drawn.

As a final observation, the three least preferred fungi are known to contain toxic compounds: *H. crustiliniforme* and *P. involutus* are known to be dangerously poisonous to man due to high concentrations of muscarine and other toxins, while *R. luteolus* is little known but reported to be toxic to slugs (S. Visser, pers. comm.). This may indicate that these fungal toxins protect the hyphae in the soil against grazing arthropods, and that the toxicity of the sporophore hyphae is secondary or even incidental. The results of Lacy (1984) show that sporophores are not protected from mycophagous *Drosophila* by secondary toxins, while the results presented above suggest that toxins may protect mycelial hyphae from grazing by *Onychiurus*.

This hypothesis predicts that fungi whose hyphae ramify through decaying wood, where few if any arthropods can penetrate, should not possess secondary toxins since the hyphae have no need of protection. A limited search, using Phillips (1981) indicated that no British wood-decaying fungi are poisonous, although many are classed as 'inedible' due to the hardness of the sporophore.

ACKNOWLEDGMENTS

This work was performed whilst in receipt of a NERC research studentship supervised by Dr J. Dighton and Dr M. B. Usher, and would not have been possible without the help and cooperation of many members of staff at ITE Merlewood, and Mr T. C. Mitchell of the Forestry Commission.

REFERENCES

Allen, S.E. (1974). *Chemical Analysis of Biological Materials*. Blackwell Scientific Publications, Oxford.

Chu-Chou, M. & Grace, L.J. (1982). Mycorrhizal fungi of *Eucalyptus* in North Island of New Zealand. *Soil Biology and Biochemistry*, 14, 133–137.

Dighton, J. & Mason, P.A. (in press). Mycorrhizal dynamics during forest tree development. *Resource Relationships of Agarics and Developmental Biology of Agarics* (Ed. by D. Moore, L.A. Casselton, D.A. Wood & J.C. Frankland).

Goto, H.E. (1961). Simple techniques for the rearing of Collembola. *Entomologist's Monthly Magazine*, 96, 138–140.

Hale, W.G. (1966). The Collembola of the Moor House National Nature Reserve, Westmorland: a moorland habitat. *Revue d'Ecologie et de Biologie du Sol*, 3, 97–122.

Harley, J.L. (1969). *The Biology of Mycorrhiza*. Leonard Hill, London.

Lacy, R.C. (1984). Predictability, toxicity, and trophic niche breadth in fungus-feeding Drosophilidae (Diptera). *Ecological Entomology*, 9, 43–54.

Marks, G.C & Foster, R.C. (1967). Succession of mycorrhizal associations on individual roots of radiata pine. *Australian Forestry*, 31, 193–200.

McMillan, J.H. (1976). Laboratory observations on the food preference of *Onychiurus armatus* (Tullb.) Gisin (Collembola, family Onychiuridae). *Revue d'Ecologie et de Biologie du Sol*, 13, 353–364.

Mosse, B. (1973). Advances in the study of vesicular-arbuscular mycorrhizas. *Annual Review of Phytopathology*, 11, 171–195.

Newell, K. (1980). *The interaction of soil arthropods and litter decaying fungi*. Ph.D. thesis, Lancaster University.

Nosek, J. (1977). Proturan synusies and niche separation in the soil. *Ecological Bulletins*, 25, 138–142.

Parkinson, D., Visser, S. & Whittaker, J.B. (1979). Effect of Collembolan grazing on fungal colonization of leaf litter. *Soil Biology and Biochemistry*, 11, 529–535.

Phillips, R. (1981). *Mushrooms and Other Fungi of Great Britain and Europe*. Pan Books, London.

Tinker, P.B. (1975). Mycorrhizal effects on plant growth. *Endomycorrhizas* (Ed by S. Sanders, B. Mosse & P.B. Tinker), pp. 353–371, Academic Press, London.

Usher, M.B. (1967). *Studies on population ecology with particular reference to soil Collembola*. Ph.D. thesis, Edinburgh University.

Usher, M.B. & Booth, R.G. (1984). A portable extractor for separating microarthropods from soil. *Pedobiologia*, 26, 17–23.

Visser, S. & Whittaker, J.B. (1977). Feeding preferences for certain litter fungi by *Onychiurus subtenuis* (Collembola). *Oikos*, 32, 320–325.

Warnock, A.J., Fitter, A.H. & Usher, M.B. (1982). The influence of a springtail *Folsomia candida* (Insecta: Collembola) on the mycorrhizal association of leek *Allium porrum* and the vesicular-arbuscular mycorrhizal endophyte *Glomus fasciculatum*. *New Phytologist*, 90, 285–292.

Zak, B. (1965). Aphids feeding on Douglas fir. *Forest Science*, 11, 410–411.

Resource quality and trophic structure in the soil system

O. W. HEAL* AND J. DIGHTON

ITE Merlewood Research Station, Grange-over-Sands, Cumbria LA11 6JU

INTRODUCTION

'In seeking ways to reduce the apparent complexity of the ecosystem to a manageable level, ecologists have adopted, with varying success, a number of different (not mutually exclusive) bases for their work, such as the trophic level concept; the cyclic flow of carbon atoms; the community concept; the unidirectional flow of energy; the generalized Lotka–Volterra equation; and several others.' Platt & Denman (1978) thus introduced an analysis of the structure of pelagic marine ecosystems. Their comment is relevant to the discussion of trophic structure in the soil system because the diversity of microflora and fauna in soil, combined with practical problems of sampling and observation on relevant scales of time and space, have inhibited soil ecologists from developing a satisfying cohesive picture of the inter-relationships between soil organisms. Whilst we have considerable information on the composition of soil populations, and on the biology of individual species, we have not managed to combine this information with that of soil and ecosystem processes–the concluding point of Coleman, Reid & Cole (1983) in their constructive review of soil biology.

Coleman *et al.* (1983) also emphasize the need to examine the application to soil ecology of theoretical aspects of ecology developed from above-ground and aquatic research. Definition of trophic levels in the classical sense of Lindeman (1942) has proved particularly difficult to translate into the trophic structure of soil systems, the plant–herbivore–carnivore chain being confounded by recycling of production within the microflora. However, some recent emphasis on the importance of size as a basis for understanding feeding relationships in ecosystems (Platt & Denman 1978; Cousins 1980) could usefully be incorporated into soil biological research. Similarly, the ideas of life-history strategies based on the definitions of *r*- and *K*-types by MacArthur & Wilson (1967), with the addition of a second dimension of adversity (*A*) or stress (Southwood 1977; Grime 1979), has begun to clarify organism inter-relationships in soil (Greenslade 1982; Heal & Ineson 1984; Pugh 1980; Swift 1984).

There are many examples of conceptual trophic models which identify a sequence of transfers of carbon, energy or nutrients between organisms. These models attempt to classify the organisms functionally as saprophages and biophages or as biotrophs, necrotrophs, and saprotrophs or as herbivores, decomposers, microbivores and carnivores (see Swift, Heal & Anderson 1979). Such models have, however,

*Present address: NERC, Polaris House, North Star Avenue, Swindon, Wiltshire SN2 1EU.

339

proved inadequate to describe the functional organization of soil organisms quantitatively except at the broadest level of resolution. They lose the finer understanding of inter-relationships described in the earlier qualitative research founded on more detailed taxonomic analysis, especially amongst fauna, of specific habitats such as decomposing logs (see Elton 1966). The problem is partly practical, since different techniques are required for microbial and faunal reseach, and partly due to the flexibility of feeding relationships, in which there are recurrent cycles of microbe feeding on microbe, and of fauna ingesting and probably assimilating both plant and microbial material.

In this paper we will not go into the details of organism inter-relationships–these are dealt with in the individual papers of this volume. Nor do we attempt to provide an all-embracing definition of trophic stucture: that is beyond our comprehension. Instead, we focus on:

  (i) the main basis of the trophic structure in soil (dead plant material) examining its variation in quality;
  (ii) how variation in the resource quality can influence the composition and inter-relationships of the organisms dependent upon it, combining the general niche model of Swift (1976) with the concept of r-, K- and A-strategies;
  (iii) identification of the particular physical constraints associated with soil, which in combination with the influence of resource quality, suggest the occurrence of size-related systems–microtrophic, mesotrophic and macrotrophic;
  (iv) examination of results from some intensive ecosystem studies as an initial test of the hypothesis of the co-existence of three trophic systems.

RESOURCE QUALITY

Plant and animal detritus entering the soil represents the initial food resource* on which the trophic structures are developed. Resource quality influences the type and rate of growth of the microflora, and hence the grazing fauna, as well as influencing palatability to the saprotrophic fauna. A suite of factors determine resource quality: the composition of carbon and energy-providing substrates, concentration of essential nutrients, the presence and concentration of modifying compounds such as polyphenols which can inhibit or stimulate resource utilization, and the physical arrangement of the resource (Swift, Heal & Anderson 1979).

Surface resource quality

The components of resource quality may be broadly defined chemically but the influence of the suite of factors is most clearly seen through the comprehensive data

---

*The term 'resource' is used for any identifiable piece of detritus. The term 'substrate' is not used because of its more specific use in biochemistry and microbiology for a chemically defined substance (Swift, Heal & Anderson 1979).

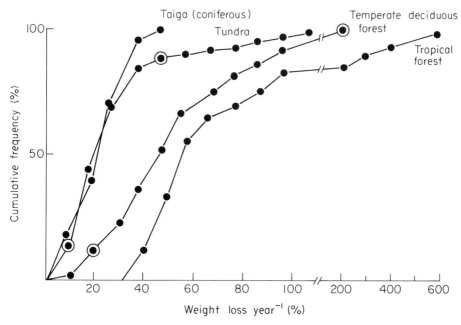

FIG. 1. Cumulative frequency curves of rates of litter and wood decomposition (% weight loss in first year) of a range of sites within a number of different biomes. In tundra and temperate deciduous woodland, minimum and maximum rates within a particular site (Moor House and Meathop Wood respectively) are encircled. Compiled from a wide range of literature.

now available on rates of litter decomposition (Fig. 1). The first year weight loss from litters integrates losses through catabolism, leaching and comminution, resulting from a combination of abiotic factors, microbial and faunal activity. The pattern is of a broad distinction between decomposition in different biomes through the influence of climate but with a wide variation within each biome reflecting mainly the variation in resource quality. The samples of resource types do not necessarily correspond to the actual proportions of different qualities within a biome; the selection has been determined by the various research workers, but in each case a wide range of species and plant parts are included. This is indicated by two examples in which the minimum and maximum values recorded within a site are shown, emphasizing the point that within a site there is a wide range of decay rates related to resource quality and that there is the opportunity for development of different trophic relationships.

Below-ground resource quality

Unfortunately, although there is a wealth of information for surface litter input from plants, we are still remarkably ignorant of both the quality and quantity of below-ground inputs and of their rates of decomposition. In a few studies concerned with root input and decomposition, the majority of work has been concerned with live roots. Resorption of soluble inorganic and organic fractions probably occurs prior to root death, as in leaves, and it is probable that the observed rates of decomposition (Table 1) are overestimates of decomposition of naturally dead root litter. As with surface litters, there is variation in root composition, for example, the size-related nitrogen content and decay rate of *Pinus resinosa* and *P. sylvestris* roots (Table 1), while *Pseudotsuga menziesii* mycorrhizal roots contain about twice the concentration of nitrogen, phosphorus and potassium of fine (<5 mm diam.) roots (Fogel & Hunt 1983).

Root exudates may also be important elements in the below-ground contribution to decomposer communities; for example, Smith (1976) calculated annual exudation from three hardwoods, in kg ha^{-1}, to include carbon 4, total nitrogen 0·8, total sulphur 1·9, phosphate 0·2, potassium 8·1, calcium 3·6 and sodium 34. The amount and proportion of photosynthate which is exuded from roots is very variable. It is affected by the plant's age and physiological state and a number of soil environmental factors. Up to 80% of translocated assimilate may be exuded as in *Pinus contorta* (Bowen 1980). The importance of this resource to the soil trophic structure is difficult to assess because of its variability and localization. As indicated by Smith (1976), exudates, containing simple carbohydrates, amino acids, organic acids and minerals, mediate the composition of the rhizoplane microflora with stimulation of gram-negative bacteria, spore germination, mycelial growth of pathogenic fungi and hatching of parasitic nematode eggs. The effects of changes in the amount, and possibly composition, of exudates to the composition of the rhizosphere microflora is shown by the low carbohydrate demand of mycorrhizal fungi from early successional stages of birch compared with the larger demand of species later in the succession (Dighton & Mason 1984).

The influence of the variety in quality and quantity of below-ground resource input on the trophic system may be (i) highly restricted to the rhizoplane and rhizosphere microflora and associated microfauna (Coleman *et al.* 1983), (ii) extended by mycorrhizal hyphae to beyond the rhizosphere with the possible development of a mesofaunal food link (Finlay, p. 319) or (iii) generally distributed in the soil through macrofaunal consumption of live or dead roots. Data on the effect of the quality of roots on macrofauna are limited but Lavelle, Sow & Schaefer (1980) showed that pulverized live roots of the grass *Loudetia simplex* supported less growth of young earthworms (*Millsonia anomala*) than comparable leaf material, corroborating the indications from the chemical and decomposition data that roots are of lower quality than above-ground parts. Growth was influenced by the state of decomposition of both leaf and root material and by the aerobic or anaerobic conditions for decomposition. Initial plant resources of high quality ensured an energy supply sufficient for both worms and microflora. With increasing humification resource quality declined leading to competition between worms and microflora for the available resources. When only complex plant residues remained, a symbiotic association between the microflora and earthworms was required to provide suitable enzyme suites to utilize the more recalcitrant resource. Although the interpretation by Lavelle *et al.* (1980) remains an

TABLE 1. Rates of weight loss of root litter: examples from published literature over a wide range of species and site conditions

	Weight loss (% 1st year^{-1})	Diameter (mm)	Bag mesh size (mm)	Position in site	N (%)	Initial lignin (%)
1. *Pinus resinosa*: live roots						
	11–16	0·5	0·4	Surface	1·0–1·3	22–25
	15–24	0·5–3·0	0·4	and	0·6–0·8	22–23
	25–48	0·5–3·0	3·0	0–15 cm	0·6–0·8	22–23
2. *P. sylvestris*: live roots						
	10–27	1–2 to 9–11	1	F+H	0·3–0·6	21–22
3. *Vaccinium vitis-idaea*: live rhizomes						
	15 ± 0·9	1–2	1	F+H	0·5	31
4. *Calluna vulgaris*: live rhizomes						
	5·3 ± 0·44	2–3	1	F+H	0·3–0·4	28
5. *C. vulgaris*: live rhizomes and below-ground stems						
	5·0 ± 2·0	c. 2–3	1	0 cm	0·7	49
	0·6 ± 0·3	c. 10	unconfined	0 cm	0·5	35
6. *Eriophorum vaginatum*: live roots						
	8–19	c. 3	1	0–23 cm	0·5	34
7. *Juncus romerianus*: live & dead (a) roots, (b) rhizomes						
	(a) 17	—	—	5 cm	—	—
	(b) 26	—	—	5 cm	—	—
8. *Spartina cynosuriodes*: live & dead roots and rhizomes						
	20	—	—	5 cm	—	—
9. *Larrea tridentata*: live roots						
	0·21–0·45 day^{-1} (over 38–141 day)	—	—	10 cm	—	—

Sources: 1. McClaugherty, Aber & Melillo (1982); 2–4, Berg (1984); 5&6, Heal, Latter & Howson (1978); 7&8, Hackney & de la Cruz (1980); 9, Comanor & Staffeldt (1978).

hypothesis, the emphasis is that the inter-relationship between fauna and micro-flora, is affected by resource quality (Lavelle 1983).

INFLUENCE OF RESOURCE QUALITY ON TROPHIC STRUCTURE

There are many papers (see Dickinson & Pugh 1974) describing field variation in microbial and faunal populations associated with different resource and substrate types, decomposition of a particular type of litter, and laboratory experiments on the influence of different resources and substrates on growth or ingestion by micro-flora and fauna. Swift (1976) provided a general model which helps to clarify the basic resource characteristics which determine the diversity and composition of the microflora (Fig. 2a). He recognized that the variety of fundamental niches available in a resource entering the decomposer system was represented by its range of component substrates and by its physical structure. As the resource decomposes, the range of substrates is reduced as labile fractions are metabolized, but there is

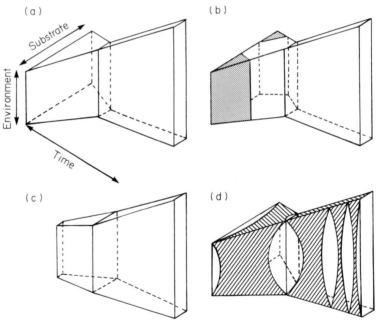

FIG. 2. Changes in the potential variety of fundamental microbial niches within primary resources of different quality during decomposition. A cross section of the volume at right angles to any point in time represents the available niche space in relation to the two major determinants, environment and resource. The time-scale is simplified to represent two stages in decomposition: the initial rapid exploitation phase characteristic of primary resources and the much slower interaction phase of the terminal resources (see text). (a) General model representing high resource quality (from Swift 1976). (b) Initial substrate availability masked by the presence of inhibitory chemicals, e.g. polyphenols. (c) Low quality resource, e.g. wood with a limited range of substrates. (d) Availability of niches limited by periods of severe climate, e.g. drought or cold.

some compensation through production of secondary substrates. The gradual physical disruption of the resource over time increases the physically determined niches available to organisms.

Swift's general model of niche availability during decomposition, may be modified by resource quality. For example, the presence of secondary compounds selected to inhibit leaf consumption by herbivores can inhibit the availability of substrates in an otherwise high quality resource (Fig 2b). A second variation is of a resource such as wood, in which substrate and physical availability is initially limited (Fig 2c). As decomposition proceeds, the range of niches expands with production of metabolites from lignin and cellulose decomposition and physical disintegration, the latter particularly associated with a phase of faunal colonization (Swift 1976). A third situation may be represented as in Fig 2d, in which severe climatic conditions such as cold or drought, restrict niche availability to a few 'windows' in time.

The general model developed by Swift (1976) critically identifies the quality of the resource as the primary factor which selects the composition and characteristics of the microflora. The recurrent provision of fresh resources for colonization and their residence time are readily related to the concepts of disturbance or durational stability inherent in the selection for species along the $r-K$ continuum of MacArthur & Wilson (1967). A second dimension of adversity or stress defined by Grime (1979) and Southwood (1977) as modifying the $r-K$ continuum can be equated with low resource quality and severe climatic conditions. The recognition of the general principles of durational stability and adversity has helped to explain patterns of fungal associations (Pugh 1980; Swift 1984) and has been extended to soil fauna (Greenslade 1982). However, it is essential to recognize that the selective effects of resource quality, representing degrees of durational stability and adversity, act not on single characteristics such as substrate utilization, but on a combination of morphological, physiological and phenological characteristics of the microflora. Given that resource quality influences the characteristics of the microflora populations, it is a logical consequence that it will influence the composition of the fauna which graze the microflora, as well as directly determining the composition of decomposer fauna. On this basis, Heal & Ineson (1984) hypothesized general patterns of development of microflora and fauna associations related to resource quality.

High resource quality

Exploitation phase

There is rapid colonization by microflora with small cell size or diffuse mycelium, and with maximum production and dispersal of propagules. Growth rates tend to be high on low molecular weight substrates, with high efficiency of conversion to biomass production. These colonizers tend to be demanding of nutrients as a result of their growth characteristics. Population growth is rapid and is often followed by a sharp decline as readily available substrates are exhausted or as fauna exploit the food source. Fauna also tend to be small, with rapid growth rates and dormant dispersal stages, such as protozoa, or mobile and longer-lived, e.g. Collembola. Consumption by microfauna tends to disrupt populations, fragment and disperse the resource and enhance the rate of decomposition at an early stage. The exploitation phase, selecting organisms with r characteristics, merges into an interaction phase in which organisms tend to be of the K type.

Interaction phase

Associated with the more resistant substrates and slower decomposition, equivalent to greater durational stability, populations in the later stages tend to have large cell size or compact mycelium and slower growth rates than in the exploitation phase. With longer generation times interactions with competitors and grazers tend to be more stable, with development of defence mechanisms such as resistant cell walls and antibiotic production. Population dynamics tend to be damped, with more density-dependent control and faunal grazing is likely to be more selective. The increasingly resistant residue from decomposition selects for a limited range of specialized microflora, the characteristics of which are similar to the associations on low quality resources.

Low resource quality

Resources dominated by combinations of high molecular weight and resistant substrates, low nutrient concentrations and inhibitory compounds have slow decay rates and select for relatively specialized microflora with slow growth rates because of the intransigent substrates. Nutrient conservation mechanisms such as translocation help to overcome deficiencies and may be associated with large size. Microflora interactions are often intraspecific, reflecting the limited range of species present. Interactions between microflora and fauna may include intimate symbiosis in response to lower resource quality. Slow-growing microflora tend to develop physical or chemical defences against grazing, resulting in specific fauna–microflora relationships. Concentration of nutrients by microflora in nutrient-poor resources can gradually provide suitable food sources for fauna, e.g. in wood, and faunal colonization can enhance physical disintegration with exposure of new surfaces for exploitation (Fig 2c).

It is expected that the actual species involved in the succession of organisms on a particular resource type will be very variable; chance plays a large part in the initial colonization. The composition of populations adjacent to a new resource and current microclimatic conditions will influence the species succession. Sequences have also been obscured by the use of samples of resources rather than following the fate of individual resource units (Swift 1976). Thus, the populations of organisms developing on a resource do not follow a simple pattern. However, logic suggests that the diverse microflora and fauna of the soil is constituted from a small-scale mosaic of resources which supports associations of organisms at varying states of development with distinctive inter-relationships. The characteristics of the organisms are constrained, however, by the physical environment of the soil, including that in the surface organic matter, as discussed in the next section.

PHYSICAL CONSTRAINTS ON TROPHIC RELATIONSHIPS

So far we have concentrated on the selection of species characteristics and population inter-relationships through variation in the quality of resource input both above- and below-ground. The feeding relationships are related to the size of the organisms, particularly amongst the fauna, which can be broadly grouped into micro-, meso- and macrofauna. As emphasized by Cousins (1980), fauna do not distinguish their prey on the basis of the past history of the prey, i.e. whether it is a herbivore, microbivore or carnivore; size is a much more important criterion. Platt & Denman (1978) suggest that the body weight of prey is typically from 0·01 to 0·1% of their predators. Within the soil, there are strong physical constraints on organism size, related to moisture characteristics, pore size and particle size. These constraints, when linked to resource quality, suggest that within the soil, three trophic systems may be recognized.

Trophic systems

Microtrophic

Largely confined within the water film in surface resources, on root surfaces and around soil particles, this is a system based on utilization by bacteria or yeasts of the more readily available carbon sources. Consumption of the microflora is by protozoa and nematodes, which may also be preyed on by protozoa and nematodes, still within the water film. The increase in weight from bacteria or yeasts to flagellates or small amoebae to predatory ciliates is probably of the order of 10^{-15}–10^{-13} to 10^{-12}–10^{-9} to 10^{-9}–10^{-7} g. The size of individual organisms will have increased considerably but only the larger ones have the capacity to break the surface of moisture films. The system is virtually contained within the moisture film, which will obviously be less inhibiting in moister habitats. Larger fauna are not commonly involved directly in this food web, although enchytraeids and some Diptera larvae may capitalize on concentrations of micro-organisms.

Mesotrophic

Largely confined to the air spaces within and between organic resources or to soil pores, this is a system based on the utilization by fungi of a wide range of substrates, including the less readily available carbon sources. The larger mass of mycelium and its extension into air spaces allows consumption by mesofauna, predominantly Collembola and mites, which have the capacity to move through, but not significantly disrupt, the physical structure of the soil or litter. The physical size of spaces tends to limit the size of individuals and predation tends to be amongst the mesofauna. Some larger predators, such as pseudoscorpions and linyphiid spiders, may exploit this food web, particularly at the soil or litter surface, but individual body size or population density are not large enough to support macropredators. Size relationships are difficult to define because of the problem of identifying the individual fungus weight, but fungus-grazing Collembola and mites are of the order of 10^{-7}–10^{-4} g, with their predators at 10^{-4}–10^{-2} g.

Macrotrophic

With a body size or shape large enough to disrupt the physical structure of the soil or litter, macrofauna such as earthworms and millipedes transcend the physical limitations imposed on microtrophs and mesotrophs. They ingest both the basic resource and the associated populations of microflora and fauna. They utilize the concentrated nutrient and carbon sources of the high quality, initial resources or the concentration which has occurred through microbial decomposition or enhance concentration and availability through gut microflora. In each case, there is a large increment in body size, with the production of size units which can support larger predators, including vertebrates. Animals with large body size are unlikely to be supported directly from the microtrophic and mesotrophic systems because of the successive energy loss through the smaller size increments associated with the transfer along these more selective food webs.

It is recognized that the distinctions between these three trophic systems are blurred, but it is suggested that the physical limitations of the soil environment, combined with the varying quality of resources derived from the vegetation, results in a small-scale mosaic of trophic associations. These associations are not identifiable through the coarse sampling regimes adopted in most soil biological research, but are more recognizable when individual resource units can be isolated (Swift 1976) and the trophic relationships followed more directly. The dissipation of energy in the transfer of material along successive size-defined links in a food web may be a major limitation to the maximum size of organism that can be produced from small individual substrates, analogous to marine microtrophic associations (Pomeroy 1984).

A CONVENTIONAL QUANTITATIVE APPROACH TO TROPHIC STRUCTURE

Although the complexity of the trophic relationships in soil systems has frustrated attempts to produce a satisfying synthesis of field data, it is worth re-examining the results of the intensive ecosystem research of the late 1960s and early 1970s to see to what extent the microtrophic, mesotrophic and macrotrophic systems are detectable, and are related to resource quality. The intensive soil fauna research at Moor House in the north Pennines was summarized by Coulson & Whittaker (1978), and has been combined with microbial information from Collins, D'Sylva & Latter (1978), to provide a summary, however speculative, of the quantitative trophic relationships in one soil system (Fig 3).

In the acid blanket peat at Moor House, fauna are virtually confined to the

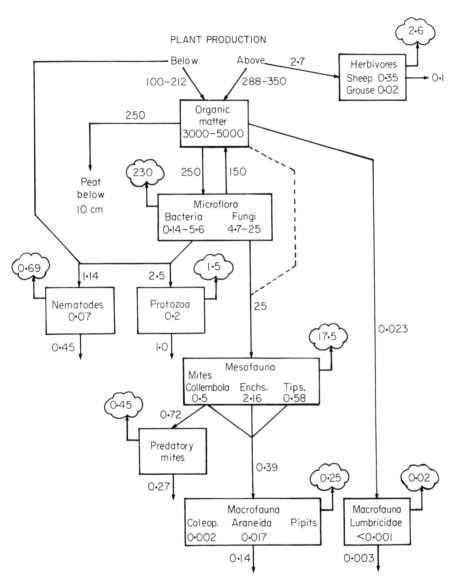

FIG. 3. A summary of the trophic structure of blanket peat (0–10 cm) at Moor House, Cumbria. Biomass in compartments is expressed as g m^{-2} and transfers as g m^{-2} year^{-1}. Litter input, accumulated peat and transfer below 10 cm are based on Jones & Gore (1978). Bacterial and fungal biomass range represent dilution to direct counts and stained to total mycelium (Collins, D'Sylva & Latter 1978). Protozoa estimates, for testate amoebae only, are from Heal (1964). Other fauna data are from Coulson & Whittaker (1978).

surface 10 cm. The basis of the trophic structure is surface and root litter input of 450–500 g m^{-2} year^{-1} from *Calluna vulgaris*, *Eriophorum vaginatum* and *Sphagnum* spp. About half of the input passes into peat below 10 cm although the time taken may vary from 15 to 150 years depending on site conditions (Jones & Gore 1978). Thus, about 250 g m^{-2} are decomposed annually in the surface 10 cm. No direct measure of microbial production is available, but after subtraction of the amount respired by the fauna, about 230 g m^{-2} year^{-1} must be respired by the microflora. Given a yield coefficient of 0·4 and the recirculation of microbial cells and hyphae produced (Heal & MacLean 1975), the best estimate of annual microbial production is of the order of 150 g m^{-2}.

Defining who eats whom in the soil fauna is hazardous and the pathways indicated in Fig. 3 must be accepted as first approximations. Despite these reservations the results indicate that about 25 g m^{-2} year^{-1} is assimilated by the soil fauna, of which about 8 g is converted into production. If it is assumed that the fauna, apart from known predators, are feeding on microflora, then the microfauna and mesofauna assimilate 4 and 25 g m^{-2} year^{-1} respectively, i.e. 3 and 17% of microflora production. A few macrofauna (lumbricids) occur on the bog, but their assimilation is estimated at only 0·02 g m^{-2} year^{-1} (Coulson & Whittaker 1978). Of the 8 g m^{-2} year^{-1} produced from the 'microbivore trophic level', 1 g m^{-2} year^{-1} (12%) is assimilated by carnivores and, as emphasized by Coulson & Whittaker (1978), subsequent carnivore levels are severely restricted.

Although information on the microfauna is particularly limited, the general pattern seems to be of dominance of the food chain sequence:

$$\text{organic matter} \rightarrow \text{microflora} \rightarrow \text{mesofauna} \rightarrow \text{mesofauna/macrofauna,}$$

the major activity being recirculation within the microflora. In contrast, below the 10 cm level on the bog, near to the water table, metazoan and probably protozoan activity is negligible. Thus, the trophic system based on the slow decomposition of the residual organic matter is one of microflora alone.

Comparison of the blanket bog with adjacent soil systems at Moor House is also revealing. The transition from blanket bog, through *Juncus squarrosus* on shallow peat and *Nardus* grassland on alluvial peaty podzols and gleys, to *Agrostis–Festuca* grasslands on brown earth, represents a sequence of increased aeration, reduced acidity, higher quality of organic input and higher rate of organic matter turnover (Rawes & Heal 1978). Associated with this gradient is an increase in assimilation by the mesofauna (25, 56, 97 and 69 g m^{-2} year^{-1} respectively) and there is a more marked increase in macrofauna assimilation from less than 1 g m^{-2} year^{-1} in the blanket bog and *Juncus* peat to about 73 g m^{-2} year^{-1} in the *Nardus* and *Festuca–Agrostis* grassland with mineral soils (Coulson & Whittaker 1978).

Increased importance of larger organisms is associated with the increase in resource quality of the litters. This continuum may be seen in relation to the Swift model (Fig. 2), with the bog community providing a narrow substrate window and with waterlogging conditions restricting the environment parameter of the model. Increased soil fertility and reduction of environmental stresses in the grassland

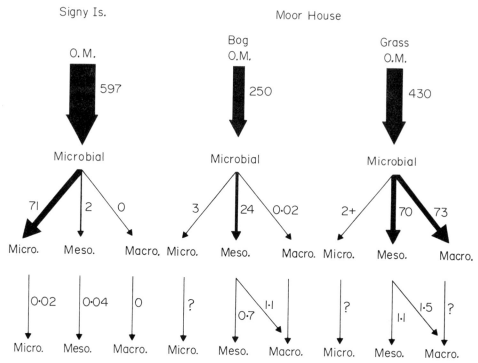

FIG. 4. The general trophic structure in the soil of three ecosystems, distinguished by variation in severity of climatic and resource quality constraints: (a) Signy Island moss bank in the Antarctic (from Davis 1980); (b) and (c) are respectively acid blanket bog and *Festuca–Agrostis* grassland on limestone, at Moor House in the north Pennines, UK (see text for details).

increase the initial resource quality window size. In the highly stressed blanket bog system, movement along the time axis decreases, rather than increases, decomposer niches on the substrate, hence residual organic matter tends to accumulate.

Quantitative information on the microfauna and microflora are limited but Fig. 4 shows a simplified summary of the trophic structure of the blanket bog compared with that of the adjacent grassland on which the nutrient constraint has been reduced. Also shown is the structure of an Antarctic moss community, an extremely simple community in terms of numbers of species, developed under severe climatic constraints (Davis 1980). The comparison of these three systems suggests differential expansion of distinct trophic structures related to variation in climate and resource quality.

The indications are that an increasing importance of fauna, and an increasing dominance of macrofauna, are related to improvements in the physico-chemical environment and in quality, rather than in quantity, of the residues of primary production. This interpretation from comparative descriptions tends to be confirmed by experiment, with particular emphasis on the nutrient quality of the plant material. Coulson & Butterfield (1978) showed that whilst the rates of decomposition of the same litters were similar on adjacent sites at Moor House, the contribution of fauna to decomposition tended to be larger on the mineral than on the peat soils. Fertilization with nitrogen increased the numbers of enchytraeids and tipulids on peats and increased the rate of decomposition through enhanced nutrient content of the plant material. Surprisingly, fertilization with phosphorus tended to reduce both rate of decomposition and soil fauna populations.

In this limited example, we see that the relative microflora and fauna contributions to decomposition processes are strongly affected by resource quality which determines the length of food chains and size of their component organisms. Cragg (1961) and Macfadyen (1963) synthesized and interpreted the limited data on soil fauna available for a few sites. Their general conclusions have been confirmed and amplified by the subsequent intensive studies on a wide range of sites (Petersen & Luxton 1982) but with a shift in emphasis. One of the conclusions of Petersen & Luxton (1982) was that 'chemical nutrient, rather than energy, availability may impose the most serious limitations on soil animal populations . . . more information is required on food substrate and space utilization before a complete understanding of niche exploitation by soil animals is attained'.

CONCLUSION

Well defined, replicated patterns of species associations with clearly organized trophic structures are not expected to occur in soil systems. The diversity of resources and organisms, microbial flexibility and chance, mitigate against this. But neither is the composition of the microbial and faunal populations random. The argument developed in this paper is that the physico-chemical features of individual resource units select for distinctive combinations of morphological, physiological and phenological characteristics of the microflora. These in turn select for particular characteristics in the associated fauna.

Activities of the microflora and fauna modify resource quality with time, resulting in a change in the biota. Within any ecosystem, the wide variety in quality of resources entering the soil system results in a variety of microflora–fauna associations. The physical structure of the resources and of the soil superimposes additional constraints on the biota and differentiates three trophic systems which are broadly separated on size. The extent to which these micro-, meso- and macrotrophic systems are developed in any ecosystem is thus determined by the combined effects of resource quality and physical environment.

REFERENCES

Berg, B. (1984). Decomposition of root litter and some factors regulating the process: long-term root litter decomposition in a Scots pine forest. *Soil Biology and Biochemistry* (in press).

Bowen, G.D. (1980). Misconceptions, concepts and approaches in rhizosphere biology. *Contemporary Microbial Ecology* (Ed. by D.C. Ellwood, J.N. Hedger, M.J. Latham, J.M. Lynch & J.H. Slater), pp. 283–304. Academic Press, London.

Coleman, D.C., Reid, C.P.P. & Cole, C.V. (1983). Biological strategies of nutrient cycling in soil systems. *Advances in Ecological Research*, **13** (Ed. by A. Macfadyen & E.D. Ford), pp. 1–55. Academic Press, New York.

Collins, V.G., D'Sylva, B.T. & Latter, P.M. (1978). Microbial populations in peat. *Production Ecology of British Moors and Montane Grasslands* (Ed. by O.W. Heal & D.F. Perkins), pp. 94–112. Springer-Verlag, Berlin.

Comanor, P.L. & Staffeldt, E.E. (1978). Decomposition of plant litter in two western North American deserts. *Nitrogen in Desert Ecosystems* (Ed. by N.E. West & J. Skujins), pp. 31–49. Dowden, Hutchinson and Ross, Stroudsburg.

Coulson, J.C. & Butterfield, J. (1978). An investigation of the biotic factors determining the rates of plant decomposition on blanket bog. *Journal of Ecology*, **66**, 631–50.

Coulson, J.C. & Whittaker, J.B. (1978). Ecology of moorland animals. *Production Ecology of British Moors and Montane Grasslands* (Ed. by O.W. Heal & D.F. Perkins), pp. 52–93. Springer-Verlag, Berlin.

Cousins, S.H. (1980). A trophic continuum derived from plant structure, animal size and a detritus cascade. *Journal of Theoretical Biology*, **82**, 607–18.

Cragg, J.B. (1961). Some aspects of the ecology of moorland animals. *Journal of Ecology*, **49**, 477–506.

Davis, R.C. (1980). Structure and function of two Antarctic terrestrial moss communities. *Ecological Monographs*, **51**, 125–143.

Dickinson, C.H. & Pugh, G.J.F. (1974). *Biology of Plant Litter Decomposition*. Academic Press, London.

Dighton, J. & Mason, P.A. (1984). Mycorrhizal dynamics in forest tree development. *Developmental Biology of Higher Fungi* (Ed. by D. Moore, L.A. Casselton, D.A. Wood & J.C. Frankland). British Mycological Society Symposium, Cambridge University Press (in press).

Elton, C.S. (1966). *The Pattern of Animal Communities*. Methuen, London.

Fogel, R. & Hunt, R. (1983). Contribution of mycorrhizae and soil fungi to nutrient cycling in a Douglas fir ecosystem. *Canadian Journal of Forest Research*, **13**, 219–32.

Greenslade, P.J.M. (1982). Selection processes in arid Australia. *Evolution of the Flora and Fauna of Arid Australia* (Ed. by W.R. Baker & P.J.M. Greenslade), pp. 125–30. Peacock, South Australia.

Grime, J.P. (1979). *Plant Strategies and Vegetation Process*. Wiley, New York.

Hackney, C.T. & de la Cruz, A.A. (1980). *In situ* decomposition of roots and rhizomes of two tidal marsh plants. *Ecology*, **61**, 226–31.

Heal, O.W. (1964). Observations on the seasonal and spatial distribution of Testacea (Protozoa: Rhizopoda) in *Sphagnum*. *Journal of Animal Ecology*, **33**, 395–412.

Heal, O.W. & Ineson, P. (1984). Carbon and energy flow in terrestrial ecosystems–relevance to microflora. *Current Perspectives in Microbial Ecology* (Ed. by M.J. Klug & C.A. Reddy), pp. 394–404. American Society for Microbiology, Washington.

Heal, O.W. & MacLean, S.F. (1975). Comparative productivity in ecosystems; secondary productivity. *Unifying Concepts in Ecology* (Ed. by W.H. van Dobben & R.H. Lowe-McConnell), pp. 89–108. Junk, The Hague.

Heal, O.W., Latter, P.M. & Howson, G. (1978). A study of the rates of decomposition of organic matter. *Production Ecology of British Moors and Montane Grasslands* (Ed. by O.W. Heal & D.F. Perkins), pp. 136–59. Springer-Verlag, Berlin.

Jones, H.E. & Gore, A.J.P. (1978). A simulation of production and decay in blanket bog. *Production Ecology of British Moors and Montane Grasslands* (Ed. by O.W. Heal & D.F. Perkins), pp. 160–86. Springer-Verlag, Berlin.

Lavelle, P. (1983). The structure of earthworm communities. *Earthworm Ecology: From Darwin to Vermiculture* (Ed. by J.E. Satchell), pp. 449–66. Chapman and Hall, London.

Lavelle, P., Sow, B. & Schaefer, R. (1980). The geophagous earthworms community in the Lamto savanna (Ivory Coast): niche partitioning and land utilization of soil nutritive resources. *Soil Biology as Related to Land Use Practices* (Ed. by D.L. Dindal), pp. 653–72. Environmental Protection Agency, Washington.

Lindeman, R.L. (1942). The trophic-dynamic aspect of ecology. *Ecology*, **23**, 399–418.

MacArthur, R.H. & Wilson, E.O. (1967). *The Theory of Island Biogeography*. Princeton Univeristy Press, Princeton.

Macfadyen, A. (1963). The contribution of the microfauna to total soil metabolism. *Soil Organisms* (Ed. by J. Doeksen & J. van der Drift), pp. 3–16. North-Holland Publishing Co, Amsterdam.

McClaugherty, C.A., Aber, J.D. & Melillo, J.M. (1982). The role of fine roots in the organic matter and nitrogen budgets of two forested ecosystems. *Ecology*, **63**, 1481–90.

Petersen, H. & Luxton, M. (1982). A comparative analysis of soil fauna populations and their role in decomposition processes. *Oikos*, **39**, 287–388.

Platt, T. & Denman, K. (1978). The structure of pelagic marine ecosystems. *Rapport et procès-verbaux des réunions. Conseil permanent international pour l'exploration de la mer*, **173**, 60–5.

Pomeroy, L.R. (1984). Significance of micro-organisms in carbon and energy flow in marine exosystems. *Current Perspectives in Microbial Ecology* (Ed. by M.J. Klug & C.A. Reddy), pp. 405–411. American Society for Microbiology, Washington, D.C.

Pugh, G.F.J. (1980). Strategies of fungal ecology. *Transactions of the British Mycological Society*, **75**, 1–14.

Rawes, M. & Heal, O.W. (1978). The blanket bog as part of a Pennine moorland. *Production Ecology of British Moors and Montane Grassland* (Ed. by O.W. Heal & D.F. Perkins), pp. 224–43. Springer-Verlag, Berlin.

Smith, W.H. (1976). Character and significance of forest tree root exudates. *Ecology*, **57**, 324–31.

Southwood, T.R.E. (1977). Habitat, the templet for ecological strategies? *Journal of Animal Ecology*, **46**, 337–65.

Swift, M.J. (1976). Species diversity and the structure of microbial communities. *The Role of Terrestrial and Aquatic Organisms in Decomposition Processes* (Ed. by J.M. Anderson & A. Macfayden), pp. 185–222. Blackwell Scientific Publications, Oxford.

Swift, M.J. (1984). Microbial diversity and decomposer niches. *Current Perspectives in Microbial Ecology* (Ed. by M.J. Klug & C.A. Reddy), pp. 8–16. American Society for Microbiology, Washington.

Swift, M.J., Heal, O.W. & Anderson, J.M. (1979). *Decomposition in Terrestrial Ecosystems*. Blackwell Scientific Publications, Oxford.

Possible roles for roots, bacteria, protozoa and fungi in supplying nitrogen to plants

MARIANNE CLARHOLM

Department of Microbiology, Swedish University of Agricultural Sciences, S-750 07 Uppsala, Sweden

SUMMARY

1 The relations between micro-organisms and soil organic matter are explored and related to nitrogen mineralization. Special emphasis is given to soil protozoa, a neglected group of micro-organisms with their main area of activity within the rhizosphere.

2 Interactions are described where the root induces a series of events which leads to mineralization of N from the soil organic matter. The first step is excretion of easily available carbon from the root-tip into the surrounding soil. The carbon enables the bacteria to multiply and so to release enough N to support their own growth. The bacterial production attracts protozoa, mainly naked amoebae. When feeding on the bacteria, the protozoa excrete part of the bacterial N as ammonium, close to the root surface where it can be taken up by the plant.

3 In a microcosm experiment, wheat plants grown for 6 weeks in sterilized soil with both bacteria and protozoa present had a 60% larger dry weight with 60% higher N content as compared to plants grown with bacteria only. Since fungi were excluded and no inorganic N added, the experiment showed that bacteria were able to mineralize N from older organic matter when supplied with a suitable energy source. Protozoa were needed to make the bacterial N available for plant uptake.

4 In another part of the same experiment repeated small additions of glucose were made to mimic an increased release of carbon from the roots. The additions resulted in a significantly higher amount of N in the shoots. When C was added to similar microcosms without plants a large amount of N was tied up in protozoan biomass. With only bacteria present very small N transformations were registered.

INTRODUCTION

The mineralization of nitrogen in soil is acknowledged to be a result of microbial activities, but little is known about the mechanisms of its release. Protozoa feeding on bacteria have been shown to mineralize N, which previously was immobilized in bacterial biomass (Coleman *et al.* 1977; Woods *et al.* 1982). Plants have also been reported to take up more N when grown in the presence of both protozoa and bacteria as compared to when grown with bacteria only (Elliott, Cole & Coleman 1979). Likewise, a series of events along the root, starting with bacterial immobilization of inorganic N initiated by the root-derived carbon, ending with root-uptake

numbers of bacteria are also reported for wild plants, compared with values for cultivated species (Woldendorp 1981).

MINERALIZATION AT THE ROOT SURFACE

Local N mineralization around the root

I would like to advance the idea that the root, through its carbon inputs, will induce a chain of events in the surrounding soil, which will lead to a local mineralization of soil organic N around the root. The main part of this N will be taken up by the plant.

Bacteria in soil are concentrated on the surfaces of organic matter (Fig. 2a) where they exist in a very low metabolic state due to restricted C availability

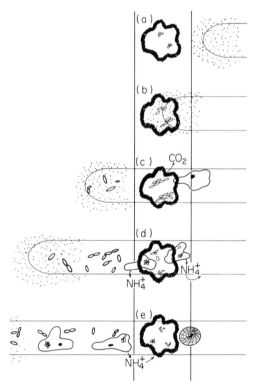

FIG. 2. Model of interactions around the root initiated by root-derived carbon to release soil organic N for plant uptake. (a) Bacteria in soil are concentrated on the surfaces of organic matter where they exist in a low metabolic state due to restricted C availability. (b) When a root grows through the soil a carbon input will take place around and just behind the tip. To any specific part of the soil, fresh root-derived C will thus be added as a pulse when the tip passes. Bacteria will start to grow. They will then release enough N from the organic matter to meet their need for growth. (c) The bacterial growth attracts protozoa. (d) When protozoa consume bacteria, one-third of the bacterial N is released as NH_4^+. (e) Part of the ammonium will be taken up by the root.

(Stotsky & Norman 1963). When a root grows through the soil a carbon input will take place around and just behind the root-tip (Rovira 1969). Thus, fresh root-derived carbon will be added as a pulse to any specific part of the soil when the tip grows through (Fig. 2b). The C input releases the bacteria from their usual shortage of C and they will instead become temporarily N-limited (Woldendorp 1981).

Utilizing the energy provided by the root, the bacteria will start to grow. Through enzymatic activities, they will release enough N from organic material to meet their needs for growth. Because of its resistance, this material cannot meet the C requirements of the bacteria, but if another energy and C source is available, it could serve as a N source. Amino-acids are the most likely form for this N transfer.

The growth of bacteria will cause a local increase of the CO_2 concentration in the soil, which in turn attracts and activates protozoa (Fig. 2c). When protozoa consume bacteria with approximately the same C/N ratio as their own, one-third of the bacterial N will be built into protozoan biomass, one-third will be excreted as bacterial cell walls and one-third as NH_4^+ (Fenchel 1982) (Fig. 2d). The ammonium will be released very close to the root, in an area where most of the soluble root-derived carbon is depleted through previous bacterial growth. It is therefore likely that at least part of the ammonium will be taken up by the root (Fig. 2e).

EXPERIMENTAL EVIDENCE AND OBSERVATIONS TO SUPPORT THE MODEL

In a microcosm experiment, wheat plants were grown in the presence and absence of protozoa in autoclaved but not leached arable soil to which a natural bacterial flora had been added. After 6 weeks, the plants grown in soil with protozoa had on average a 60% larger dry weight (Clarholm 1984b). For plants grown with bacteria only, the roots made up 60 and 62%, respectively, of the total plant weight with and without C additions. With protozoa present, the equivalent figures were 53 and 42%. The lower S/R ratios found in the presence of grazers indicate an improved nutrient supply to the roots (Davidson 1969), which became even better after the C additions. The increase in N uptake with grazers present was on average 60% for the treatments with and without C additions (Table 2). The largest N increase was recorded for the shoots which contained significantly more N as compared to the shoots of plants without C additions. Since the fungi were carefully excluded and no inorganic N added, the experiment showed that bacteria can mineralize N from the soil organic matter and that the presence of protozoa makes more N available for plant uptake.

The observation that N uptake by annual crops is nearly finished at the tillering stage (Gregory, Crawford & McGovan 1979a) could be related to observations of bacteria and naked amoebae in the rhizosphere of white mustard over the whole period (Table 1 in Darbyshire & Greaves (1967), shown diagrammatically in Fig. 1 of Clarholm (1984a)). After low initial values, large numbers of amoebae were

TABLE 2. Inorganic N and N content of three wheat plants plus micro-organisms grown for 6 weeks in microcosms which held 35 g soil (d.w.) and contained 0·68 mg inorganic N after autoclaving. The N content of the seeds (3·24 mg) has been subtracted from the figures given for the plant N. The observable change was calculated by addition of all biomass values and subtraction of the 0·68 mg present at start. The C and N additions were made twice a week (Clarholm 1984b).

Treatments	Additions	Inorganic N	Bacterial N	Protozoan N	Plant N −seed N	Total observable change in biomass N plus inorganic N
Bacteria	none	0·17	0·36			−0·15
	14·4 mg C	0·26	0·53			0·11
Protozoa	none	0·24	0·28	0·23		−0·07
Bacteria	14·4 mg C	0·24	0·36	4·66		4·58
Wheat	none	0·15	0·45		1·61	1·53
Bacteria	14·4 mg C	0·01	0·70		1·78	1·81
Wheat	none	0·22	0·44	0·16	2·55	2·69
Protozoa	14·4 mg C	0·00	0·56	0·18	3·04	3·10
Bacteria						

recorded over the 20 days preceding the onset of flowering: 5 days later bacterial numbers were drastically decreased and the numbers of amoebae diminished quickly. There are thus indications of high microbial activities around the roots over the period of the largest N uptake. The drastic reduction of micro-organisms was concurrent with the time when more C is directed to above-ground parts to support seed production, with the seeds largely being supplied by N through translocation. These observations indicate that the size and timing of the input of root-derived C is at least partly controlled by the plant and related to N uptake.

POSSIBLE SIZE AND AVAILABILITY OF ROOT-DERIVED N

It is difficult to evaluate the possible impact of the root-derived carbon on mineralization, but as a first attempt one could assume that *all* the root-derived carbon is utilized by the bacteria and that *the whole* bacterial production is consumed by protozoa. Input by the roots varies between 7·5 and 55 kg organic matter $ha^{-1} d^{-1}$ (Woldendorp 1981). Assuming 42% C in the fresh plant material, a bacterial growth efficiency of 40%, a bacterial C/N ratio of 10 and that one-third of the bacterial N is released as ammonium, together with two-thirds of the one-third of the bacterial N built into protozoan biomass: then, with a calculated uptake of N for 60 days, *c.* 4–30 kg ha^{-1} could be released. The lowest C input value is conservative, indicating that 15–25 kg ha^{-1} could be a reasonable figure for a local N addition induced by the root at the right time and place.

CONCLUSIONS

The following statements should be considered in future research on N mineralization:

(i) A vegetation cover is the normal state for an undisturbed fertile soil, and roots have evolved accompanied by micro-organisms.

(ii) The size and cause of mineralization of N is not the same in a vegetated soil as in a soil under fallow. The shift in microbial activities could be a result either of competition or of antagonism involving both rhizosphere micro-organisms and the root.

(iii) The ability of micro-organisms to liberate N from different substrates is more important than their taxonomic status (fungi, bacteria, etc.).

(iv) The transport mechanisms for N in the soil should be further studied and related to plant uptake of N over time.

(v) All N mineralized in soil is probably not equally available for plant uptake. Possibly there is a local mineralization close to the root, which could be more important in providing the plant with N than N mineralization outside the rhizosphere.

ACKNOWLEDGMENTS

Thanks are due to T. Lindberg, S. L. Jansson, K. Paustian and J. Schnürer for valuable discussions, criticism and suggestions on the form and contents of this paper, and to A. Fitter for correcting the language.

The work is a part of the project 'Ecology of arable land–the role of organisms in nitrogen cycling' financed by Swedish Council for Planning and Co-ordination of Research, Swedish Council for Forestry and Agricultural Research, Swedish Environmental Protection Board and Swedish Natural Science Council.

REFERENCES

Anderson, R.V., Coleman, D.C. & Cole, C.V. (1981). Effects of saprophytic grazing on net mineralization. *Ecological Bulletins* (Stockholm), **33**, 201–215.

Anderson, J.P.E. & Domsch, K.H. (1975). Measurements of bacterial and fungal contributions to respiration of selected agricultural and forest soils. *Canadian Journal of Microbiology*, **21**, 314–322.

Baldwin, J.P. (1976). Competition for plant nutrients in soil; a theoretical approach. *Journal of Agricultural Sciences*, **87**, 341–356.

Barber, S.A. (1977). Nutrients in soil and their flow to plant roots. *The Belowground Ecosystem: A Synthesis of Plant-Associated Processes* (Ed. by J.K. Marshall), pp. 161–170. Range Science Department Science Series No. 26, Colorado State University, Fort Collins.

Bergström, L. (1982). Leaching measurements in the Kjettslinge project. *Ecology of Arable Land–The Role of Organisms in Nitrogen Cycling. Progress Report 1981* (Ed. by T. Rosswall), pp. 202–209 Swedish University of Agricultural Sciences, Uppsala.

Campbell, C.A. (1978). Soil organic carbon, nitrogen and fertility. *Soil Organic Matter* (Ed. by M. Schnitzer & S.V. Kahn), pp. 173–271. Elsevier Scientific Publishing Company, Amsterdam.

Chapin, F.S. (1980). The mineral nutrition of wild plants. *Annual Review of Ecology and Systematics*, **11**, 223–260.

Clarholm, M. (1981). Protozoan grazing of bacteria in soil–impact and importance. *Microbial Ecology*, **7**, 343–350.

Clarholm, M. (1984a). Heterotrophic, free-living protozoa: neglected micro-organisms with an important task in regulating microbial populations. *Current Perspectives in Microbial Ecology* (Ed. by M.J. Klug & C.A. Reddy), pp. 321–326. American Society for Microbiology, Washington, D.C.

Clarholm, M. (1984b). Interactions of bacteria, protozoa and plants leading to mineralization of soil nitrogen. *Soil Biology and Biochemistry* (in press).

Coleman, D.C., Cole, C.V., Anderson, R.V., Blaha, M., Campion, M.K., Clarholm,, M., Elliott, E.T., Hunt, H.W., Shaefer, B. & Sinclair, J. (1977). An analysis of rhizosphere-saprophage interactions in terrestrial systems. *Ecological Bulletins (Stockholm)*, **25**, 299–309.

Darbyshire, J.F. & Greaves, M.P. (1967). Protozoa and bacteria in the rhizosphere of *Sinapis alba* L. and *Lolium perenne* L. *Canadian Journal of Microbiology*, **13**, 1057–1068.

Davidson, R.L. (1969). Effect of soil nutrients and moisture on root/shoot ratio in *Lolium perenne* L. and *Trifolium repens* L. *Annals of Botany*, **33**, 571–577.

Elliott, E.T. & Coleman, D.C. (1977). Soil protozoan dynamics in a short-grass prairie. *Soil Biology and Biochemistry*, **9**, 113–118.

Elliott, E.T., Coleman, D.C. & Cole, C.V. (1979). The influence of amoebae on the uptake of nitrogen in gnotobiotic soil. *The Soil–Root Interface* (Ed. by J.L. Harley & R. Scott Russell), pp. 221–230. Academic Press, London.

Fenchel, T. (1982). Ecology of heterotrophic microflagellates, II Bioenergetics and growth. *Marine Ecology Progress Series*, **8**, 225–231.

Frankland, J.C. (1982). Biomass and nutrient cycling by decomposer basidiomycetes. *Decomposer Basidiomycetes–Their Biology and Ecology* (Ed. by J.C. Frankland, J.N. Hedger & M.J.Swift), pp. 241–262. Cambridge University Press, Cambridge.

Gregory, P.J., Crawford, D.V. & McGowan, M. (1979a). Nutrient relations of winter wheat, 1. Accumulation and distribution of Na, K, Ca, Mg, P, S and N. *Journal of Agricultural Sciences (Cambridge)*, **93**, 485–494.

Gregory, P.J., Crawford, D.V. & McGowan, M. (1979b). Nutrient relations of winter wheat, 2. Movements of nutrients to the root and their uptake. *Journal of Agricultural Sciences (Cambridge)*, **53**, 495–504.

Huntjens, J.L.M. (1972). *Immobilization and mineralization of nitrogen in pasture soil*. Ph.D. Thesis, Wageningen.

Jansson, S.L. (1971). Use of ^{15}N in studies of soil nitrogen. *Soil Biochemistry*, Vol. 2 (Ed. by A.D. McLaren & J. Skujins), pp. 129–166. Marcel Dekker Inc., New York.

Jenkinson, D.S. (1977). Studies of the decomposition of plant material in soil. V. The effects of plant cover and soil type on the loss of carbon from ^{14}C-labelled ryegrass decomposing under field conditions. *Journal of Soil Science*, **28**, 424–434.

Lofs-Holmin, A. & Boström, U. (1982). Role of earthworms in carbon and nitrogen cycling. *Ecology of Arable Land–The Role of Organisms in Nitrogen Cycling. Progress Report 1981* (Ed. by T. Rosswall), pp. 160–166. Swedish University of Agricultural Sciences, Uppsala.

Nye, P.H. (1977). The rate-limiting step in plant nutrient absorption from soil. *Soil Science*, **623**, 292–297.

Nye, P.H. & Tinker, P.B. (1977). *Solute Movement in the Soil–Root System*. University of California Press, Berkeley.

Persson, T., Bååth, E., Clarholm, M., Lundkvist, H., Söderström, B. & Sohlenius, B. (1980). Trophic structure, biomass dynamics and carbon metabolism in a Scots pine forest. *Ecological Bulletins (Stockholm)*, **32**, 519–559.

Prenzel, J. (1979). Mass flow transport towards the root systems of two forest ecosystems in relation to the uptake by the above ground tree parts. *The Soil–Root Interface* (Ed. by J.L. Harley & R. Scott Russell), p. 431. Academic Press, London.

Reid, J.B. & Goss, M.J. (1982). Suppression of decomposition of ^{14}C-labelled plant roots in the presence of living roots of maize and perennial ryegrass. *Journal of Soil Science*, **33**, 387–395.

Rosswall, T. & Paustian, K. (1984). Cycling of nitrogen in modern agricultural systems. *Plant and Soil*, **76**, 3–21.

Rovira, A.D. (1969). Plant root exudates. *Botanical Review*, **35**, 35–55.

Schnürer, J., Clarholm, M. & Rosswall, T. (1982). Microorganisms. *Ecology of Arable Land–The Role of Organisms in Nitrogen Cycling. Progress Report 1981* (Ed. by T. Rosswall), pp. 77–78. Swedish University of Agricultural Sciences, Uppsala.

Shields, J.A., Paul, E.A., Lowe, W.E. & Parkinson, D. (1973). Turnover of microbial tissue in soil under field conditions. *Soil Biology and Biochemistry*, **5**, 753–764.

Sparling, G.P., Cheshire, M.V. & Mundie, M. (1982). Effect of barley plants on the decomposition of ^{14}C-labelled soil organic matter. *Journal of Soil Science*, **33**, 98–100.

Stanford G. (1982). Assessment of soil nitrogen availability. *Nitrogen in Agricultural Soils* (Ed. by F.J. Stevenson), pp. 651–688. American Society of Agronomy, Maidson.

Stotsky, G. & Norman, A.G. (1963). Factors limiting microbial activities in soil, III. Supplementary substrate additions. *Canadian Journal of Microbiology*, **10**, 143–147.

Vancura, V. & Kunc, F. (1977). The effect of streptomycin and Actidione on respiration in the rhizosphere and non-rhizosphere. *Zentralblatt für Bakteriologie Abteil II*, **132**, 472–478.

Van Vuurde, J.W.L. & Schippers, B. (1980). Bacterial colonization of seminal wheat roots. *Soil Biology and Biochemistry,* **12**, 559–565.

Woods, L., Elliott, E.T., Anderson, R.V. & Coleman, D.C. (1982). Nitrogen transformations in soil as affected by bacterial-microfaunal interactions. *Soil Biology and Biochemistry*, **14**, 93–98.

Woldendorp, J.W. (1986). Nutrients in the rhizosphere. *Agricultural Yield Potentials in Continental Climates.* Proceedings of the 16th International Potash Institute, pp. 99–126. Berne.

Effects of collembolan grazing on nitrogen dynamics in a coniferous forest

H. A. VERHOEF AND R. G. M. DE GOEDE

Department of Biology, Free University, De Boelelaan 1087, NL-1081 HV Amsterdam, The Netherlands

SUMMARY

1 In field and laboratory studies the effects of the hemiedaphic collembolan *Tomocerus minor* on the nitrogen dynamics in a *Pinus nigra* forest have been examined.

2 Exclusion experiments in the field showed that during a 3 month winter period *T. minor* had a significant effect on the nitrogen concentration in the $A_{00}F$ soil layer

3 In the laboratory *T. minor* increased the existing nitrogen fluxes in the A_{000} and the $A_{00}F_1$ layers (immobilization and release respectively).

4 These effects were influenced by temperature, inoculation with the soil fungus *Verticillium bulbillosum*, and density of *T. minor*.

INTRODUCTION

In many studies on the role of soil animals in decomposition processes, attention has been focused on the indirect effect of soil fauna on nutrient mobilization and mineralization (see review by Anderson, Coleman & Cole 1981). Effects of the grazing of different functional groups of soil animals—divided by body width into micro-, meso- and macrofaunal elements—on the release of nutrients have been investigated in experimental soil systems (Bååth *et al.* 1978; Cole *et al.* 1978; Anderson *et al.* 1981). In a comparative experimental study, Anderson, Ineson & Huish (1983) found for the various representatives of the different faunal groups an overall correlation between increased ammonium-N mobilization and faunal biomass. However, the fact that there is a non-linear relation between the grazing intensity of the millipede *Glomeris marginata* and nitrogen fluxes shows that nutrient mobilization is not simply a function of size, biomass or activity of the grazing animals (Anderson *et al.* 1983).

This conclusion is sustained by laboratory studies on the effects of collembolan grazing on fungal standing crop. These effects on fungal growth appeared to be greatly influenced by the duration of the experiments, and the local and temporal density changes of the Collembola (Hanlon & Anderson 1979; Visser, Whittaker & Parkinson 1981; Bengtsson & Rundgren 1983; Ineson, Leonard & Anderson 1982). Therefore, it is necessary to perform experimental field work at more natural densities.

In the studies on mineral nitrogen mobilization in forest soils, temperature

367

appeared to be a major determinant (Witkamp & Frank 1970). For ammonium-N release in deciduous litter, an optimal temperature of 15 °C was found by Anderson *et al.* (1983).

The magnitude of nitrogen mobilization depends on the nitrogen concentrations in the resources exploited, as was found by comparing the effects of grazing by the millipede *Glomeris* in different organic soil layers of a mature oak-woodland. The effect of the millipede appeared to be an amplification of the existing fluxes in the different layers (Anderson *et al.* 1983).

Much of the experimental work was done in deciduous forest soils, which are rich in nutrients and animals. Coniferous forest soils have a relatively low faunal biomass, dominated by micro-arthropods, whose impact on nutrient cycling appears to be important. The soil invertebrates of a Swedish Scots pine forest contribute, depending on their assimilation efficiency, 10–50% of the total net nitrogen mineralization (Persson 1983).

In the present study the effects of the dominant collembolan *Tomocerus minor* (Lubbock) on the nitrogen dynamics in a relatively young *Pinus nigra* forest were analysed. Exclusion experiments were carried out in the field. In the laboratory the effects of (i) addition of soil fungi and *T. minor* in different densities, and (ii) temperature, on the nitrogen concentrations in different soil layers were investigated, using experimental soil systems.

MATERIALS AND METHODS

Field study

Site description

The 25-year-old pine stand studied is located near Dronten, The Netherlands, on an alluvial sand layer. The forest is exclusively composed of *Pinus nigra* (Arn.) var. *austriaca* A. et G. There is no undergrowth and the trees are even-aged and equidistant. The soil profile is a podsol with mor humus. The pH values, determined in a soil/water mixture (1:5), were 4·5 for the A_{000} layer, 4·8 for the $A_{00}F$ and 6·4 for the sand layer. The temperature in the $A_{00}F$ layer during the 3 month field experiment (3 December 1982–4 March 1983) ranged from $-2°$ to 10°C.

Exclusion experiment

The experiment was carried out in two sets of ten plots each $1 \times 0·5$ m (Fig. 1). The bottomless plots were constructed of stainless steel plates. The whole set-up was prepared 2 years before the experiment started, so the area could recover from the disturbing construction activities.

From the twenty plots, ten were chosen randomly for removal of animals. The litter was sieved on the location and the densities of *T. minor* of all twenty plots

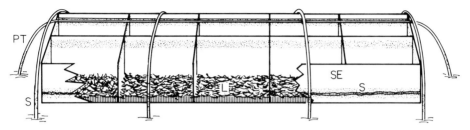

Fig. 1. Set-up of one of the two sets of ten plots of the field experiment. L, litter; PT, plastic tubes; S, strip with glue; SE, stainless steel enclosures.

were established. The litter layers were put back into the plots. *T. minor* was reintroduced into the 'untreated' plots (together with the other soil animals). In the treated plots all soil animals without *T. minor* were put back.

Strips with glue were placed on the inside and outside of the plots, to prevent migration into and out of the plots. Then a gauze net was spread over the plastic tube-frame (see Fig. 1) to prevent animals from falling out of the trees into the plots. As a consequence of this net the input of fresh needles was stopped.

Three soil samples were taken from each plot, and were divided into the A_{000} and $A_{00}F$ layers. At the end of the experiment three more soil samples were taken next to the initial samples. All samples were stored at $-20°C$, until sample preparation took place.

To establish the densities of *T. minor* at the end of the experiment, all litter from the twenty plots was collected, transported to the laboratory and extracted in Tullgren funnels.

Chemical analyses

The frozen soil samples were dried at $45°C$ and cut into $1–2$ cm fragments. The sand fraction was sieved out of the samples. Then they were ground in an electric mill and sieved. The 0.4 mm fraction was kept at $45°C$. To establish the nitrogen and carbon concentrations, these samples, with a minimum dry wt of 1 mg, were analysed in a Carlo-Erba Elemental Analyser 1106.

Laboratory study

Experimental soil systems

The experimental soils were put in perspex cylindrical jars (height, 7.4 cm; diameter, 5.6 cm) with a plaster floor, which was in contact with a continuous water supply. The jars were covered with gauze lids.

At the start of the field experiment (3 December 1982) the soil layers A_{000}, $A_{00}F_1$ and $A_{00}F_2$, were collected near the location of the exclusion plots and stored at $5°C$ until the start of the experiment. Before the jars were filled with either A_{000}, or

$A_{00}F_1$ or $A_{00}F_2$, all faunal elements were removed from the litter. Each treatment had five replicates. The needles from the A_{000} layer were fragmented into 3 cm lengths.

For the fungal treatment experiment the soil layers were inoculated with 12 ml spore suspension (concentration, 9.1×10^6 spores ml^{-1}) of a stock culture of *Verticillium bulbillosum* (W. Gams and Malla). This soil-borne hyphomycete is a common fungus in coniferous forests and appears to be appropriate food for the collembolan *Onychiurus armatus* (Tullb.) (Bengtsson, Gunnarsson & Rundgren 1983). The experiments were performed at 19°C.

Grazing intensity was studied by addition of different numbers of *T. minor* (15 and 50 respectively) to the jars. *T. minor* was collected from the field plots. This was done 2 weeks after the filling of the jars with the soil layers to permit establishment of a microflora (see Anderson *et al.* 1983). During these 2 weeks *T. minor* was acclimatized at 19°C. Only adult specimens were selected. The experiments were performed at 19°C.

For the *T. minor* × *V. bulbillosum* interaction, addition of either *V. bulbillosum* or 15 *T. minor* was compared with simultaneous addition of *V. bulbillosum* and 15 *T. minor*. The experiments were carried out at 19°C. Influence of temperature was measured by carrying out experiments at 13°, 19° and 25°C. This was done with and without 15 *T. minor*, to study the *T. minor* × temperature interaction. Every 3 weeks the numbers of *T. minor* were counted in all of the jars. Dead animals were replaced and all juveniles were removed.

Chemical analyses

Samples from the three soil layers were dried at the start and at the end of the experiments. The total duration of the experiments without animals was 128 days, and that of the experiments with animals was 114 days. The measurements of nitrogen and carbon concentrations were as described for the field experiment.

Analysis of data

The numbers of *T. minor* in the field experiment were log transformed and are given as geometric means. Three-way, two-way and one-way analyses of variance were applied to the results of the laboratory experiments.

RESULTS

Field data

Nitrogen dynamics during a 3 month winter period

At the start of the field experiment the nitrogen concentration of the A_{000} layer was lower than that of the deeper $A_{00}F$ layer (Table 1). There was a similar difference at the end of the experiment although the nitrogen concentration increased in both

TABLE 1. Nitrogen concentration and carbon/nitrogen ratio for two layers of *Pinus nigra* needle litter at the start and the end of the field experiment. The number of samples is *n*, and means are quoted with their standard errors

		Soil layer			
		A_{000}		$A_{00}F$	
Date	*n*	N conc. (mg g^{-1})	C/N	N conc. (mg g^{-1})	C/N
3 Dec 1982	20	$8 \cdot 72 \pm 0 \cdot 128$	$57 \cdot 7 \pm 2 \cdot 96$	$12 \cdot 67 \pm 0 \cdot 112$	$32 \cdot 1 \pm 0 \cdot 73$
4 March 1983	20	$11 \cdot 64 \pm 0 \cdot 108$	$45 \cdot 4 \pm 1 \cdot 47$	$15 \cdot 25 \pm 0 \cdot 095$	$31 \cdot 4 \pm 0 \cdot 63$

layers. The C/N ratio of the A_{000} layer was higher than that of the $A_{00}F$ layer. In the A_{000} layer it decreased, whereas it stayed constant in the $A_{00}F$ layer.

Effects of Tomocerus minor *exclusion*

The mean density of *T. minor* for all plots at the start was 4169 m^{-2} (Table 2). After 3 months the numbers of *T. minor* had not changed in the untreated plots. On the other hand in the treated plots lower numbers ($P<0\cdot001$) were found. It may be assumed that immediately after the removal of animals this number was even lower.

Comparison of the plots showed that the initial nitrogen concentrations of the untreated plots had a larger variance than that of the treated plots. We therefore selected plots with comparable start concentrations. The removal of animals had no effect in the A_{000} layer ($F_{s1,11} = 0\cdot02$; NS). However, in the $A_{00}F$ layer a significant difference was found ($F_{s1,10} = 4\cdot98$; $P<0\cdot05$). In the untreated plots the nitrogen concentration increased by a factor of $2\cdot3$ over the treated plots.

TABLE 2. Densities of *T. minor* in the experimental plots at the start and the end of the field experiment. The untreated plots had their numbers of *T. minor* replaced prior to the experiment, whereas the treated plots had no replenishment with *T. minor*. The number of samples is *n*, and geometric means are quoted with their 95% confidence limits

		Density of *T. minor* (geometric means m^{-2})		
Date	*n*	Untreated plots		Treated plots
3 Dec 1982	20		4169	
			(2630−6607)	
4 March 1983	10	3236		1047
		(2455−4266)		(794−1380)

Laboratory data

Effects of fungal treatment on nitrogen concentration

The initial nitrogen concentrations of the three soil layers were 0·9% for the A_{000} layer, 1·7% for the $A_{00}F_1$ layer and 1·3% for the $A_{00}F_2$ layer (Fig. 2a). At the end of

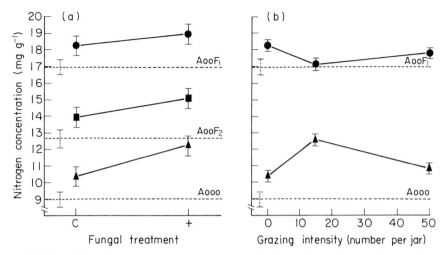

FIG. 2. (a) Effect of fungal treatment on nitrogen concentration in the three soil layers. C, control; +, inoculation with *Verticillium bulbillosum*; initial concentrations (- - - -); concentration in A_{000} (▲), $A_{00}F_1$ (●) and $A_{00}F_2$ (■): $n = 5$, and bars represent one standard error. (b) Effect of grazing intensity of *T. minor* on nitrogen concentration in two soil layers. Initial concentrations (- - - -); concentration in A_{000} (▲) and $A_{00}F_1$ (●); $n = 5$, and means ± standard error are plotted.

the experiment the concentrations had increased slightly in all three layers. The only significant effect on nitrogen concentration of the inoculation with *Verticillium bulbillosum* was found in layer A_{000} (Fig. 2a). In this layer a clear increase of fungal hyphae was also shown.

Effects of grazing intensity on nitrogen concentration

The effects of the addition of different numbers of *T. minor* to the systems could only be estimated for the A_{000} and the $A_{00}F_1$ layer, as in the rather dense and wet $A_{00}F_2$ layer mortality after 1 week was about 67% for both densities. In A_{000} mean mortality was 14% and 29%, and in $A_{00}F_1$ 24% and 35%, for the 15 and 50 densities respectively.

Two-way analysis of variance of all data shows that the nitrogen changes in the two layers depended on the grazing of *T. minor* ($F_{s4,36} = 3 \cdot 571$, $P < 0 \cdot 05$). In the A_{000} layer, at a density of 15 *T. minor*, there was a significant increase of nitrogen concentration compared with the control (Fig. 2b). At a density of 50 *T. minor* the concentration was not significantly different from the control. Thus, the relation between density and change in nitrogen concentration is not linear. In the $A_{00}F_1$ layer, at a density of 15 animals, there was a small decrease in nitrogen concentration, whereas with 50 animals there was an increase.

Interaction between Tomocerus minor *and* Verticillium bulbillosum

Three-way analysis of variance gives a significant interaction between soil layer, *T. minor* and *V. bulbillosum* ($F_{s1.32} = 20 \cdot 551$, $P < 0 \cdot 001$). Furthermore, the effect

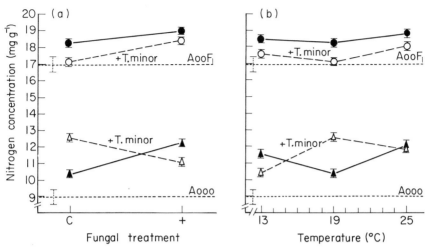

FIG. 3. (a) Effect of addition of 15 *T. minor* and fungi on nitrogen concentration in two soil layers; C, control; +, inoculation with *V. bulbillosum*; initial concentrations (----); concentration in A_{000} with ($\triangle$) and without ($\blacktriangle$) *T. minor*; in $A_{00}F_1$ with ($\bigcirc$) and without ($\bullet$) *T. minor*. All points are means of five samples, plotted with 1 standard error. (b) Effect of temperature and addition of 15 *T. minor* on nitrogen concentration in two soil layers: initial concentrations (----); concentration in A_{000} with ($\triangle$) and without ($\blacktriangle$) *T. minor*; in $A_{00}F_1$ with ($\bigcirc$) and without ($\bullet$) *T. minor*. All points are means of five samples, plotted with 1 standard error.

of fungal inoculation on the nitrogen concentration depended on the presence of *T. minor* ($F_{s1,32}$ = 10·026, P<0·01). Figure 3a shows the effects of simultaneous addition of both animals and fungi. In the A_{000} layer it can be seen that the positive reaction due to addition of fungi or animals was reduced by simultaneous addition of animals and fungi. In the $A_{00}F_1$ layer the significant effect of addition of animals was absent when animals and fungi were added simultaneously.

Effects of temperature on nitrogen concentration

There was a significant overall effect of temperature on nitrogen concentration (analysis of variance; $F_{s2,72}$ = 13·118, P<0·001). In Fig. 3b it can be seen that in the A_{000} layer the increase in nitrogen concentration was significantly higher at 13 °C and 25 °C than at 19 °C. In the $A_{00}F_1$ layer there was no effect of temperature on nitrogen concentration.

Interaction between Tomocerus minor *and temperature*

Three-way analysis of variance gives a significant interaction between soil layer, temperature and *T. minor* ($F_{s4,72}$=3·912, P<0·01). In the A_{000} layer (Fig. 3b) the effects of 15 *T. minor* at 19°C and 25°C were significantly greater than the effect at 13°C. In the $A_{00}F_1$ layer the small decrease in the presence of *T. minor* occurred at all three temperatures.

DISCUSSION

In the present research the nitrogen dynamics of *Pinus nigra* litter has been studied over a period of 3 months, a short period considering that complete coniferous litter decomposition takes several years (Berg & Söderström 1979). However, by studying decomposition in different (upper and lower) litter layers, it was possible to measure nitrogen dynamics during successive stages of decomposition.

In the present study the C/N value decreased from 57·7 to 45·4 in the A_{000} layer, and remained constant (approximately 31·5) in the $A_{00}F$ layer. Further, qualitative observation showed more fungi in the $A_{00}F$ layer than in the A_{000} layer. According to Berg & Staaf (1981) three phases may be distinguished in the nitrogen dynamics of *Pinus* litter: a rather short leaching phase (leaching was hardly measurable in the present research (unpublished data) and will not be considered), an immobilization phase (characterized by large C/N values which decrease with time and low initial fungal biomass) and a release phase (characterized by a low and constant C/N value and large fungal biomass). Thus, it may be concluded that in the A_{000} layer immobilization takes place and in the $A_{00}F$ layer release of nitrogen is occurring.

In the absence of *T. minor*, the results show that immobilization in layer A_{000} can be influenced by the addition of soil fungi. In the $A_{00}F_{1,2}$ layers, which are rich in fungi, addition of fungi had no effect. This may be due to the stabilized situation of the fungal community.

In the absence of *T. minor* immobilization was higher at 13°C and 25°C than at 19°C in the A_{000} layer. This may be caused by the fact that fungi common in *Pinus* litter, such as basidiomycetes and *Penicillium* spp., have their optimal temperatures at about 13°C and 25°C, respectively (C. W. Gams, pers. comm.). In the $A_{00}F_1$ layer temperature had no effect, as was found for nitrogen mineralization in deciduous litter (Anderson *et al.* 1983).

In the presence of *T. minor* immobilization in the A_{000} layer increased and in the $A_{00}F_1$ layer there were signs that release increased. It can be concluded that *T. minor* amplifies nitrogen fluxes in the different soil layers (cf. Anderson *et al.* 1983). The effects, however, depend on the density of *T. minor*: higher densities tend to decrease the effect, as was also found for millipedes (Anderson *et al.* 1983). This may be caused by the fact that the grazing rate of *T. minor* at higher densities exceeds the production of fungal hyphae leading to a lower fungal standing crop.

In the A_{000} layer the effect of *T. minor* were greater at 19°C and 25°C than at 13°C. The differences might be caused by an increased grazing rate at higher temperatures. In the $A_{00}F_1$ layer, temperature did not influence the effects of *T. minor*. No explanation can be given for the absence of temperature effects in this layer.

In the field experiment there was no effect of exclusion of *T. minor* in the A_{000} layer. This can be explained by the fact that during the drought of the winter period, the drought-sensitive *T. minor* moves down to deeper soil layers (Verhoef 1981; Verhoef & Witteveen 1980). In the $A_{00}F$ layer, which is supposed to be in the release phase, the results indicate that the effect of *T. minor* is inhibitory in the untreated plots, compared to the treated ones. This inhibition may be caused by the densities of *T. minor* in the untreated plots, which were relatively high, probably due to the mild winter (cf. van Straalen 1983). As mentioned before, high densities may have a negative effect.

Finally, it can be concluded that *T. minor* influences the nitrogen dynamics in *Pinus nigra* litter. Its effects depend strongly on the existing fluxes in the soil layers, its own density and temperature. In future research moisture changes, which influence greatly both the distribution and local density of the Collembola (Verhoef & Nagelkerke 1977; Verhoef & van Selm 1983), will be examined for their effects on the nitrogen dynamics in *Pinus* litter.

ACKNOWLEDGMENTS

The authors are grateful to Prof.dr E. N. G. Joosse and Prof.dr H. H. Boer for helpful comments on the manuscript, to Dr N. M. van Straalen for statistical advice, to Mr J. H. van Meerendonk and Mr C. S. Verhoef for technical assistance and to Miss M. S. Benninga for typing the manuscript.

REFERENCES

Anderson, J.M., Ineson, P. & Huish, S.A. (1983). The effects of animal feeding activities on element release from deciduous forest litter and soil organic matter. *Proceedings VIII International Soil Zoology Colloquium* (Ed. P. Lebrun), pp. 87–99. Dieu-Brichart Louvain-la-Neuve.

Anderson, R.V., Coleman, D.C. & Cole, C.V. (1981). Effects of saprotrophic grazing on net mineralization. *Terrestrial Nitrogen Cycles* (Ed. by F.E. Clark & T. Rosswall). *Ecological Bulletins (Stockholm)*, **33**, 201–16.

Bååth, E., Lohm, U., Lundgren, B., Rosswall, T., Söderström, B., Sohlenius, B. & Wirèn, A. (1978). The effect of nitrogen and carbon supply on the development of soil organism populations and pine seedlings: a microcosm experiment. *Oikos*, **31**, 153–56.

Bengtsson, G. & Rundgren, S. (1983). Respiration and growth of a fungus, *Mortierella isabellina*, in response to grazing by *Onychiurus armatus* (Collembola). *Soil Biology & Biochemistry*, **15**, 469–73.

Bengtsson, G., Gunnarsson, T. & Rundgren, S. (1983). Growth changes caused by metal uptake in a population of *Onychiurus armatus* (Tullb.) (Collembola) feeding on metal polluted fungi. *Oikos*, **40**, 216–25.

Berg, B. & Söderström, B. (1979). Fungal biomass and nitrogen in decomposing Scots pine needle litter. *Soil Biology & Biochemistry*, **11**, 339–41.

Berg, B. & Staaf, H. (1981). Leaching, accumulation and release of nitrogen in decomposing forest litter. *Terrestrial Nitrogen Cycles* (Ed. by F.E. Clark & T. Rosswall). *Ecological Bulletins (Stockholm)*, **33**, 163–78.

Cole, C.V., Elliot, E.T., Hunt, H.W. & Coleman, D.C. (1978). Trophic interactions in soils as they affect energy and nutrient dynamics, V. Phosphorus transformations. *Microbial Ecology*, **4**, 381–87.

Hanlon, R.D.G. & Anderson, J.M. (1979). The effects of Collembola grazing on microbial activity in decomposing leaf litter. *Oecologia (Berlin)*, **38**, 93–99.

Ineson, P., Leonard, M.A. & Anderson, J.M. (1982). Effect of collembolan grazing upon nitrogen and cation leaching from decomposing leaf litter. *Soil Biology & Biochemistry*, **14**, 601–05.

Persson, T. (1983). Influence of soil animals on nitrogen mineralization in the northern Scots pine forest. *Proceedings VIII International Soil Zoology Colloquium* (Ed. P. Lebrun), pp. 117–26. Dieu-Brichart, Louvain-la-Neuve.

Straalen van, N.M. (1983). Vergelijkende demografie van springstaarten. Thesis. Free University, Amsterdam 206 pp.

Verhoef, H.A. (1981). Water balance in Collembola and its relation to habitat selection; water content, haemolymph osmotic pressure and transpiration during an instar. *Journal of Insect Physiology*, **27**, 755–60.

Verhoef, H.A. & Nagelkerke, C.J. (1977). Formation and ecological significance of aggregations in Collembola; an experimental study. *Oecologia (Berlin)*, **31**, 215–26.

Verhoef, H.A. & van Selm, A.J. (1983). Distribution and population dynamics of Collembola in relation to soil moisture. *Holarctic Ecology*, **6**, 387–94.

Verhoef, H.A. & Witteveen, J. (1980). Water balance in Collembola and its relation to habitat selection; cuticular water loss and water uptake. *Journal of Insect Physiology*, **26**, 201–08.

Visser, S., Whittaker, J.B. & Parkinson, D. (1981). Effects of collembolan grazing on nutrient release and respiration of a leaf litter inhabiting fungus. *Soil Biology & Biochemistry*, **13**, 215–18.

Witkamp, M. & Frank, M.L. (1970). Effects of temperature, rainfall, and fauna on transfer of Cesium-137, K, Mg and mass in consumer-decomposer microcosms. *Ecology*, **51**, 465–74.

Interactions of invertebrates, micro-organisms and tree roots in nitrogen and mineral element fluxes in deciduous woodland soils

J. M. ANDERSON, S. A. HUISH, P. INESON*, M. A. LEONARD
AND P. R. SPLATT

Wolfson Ecology Laboratory, Department of Biological Sciences, University of Exeter, Exeter EX4 4PS

SUMMARY

1 There is increasing evidence that soil invertebrates have quantitatively important roles in the nutrient flux pathways of forest soils, particularly in the mobilization and mineralization of nitrogen through direct and indirect effects.

2 The direct effects are simple transfers of elements through food webs which can be estimated, with variable difficulty, from population data. Indirect effects involve feedbacks on microbial populations and activities, and are poorly quantified.

3 Short-term, indirect effects of macrofaunal feeding activities result in enhanced nitrification and ammonification in forest litters and soil organic matter.

4 A model is proposed which attempts to quantify and predict nitrogen mineralization through interactions of micro-organisms and invertebrate saprotrophs.

5 Field experiments using microcosms and small lysimeters, both with integrated tree root systems, demonstrate further interactive effects of animals on nitrogen fluxes during wetting and drying cycles. There are also indications that animals may facilitate mineral uptake by roots.

INTRODUCTION

The organic nitrogen pools within litter and soil organic matter (SOM) are the main sources of nitrogen which sustain primary productivity in natural forests or unfertilized forestry systems (Aber & Melillo 1980). The timing and extent of the mobilization of this nitrogen (net mineralization) in relation to plant requirements is a key process governing the successional development and the climax productivity of the system (Clark & Rosswall 1981; Melillo & Gosz 1983). In addition, the response of the nitrogen cycle to disturbance, including various aspects of management practices, is to a large extent determined by the capacity of the litter and soil to retain or release nitrogen (Miller 1979; Swank & Waide 1980; Heal, Swift & Anderson 1982; Vitousek *et al.* 1982).

  Net mineralization represents the balance between the key processes of gross nitrogen mobilization (predominantly as ammonium but also as simple organic forms, such as amino acids, which are available for root mycorrhizal uptake) and

---

*Institute of Terrestrial Ecology, Merlewood Research Station, Grange-over-Sands, Cumbria, LA11 6JU.

immobilization in decomposer biomass or through absorption of ammonium onto soil (Nömmik 1981; Kudeyarov 1981). The availability of nitrogen for root uptake may also be influenced by the balance between rates of nitrification and losses from the system as a consequence of denitrification and leaching. Most terrestrial ecosystems are highly conservative of nitrogen–especially forests with acid, organic soils–and losses of nitrogen are small compared with the internal fluxes. Therefore, the net mineralized nitrogen potentially available to higher plants represents the fraction of gross nitrogen mineralization which exceeds microbial demand (Runge 1971) and ammonium fixation. These latter two processes will be inversely related to the availability of nitrogen (and phosphorus) in the system as a consequence of nitrogen limitations on decomposition processes and the accumulation of large SOM pools. The processes which govern the relationships between net mineralization, gross mineralization and the nitrogen pool in SOM are therefore critical parameters of ecosystem functioning in forest soils.

At a systems level the operation of these processes and their correlation with other ecological variables and physical environmental parameters are increasingly well documented, but there has been little explicit recognition of the complexity of mechanisms which regulate nutrient fluxes between litter, soil and plant roots (Frizzel & van Veen 1982), particularly in forest ecosystems. Specifically, the role of soil fauna in nitrogen flux pathways has received scant attention and the rationale for nitrogen mineralization processes is usually based almost entirely on microbial processes. However, there is mounting evidence that soil animals significantly disrupt the time course of microbial processes and enhance the turnover of microbial populations through direct and indirect effects (Anderson & Ineson 1984).

DIRECT AND INDIRECT EFFECTS OF ANIMALS ON NITROGEN MINERALIZATION

The direct effects of invertebrates on fungi and bacteria involve the release of elements through trophic transfers and population processes: feeding, excretion, and the turnover of secondary production. The indirect effects include feedbacks to lower trophic levels: for example, the effects of litter comminution and grazing on microbial activity and the functional organization of bacterial and fungal communities (Anderson & Ineson 1983), as well as predator/prey interactions. The central problem is quantifying both the direct and indirect effects of soil invertebrates on microbial processes in the context of ecosystem-level nutrient fluxes.

The direct contribution of invertebrates to nutrient fluxes can be quantified using a budgetary approach where sufficient information on the biology and population ecology of the decomposer community is available. Estimates of nitrogen turnover by invertebrate populations have been recently reviewed by Anderson, Coleman & Cole (1981), Seastedt (1984) and Anderson & Ineson (1984). Results from studies in a wide range of forest types suggest that in temperate forest soils with a mean annual invertebrate biomass of about 5–10 g (dry wt) m$^-$ (i.e. 20–30 g fresh wt) the annual turnover of nitrogen and other minerals by fauna can equal or exceed inputs to the decomposer system. In base-rich, deciduous woodland soils with an earthworm biomass of 50–100 g (fresh wt) m^{-2}, or even more (Satchell 1983), the annual nitrogen flux through the worm population may be several times the 30–70 kg ha^{-1} year^{-1} contained in leaf fall (Satchell 1963). At the other extreme, in an acid pine forest soil in Sweden, Persson (1983) has estimated that the fauna, with a mean annual biomass of only 1–7 g (dry wt) m^{-2} (in comparison with 120 g m^{-2} fungi and 39 g m^{-2} bacteria) contributed between 10 and 49% of annual nitrogen mineralization, of which 70% was excretion by bacterivores and fungivores.

We have demonstrated significant indirect effects of macrofauna on nitrogen release from forest litter and SOM in the laboratory (Anderson, Ineson & Huish 1983; Anderson & Ineson 1984). For example, millipedes feeding on oak leaf litter with a C:N ratio of 100 over a period of 15 weeks effected the release of 9·7% of the nitrogen capital in the leaves compared with 3% from controls without animals. Experiments using ^{15}N labelled millipedes and leaf litter showed that only 7·5% of the mobilized nitrogen was attributable to the excretion of the animals (Anderson & Ineson 1984). The indirect nature of these effects, through interactions with bacteria and fungi in the litter, is also demonstrated by the fact that the nitrogen release does not occur in a step-wise manner when the animals are added or removed from the experimental system, but takes several weeks to build up or decline (Anderson & Ineson 1984). The enhancement of nitrification by animal feeding activities is also indicative of indirect effects on nitrogen mineralization (since invertebrates do not excrete nitrate).

We review here a series of experiments which attempt to validate and quantify these effects on nitrogen mineralization both in the laboratory and in the field. The emphasis in this paper is on the general approach since detailed discussion of results cannot be made here. We firstly consider the variables influencing animal-mediated nutrient fluxes and propose a series of models quantifying animal and microbial effects. We then consider the dynamics of nitrogen fluxes in field experiments with tree root systems, using microcosms and lysimeters to demonstrate animal effects at the ecosystem level.

PROCESS VARIABLES OF ANIMAL-MEDIATED MINERAL NITROGEN FLUXES

The process variables determining nutrient flux rates can be considered within a framework of resource type, the qualitative and quantitative characteristics of the organism community and physical environmental parameters (Swift, Heal & Anderson 1979).

Resource type

As a first step towards quantifying indirect animal effects we added millipedes, *Glomeris marginata*, to a range of leaf litter (L), fermentation layer (F) and humus

TABLE 1. Effects of feeding by *Glomeris* on nitrogen leaching from different organic soil layers of three oak woodlands in Devon

Soil layer	Stoke Wood		Perridge Wood		Brook Wood	
	Control	Animals	Control	Animals	Control	Animals
Litter layer (L)						
NH$_4$–N	18·8 ± 5·7***	183·3± 10·3	23·7 ± 3·2*	251·1 ± 81·3	38·0 ± 19·5***	591·3 ± 69·6
NO$_3$–N	1·6 ± 0·4*	3·5 ± 0·7	1·7 ± 0·4	3·7 ± 1·0	1·1 ± 0·5	5·5 ± 1·5
Fermentation layer (F)						
NH$_4$–N	37·2 ± 8·4***	259·3 ± 24·3	217·9 ± 50·7**	513·9 ± 48·2	716·4 ± 194·4*	1333·5 ± 90·0
NO$_3$–N	46·8 ± 28·2	16·4 ± 4·4	12·7 ± 9·5	4·4 ± 0·6	17·6 ± 5·2	61·2 ± 31·6
Humus layer (H)						
NH$_4$–N	15·3 ± 2·1*	188·7 ± 50·7	131·0 ± 7·2***	214·7 ± 4·1	107·1 ± 6·5**	188·8 ± 17·5
NO$_3$–N	116·1 ± 0·5***	384·5 ± 50·3	2·3 ± 0·3	2·1 ± 0·2	13·9 ± 3·2	9·5 ± 2·0

Results are expressed as mean cumulative concentrations of mineral nitrogen (μg g dry litter^{-1} ± SE; $n = 3$) mobilized as ammonium-N or nitrate-N over a period of 5 weeks after the addition of four animals to the animal treatments. Values for controls (without animals) are shown for the same period of time. The experiments were incubated at 15°C. Significant differences between animal treatments and controls are shown as *** ($P < 0·001$), ** ($0·001 < P < 0·01$) and * ($0·01 < P < 0·05$) (Anderson & Ineson 1984).

materials (H) to investigate the effects of resource type on animal mediated nitrogen fluxes (Anderson & Ineson 1984). Samples were collected from three oak–beech woodlands (*Quercus robur* with small *Fagus sylvatica*) in May 1982. Materials were air dried, lightly crushed and then sieved to prepare a 5–20 mm fraction of litter and a 1–3 mm fraction of humus. Aliquots of 1·5 g litter and 3 g humus were added to microcosm chambers (Anderson & Ineson 1982), rehydrated with distilled water and inoculated with a suspension of fresh litter or humus. The chambers were then incubated at 15°C for 5 weeks before animals, four *Glomeris marginata* weighing 0·4–0·5 g (fresh wt), were added to chambers designated as animal treatments. The chambers were irrigated every 7 days with distilled water and the leachates analysed for mineral-N as nitrate and ammonium.

The results summarized in Table 1 show that the animals significantly enhanced ammonification in all sites and resource types, though the magnitude of the response is variable both within and between sites. The effects on nitrification are more variable and insignificant in most cases except for the Stoke Woods humus. This H-layer material showed a characteristic dominance of nitrate-N in controls which was maintained in animal treatments although the effects of animals were larger on the ammonium-N fluxes (enhanced twelve times) than on the nitrate-N fluxes (enhanced three times).

Losses of ammonium through nitrification complicate interpretation and therefore we will briefly consider these results in terms of total mineral nitrogen. Animal feeding activities increased mineral-N losses from the L-layer materials by ten to fifteen times control levels but only by two to three times in the F-layers and by even less (1·6 times) in the H-layer materials from Perridge and Brook Wood. The humus from Stoke Woods showed total mineral-N losses over four times control levels in animal treatments.

We interpret these differences in control and animal-mediated fluxes, both within and between sites, in terms of the reduced availability of nitrogen to microorganisms with the advancing stages of decomposition (Hayes & Swift 1978). Microbial immobilization of nitrogen is generally considered characteristic of the early stages of decomposition, especially in low quality resources, but this is disrupted by the feeding activities of animals in the litter-layers. The animals may therefore reduce nitrogen transfers to the more intractable humus pool and mobilize nitrogen in close proximity to the root/mycorrhizal network which is characteristically located at the litter/humus interface in these woodlands. In the Stoke Woods site, the higher availability of nitrogen in the humus material is reflected by the larger animal effects on nitrogen fluxes than in humus from the other two sites. The level of nitrification which is found in this site is not reflected by differences between Stoke Woods and similar woodlands in the area in terms of soil pH, total nitrogen content of the humus (ranging from 1·3 to 1·5%), parent minerals or vegetation type (Anderson *et al.* 1985). The scale of animal effects on humus nitrogen mineralization does, however, differentiate this site from the others and we are investigating the use of this response as an index of nitrogen availability.

Organisms

Our approach in these studies has been to try and find a level at which to compartmentalize animal–microbial interactions of individual species or organism groups. As a first step towards this abstraction, a necessary requisite for a simple model, we have investigated the conditioning of litter by different species of fungi (Anderson & Ineson 1984) and the activities of different faunal groups in mobilizing nitrogen (Anderson, Ineson & Huish 1983; Anderson & Ineson 1984).

Freshly fallen oak litter was collected from litter traps, autoclaved and then inoculated separately with six species of fungi, including four basidiomycetes, and incubated for several weeks before millipedes were added. The results showed that the mineral nitrogen mobilized by the animals (N_a) was related to the nitrogen mineralization by the controls (N_c) according to the function

$$N_a = 9 \cdot 3 \ N_c - 124 \cdot 2 \ (r = 0 \cdot 74, f = 16, P < 0 \cdot 001)$$

where N_a and N_c are in μg g leaves^{-1}. This function is specific to this resource type but suggests that effects of the animals transcend the taxonomic identity of the fungus. We have no indications from all the other experiments we have carried out using mixed inocula from homogenized litter to doubt the general validity of these conclusions, but the effects of fungal species conditioning leaf litter on animal feeding activities requires further study.

A similar approach was used to investigate the roles of different invertebrate groups on nitrogen mineralization (Anderson, Ineson & Huish 1983). Enchytraeids, collembola, earthworms and a range of millipede groups were added to oak leaf litter in numbers chosen to demonstrate the maximum likely effects which might occur within aggregated field populations. It was found that the overall effect of animal groups on ammonium-N release in mg g litter^{-1} over a 6-week period related linearly to biomass (B) acording to the function

$$N = 1 \cdot 6 \ B \ (r = 0 \cdot 86, f = 22, P < 0 \cdot 001)$$

where B is expressed in mg fresh wt of animal. This relationship, although again resource specific, suggests that the taxonomic identity of the animals is not as significant a variable as their biomass and activity. We acknowledge that these relationships need further investigation but in the light of these results we have chosen the millipede *Glomeris marginata* as a tool for modelling the indirect effect of litter-feeding animals on nitrogen mineralization.

ANREG: nitrogen mineralization as a function of temperature and animal biomass

A regression model has been developed quantifying and predicting the effects of soil and litter macrofauna on nitrogen mineralization. Leaf litter (F) and humus (H) materials collected from three oak–beech woodlands in Devon (Perridge, Stoke and Hillersdon Woods) were incubated in laboratory microcosms, with and without animals, at three temperatures (5, 10 and 15°C) which are representative of the mean monthly F–H layer temperatures recorded in the field (Fig. 2c) when soil fauna are active. At higher temperatures moisture usually limits animals and microbial activities, especially in the litter layers, as considered below. Three animal treatments were used which were considered to be high but not unrealistic of biomass (fresh wt) to resource (dry wt) quotients found in the field: 0.1, 0.2 and 0.3 g animals g litter^{-1} and 0.03, 0.07 and 0.1 g animals g humus^{-1}. The methods are essentially the same as those outlined above for resource types.

Nitrogen mineralized as nitrate-N or ammonium-N (N_A in μg N g resource^{-1}) from animal treatments (B as defined above and including controls as zero biomass) and at the three temperatures ($T°C$) was monitored weekly over a period of 8 weeks after the animals were added. Periods of 5, 6 and 7 weeks were analysed using the following multiple regression (which we term ANREG):

$$N_A = a + bT + cB + dBT$$

Nitrogen mineralized over 6 weeks showed the best fit to ANREG for all sites, resource types and forms of mineral nitrogen. For the purposes of the present discussion we will only consider ANREG functions for total nitrogen in Perridge and Stoke Woods to illustrate the model. Details are published by Anderson *et al.* (1985).

Equations for the litter are ($\pm$ SE for partial regression coefficients):

Perridge: $N_A = 14.3(\pm38.3) + 31.6(\pm3.5)T - 135(\pm102)B + 47.8(\pm9.5) BT$
$$R^2\% = 94.2, F_{3,32} = 191.7, P < 0.01$$
Stoke: $N_A = 171(\pm108.7) + 41.4(\pm10.1)T + 504(\pm191)B + 74.6(\pm26.9) BT$
$$R^2\% = 87.6, F_{3,32} = 158.7, P<0.01$$

and for the humus:

Perridge: $N_A = 26.5(\pm3.8) + 4.2(\pm0.4)T + 15.0(\pm10.3)B - 0.07(\pm0.9) BT$
$$R^2\% = 92.1, F_{3,32} = 158.7, P < 0.01$$
Stoke: $N_A = 165(\pm28) + 8.0(\pm2.6)T - 115(\pm74.7)B + 21.7(\pm6.9) BT$
$$R^2\% = 75.6, F_{3,32} = 37.1, P<0.01$$

The resource quality differences between the litter and humus materials are reflected in the partial regression coefficients for temperature and biomass which vary, significantly in most cases, between sites as well as resources.

There is generally poor agreement between levels of nitrogen mineralization predicted by ANREG for Stoke and Perridge Woods and the values shown in Table 1. Specifically, that ANREG tends to overestimate nitrogen mineralization from the controls (run with zero biomass) and underestimate animal effects (the sum of B and BT functions). These differences will not be discussed in detail because their magnitude and significance is less important here than the value emphasizing the uncontrolled parameters which may preclude comparisons of these experiments.

It can be seen from Table 1 that surface leaf litter (L) showed low levels of nitrogen mineralization by controls but this was enhanced ten to fifteen times by the animals. As decomposition progresses and the material becomes classifiable as F-layer, microbial mineralization of nitrogen increases but the scale of the indirect animal effects generally decreases. Finally, in the H-layers of these woodland types the nitrogen in humus usually becomes more intractable to animal and microbial activities (Stoke Woods is the main exception to this pattern which we have encountered). The partitioning of these materials according to time–depth relationships is therefore critical for experimental work since the combination of resource types with different net nitrogen mineralization potentials will influence the interpretation of animal and microbial effects. But the relationship between the decomposition time-series and soil horizons is open to question for the reasons outlined by Swift *et al.* (1979) in formulating a cascade model of decomposition processes. Particulate materials and soluble organic matter will be separated from the parent litter-class by comminution and leaching and move down the profile to depths determined by the resource characteristics, animal activities, soil structure and processes of transport. These cascade processes have not been quantified in any soils but the overall magnitude and their seasonal pattern must be a fundamental characteristic of different forest types.

ANREG was not determined for L-layer materials because previous experiments had led us to believe that the seasonal change in net nitrogen mineralization potential of this litter was more variable than materials from the F- and H-layers. We now feel that small differences in the combination of these resources, whether a feature of sampling procedure or through seasonal variations in animal activities, can influence the outcome and interpretation of these experiments. ANREG, however, offers a means of quantifying these effects including, for the first time, a measure of the indirect effects of macrofauna on nitrogen mineralization. Thus, on the basis of the present model we predict that litter-feeding animals will significantly contribute to nitrogen mineralization in forest soils where the biomass/resource quotient B is greater than 0·1 (Anderson *et al.* 1985). Under these conditions net nitrogen mineralization may be occurring in the field where laboratory incubations, without animals, show nitrogen immobilization. As an intermediate step towards investigating these effects in the field, and for understanding the functioning of complex systems, subject to extremes of temperature and precipitation, we established tree roots in the microcosms to investigate animal and root interactions in mineral element fluxes.

ANIMAL AND ROOT EFFECTS IN MICROCOSMS

The standard experimental chambers used in laboratory studies (Anderson & Ineson 1982) were set up in the Perridge field site with root systems established in a small quantity of F-layer material beneath the inner chambers containing the experimental material. Oak tree root systems, with extensive ectomycorrhizal development, were carefully dissected from the surface humus layers and 10–15 cm of the attached root network were introduced through a slit in the outer chambers which were then sealed with silicone grease. The chambers were estab-

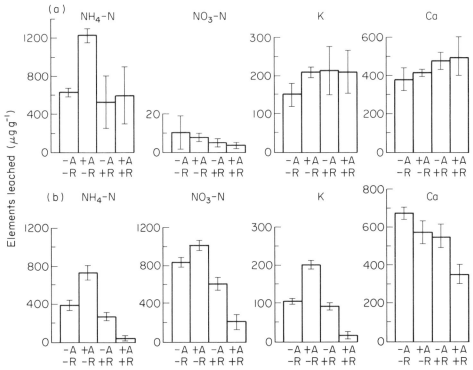

FIG. 1. Nitrogen, potassium and calcium losses from leaf litter and humus in field experiments using microcosms with and without animals and tree roots. Values are cumulative (μg g^{-1} material $\pm$SE) over 6 weeks for the Perridge F-layer material (a) and over 4 weeks for the Stoke Woods H layer materials (b) in experiments carried out sequentially using the same undisturbed rooting systems (for further details see text).

lished in early May using Perridge F-layer material and the animals introduced 8 weeks later when the root/mycorrhizal systems appeared healthy and showed growth. Leachate concentrations of mineral nitrogen, potassium and calcium were monitored for 6 weeks. In mid-August the inner chambers were then replaced with units containing humus material from Stoke Woods which had been maintained in the laboratory at 15°C, with and without animals, for several weeks before being set up in the field microcosms. The results of these two experiments are shown in Fig. 1.

Losses of nitrogen as nitrate were negligible from all the Perridge F-layer treatments (Fig. 1a) while treatments with animals showed nearly twice the cumulative ammonium-N losses of treatments without animals. The treatments with roots, but without animals, showed very similar ammonium-N losses to the non-rooted treatments suggesting negligible uptake of mineral nitrogen by the roots. But the differences between nitrogen losses from the rooted systems with and without animals suggest that there has been a transfer of all the animal-mobilized nitrogen to the roots. Potassium losses show a 30% increase attributable to animals in the non-rooted systems but both of the treatments with roots showed similar increases even when animals were absent. Calcium losses are also higher in the rooted treatments, and there are no significant animal effects.

The apparent facilitation by animals of ammonium uptake by roots is a consistent feature of similar experiments using root implants in microcosms but one which we are unable to explain at the present time. The increased potassium and calcium losses from the rooted systems are more readily attributable to root exudates. Smith (1976) showed that potassium and calcium were the dominant cations in exudates from unsuberized roots of *Betula*, *Fagus* and *Acer* species. Cumulative values over 14 days ranging from 4·2 to 12·8 μg K mg root^{-1} and 2·0 to 4·8 μg Ca mg root^{-1} are compatible with the scale of effects noted here.

The dynamics of nitrogen, potassium and calcium fluxes are totally different when the inner chambers containing Perridge litter were replaced with the Stoke Woods humus layer material (Fig. 1b). It should be noted that the same rooting systems were involved and the roots were not disturbed by the sample changes. There were also no major changes in soil temperatures over the period of the two experiments. The transition in the dynamics of the system was abrupt when the Stoke Woods material was added to the systems and the predominantly ammonium-N fluxes of the F-layer material was replaced with the strongly nitrifying H-layer material. Approximately 60% of the mineral nitrogen released from the humus was in the form of nitrate.

Interpretation of the ammonium-N fluxes, with and without animals, is complicated by losses of ammonium through nitrification. In terms of total nitrogen, treatments without animals and roots showed losses of 1·2 mg compared with 1·8 mg in the presence of animals; an increase of 50% net mineralization through animal effects. Total nitrogen losses from the rooted and non-rooted treatments, without animals, were not significantly different (1·1 mg g^{-1} and 1·2 mg g^{-1} respectively). Only 0·2 mg was recovered from the rooted treatments with animals. This suggests that 80% of the animal-mobilized nitrogen was taken up by the roots, half in the form of nitrate, and that animals apparently facilitated this uptake of mineral nitrogen as in the earlier experiment.

There are reports that nitrate inhibits mycorrhizal infection in natural soils, though attempts to differentiate between the effects of pH and other associated soil parameters have not been very satisfactory (Alexander 1983). Cole (1981) assumes ammonium to be the preferred form of nitrogen by forest trees since it is the form the mycorrhizas are most likely to encounter. If this is correct then naturally-occurring mycorrhizas might not be expected to take up nitrate readily. This is suggested by the work of Bledsoe (1976) on *Pseudotsuga* grown in culture, and excised beech mycorrhizas also show negligible uptake of nitrate from solution (Smith 1972). Field evidence is generally lacking though Haines (1977) has shown a low uptake of nitrate applied to the soil of a mixed pine–oak woodland. However, it is increasingly recognized that nitrate may represent a significant component of mineral nitrogen in forest soils and that the correlations between nitrification and soil pH is a dogma which does not survive close examination (Robertson 1982).

The high nitrification rate of the Stoke Woods humus material is an example of this phenomenon and the facultative ability of mycorrhizas to take up nitrate, as we believe we have demonstrated, is entirely consistent with nitrogen conservation in forest soils which show periodic pulses of nitrification.

The utilization of nitrate by mycorrhizas has important implications for the mineral cation nutrition of plants since the nitrate can act as a 'carrier' (Bledsoe 1976; Raven & Smith 1976; Kirkby 1981). Pulses of nitrification in these forest soils may therefore facilitate cation uptake, and the results presented in Fig. 1b support this hypothesis. In contrast to the negligible root effects on potassium and calcium losses from the ammonifying F-layer material, comparison of the animal treatments, with and without roots, in the H-layer material suggests that the roots have taken up 84% of the potassium and 34% of the calcium. We are unable to account for the reduction of potassium and calcium losses from the rooted systems except by root uptake. We have previously shown (Anderson & Ineson 1984) that it is possible to predict calcium (in μg g humus^{-1}) mobilization by animal grazing in humus from the relative nitrification index (RNI) defined as the nitrate concentration divided by total mineral nitrogen concentration, according to the function

$$Ca = 11 \cdot 5 \, RNI + 3 \cdot 6 \quad (r = 0 \cdot 90, \, df = 19, \, P < 0 \cdot 001)$$

The mechanism for this effect is believed to be the mobilization of calcium from exchange sites on SOM by hydrogen ions released through enhanced nitrification. Thus, in the absence of root uptake, and in a strongly nitrifying humus, the two animal treatments should show enhanced calcium (and potassium) losses. Clearly we must consider the ionic balances of these systems in detail before reaching any firm conclusions, but we discount denitrification as balancing these effects, though it has yet to be measured for these systems, since it would have to occur differentially in the treatments with roots and animals.

ANIMAL AND ROOT EFFECTS IN FIELD LYSIMETERS

Twelve $0 \cdot 5 \times 0 \cdot 5$ m by 30 cm deep replicate lysimeters were established in October 1982 in the Perridge oak woodland site. The lysimeters were set up as large analogues of the field microcosms with mean litter and SOM standing crops for the site. Large samples of bulked materials were air dried to extract the fauna and aliquots, equivalent to 251 g dry wt m^{-2} humus, were rehydrated before adding to the lysimeters. Extensive tree root systems were then introduced through ports in the sides of the trays, which were sealed around the roots, and all twelve units were covered with $1 \cdot 5$ mm mesh netting to exclude macrofauna. Periodic checks were made of all treatments up to April 1983, and thereafter of controls, to locate and remove small earthworms which entered the lysimeters. Leaf litter fall, precipitation and soil temperatures were monitored on the site and mineral concentrations in throughfall and lysimeter leachates measured every week.

Interpretation of results is complicated since the detection of differences between treatments depends upon rainfall; but the intensity and duration of rain, and

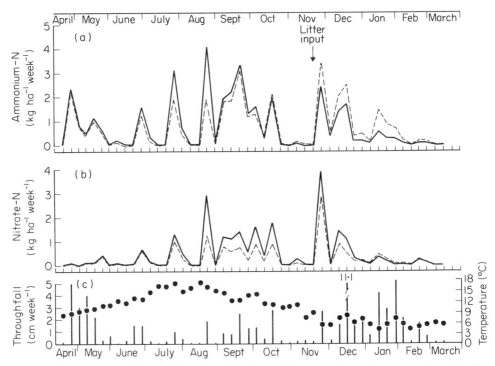

FIG. 2. Nitrogen losses as (a) ammonium-N and (b) nitrate-N from small lysimeters in an oak woodland with (——) and without (– – –) litter-feeding animals. The animals were added in April to established systems; leaf litter aliquots were added in November as indicated. Throughfall volumes (bars) and mean humus temperatures (●) measured at 5 cm below the surface litter (c).

the different effects of roots when the trees are bare or in leaf, affect leachate volumes and concentrations of elements. We will therefore briefly consider nitrate and ammonium fluxes in the treatments with and without animals, but without roots (Fig. 2).

In April 1983, 3·5 g (fresh wt) of millipedes, woodlice and earthworms (equivalent to 14 g m^{-2}) were added to three lysimeters with roots and three lysimeters without roots. Hereafter the discussions of animal treatments refer to these six units; the development of micro- and mesofauna populations following rewetting was assumed to be similar in all treatments.

No significant animal effects were recorded for 14 weeks after the addition of animals, but after that ammonium and nitrate concentrations in leachates increased relative to treatments without animals and were significantly higher in most samples taken from the beginning of July to mid-November. During this period the treatments without animals lost the equivalent (cumulative mean ± SE) of 17·3 ± 0·5 kg ha^{-1} ammonium-N and 8·6 ± 0·4 kg ha^{-1} nitrate-N, compared with 19·5 ± 0·6 kg ha^{-1} ammonium-N and 15·4 ± 1·5 kg ha^{-1} nitrate-N from animal treatments, a significant difference ($P < 0·05$) of 8·9 kg ha^{-1} total-N attributable to the presence of litter-feeding animals.

During July and August there were three periods of high soil temperatures and no rainfall which were followed by large nitrogen losses from the lysimeters when the soil was rewetted: up to 2·2 kg ha⁻ ammonium-N and 1·6 kg ha⁻ nitrate-N greater from animal treatments than controls during the course of 1 week.

The enhancement of nitrogen mineralization in the controls without animals can be interpreted in terms of wetting and drying effects on soils, though the mechanisms are not well understood (Van Veen, Ladd & Frissel 1984). The mineral nitrogen flush is thought to be the consequence of increased microbial activity resulting from the mobilization of organic substrates from SOM or the lysis of microbial biomass (Jager & Bruins 1975; McGill *et al.* 1981; Jenkinson & Ladd 1981). There are few experiments which can be used to interpret the animal effects in our systems. Witkamp & Frank (1970) observed that upon rewetting leaf litter the animals (millipedes) continued feeding at the same rate as before drying but there was a characteristic flush of microbial activity followed by nutrient immobilization during the subsequent phase of microbial growth. It was concluded, in general terms, that nutrient mobilization through wetting and drying cycles would be more significant than animal effects at high temperatures although analysis was not carried out for nitrogen mineralization and immobilization in these experiments.

We have shown in the laboratory that the effects of animals on nitrogen mineralization do not occur in a step-wise manner when the animals are added, but take time to build up, and continue for a period of weeks after the animals are removed (Anderson & Ineson 1984). We have also shown (J. M. Anderson, unpubl.) that drying of litter or humus, followed by short wet periods of a week or so, in insufficient to elicit significant animal enhancement of nitrogen mineralization and the declining differences between animal treatments and controls, following each dry–wet cycle, are equivalent to removing the animals completely. thus, wetting and drying events inhibit continuing animal enhancement of nitrogen mineralization but enhance the differences between treatments through mobilization of some labile nitrogen pool formed by animal feeding activities, a phenomenon demonstrated by these field results.

Leaf litter (300 g m⁻² and equivalent to inputs of 30 kg N ha⁻¹ year⁻¹) was added to the lysimeters in late November and during the following week of heavy rain total inorganic nitrogen losses amounted to 9·6 ± 0·2 kg ha⁻¹ from treatments without animals and 6·4 ± 0·2 kg ha⁻¹ from treatments with animals (Fig. 2). Since this period was preceded by, and followed by, weeks without rainfall, interpretation of this effect is difficult but the significant difference ($P<0·01$) of 3·2 kg ha⁻¹ week⁻¹ between treatments is the reverse trend in terms of the faunal effects expected from the wetting and drying events considered above. This points to further interactive effects of animals in soil processes resulting in nitrogen conservation during the winter period. The total nitrogen losses from the rooted lysimeters were about 30% of those from the unrooted systems. Furthermore, nitrogen losses from treatments with roots and animals were generally 12–16% lower than from treatments containing roots alone. This represents a cumulative difference of approximately 3·5 kg N ha⁻¹ year⁻¹ and substantiates the animal enhancement of root effects demonstrated in the field microcosms.

Melillo, J.M. & Gosz, J.R. (1983). Interactions of biogeochemical cycles in forest ecosystems. *The Major Biogeochemical Cycles and their Interactions* (Ed. by B. Bolin & R.B. Cook), pp. 177–222. John Wiley, Chichester.

Miller, H.G. (1979). The nutrient budgets of even-aged forests. *The Ecology of Even-Aged Forest Plantations* (Ed. by E.D. Ford, D.C. Malcolm & J. Atterson), pp. 221–56. Institute of Terrestrial Ecology, Cambridge.

Nommik, H. (1981). Fixation and biological availability of ammonium on soil clay minerals. *Terrestrial Nitrogen Cycles* (Ed. by F.E. Clark & T. Rosswall), *Ecological Bulletins (Stockholm)*, **33**, 273–80.

Persson, T. (1983). Influence of soil animals on nitrogen mineralization in a northern Scots pine forest. *New Trends in Soil Biology* (Ed. by Ph. Lebrun, H.M. Andre, A. de Medts, C. Gregoire-Wibo & G. Wauthy), pp. 117–126. Dieu–Brichart, Louvain-la-Neuve.

Raven, J.A. & Smith, F.A. (1976). Nitrogen assimilation and transport in vascular land plants in relation to intercellular pH regulation. *New Phytologist*, **76**, 415–431.

Raw, F. (1967). Arthropoda (except Acari and Collembola). *Soil Biology* (Ed. by A. Burges & F. Raw), pp. 323–62. Academic Press, London.

Robertson, G.P. (1982). Nitrification in forested ecosystems. *Philosophical Transactions of The Royal Society of London, Series B*, **296**, 445–57.

Runge, M. (1971). Investigations of the content and the production of mineral nitrogen in soils. *Integrated Experimental Ecology: Methods and Results of Ecosystem Research in the German Solling Project*, pp. 191–202. Springer-Verlag, Berlin.

Satchell, J.E., (1963). Nitrogen turnover by a woodland population of *Lumbricus terrestris*. *Soil Organisms* (Ed. by J. Doeksen & J. van der Drift), pp. 60–66. North-Holland, Amsterdam.

Satchell, J. (1983). Earthworm ecology in forest soils. *Earthworm Ecology: Darwin to Vermiculture* (Ed. by J.E. Satchell), pp. 161–170. Chapman and Hall, London.

Seastedt, T.R. (1984). The role of microarthropods in decomposition and mineralization processes. *Annual Reviews of Entomology*, **29**, 25–46.

Smith, F.A. (1972). A comparison of the uptake of nitrate, chloride and phosphate by excised beech mycorrhizas. *New Phytologist*, **71**, 875–82.

Smith, W.H. (1976). Character and significance of forest tree root exudates. *Ecology*, **57**, 324–31.

Sollins, P., Spycher, G. & Glassman, C.A. (1984). Net nitrogen mineralization from light- and heavy-fraction forest soil organic matter. *Soil Biology and Biochemistry*, **16**, 31–37.

Swank, W.T. & Waide, J.B. (1980). Interpretation of nutrient cycling research in the management context: evaluating potential effects of alternative management strategies on site productivity. *Forests: Fresh Perspectives From Ecosystem Analysis* (Ed. by R. Waring), pp. 137–58. Oregon State University Press, Corvallis.

Swift, M.J., Heal, O.W. & Anderson, J. (1979). *Decomposition in Terrestrial Ecosystems*. Blackwell Scientific Publications, Oxford.

Van Veen, J.A., Ladd, J.M. & Frissel, M.J. (1984). Modelling C and N turnover through microbial biomass in soil. *Plant and Soil*, **76**, 257–74.

Vitousek, P.M., Gosz, J.R., Gries, C.C., Melillo, J.M. & Reiners, W.A. (1982). A comparative analysis of potential nitrification and nitrate mobility in forest ecosystems. *Ecology*, **52**, 155–77.

Witkamp, M. & Frank, M.L. (1970). Effect of temperature, rainfall and fauna on transfer of Cesium-137, K, Mg and mass in consumer–decomposer microcosms. *Ecology*, **51**, 465–74.

Role of invertebrates in the decomposition of *Salix* litter in reclaimed cutover peat

J. P. CURRY, MARY KELLY AND T. BOLGER*

Department of Agricultural Zoology and Genetics, University College, Belfield, Dublin 4

INTRODUCTION

Invertebrate respiration normally accounts for 1–20% of total heterotrophic metabolism in most terrestrial systems (Persson *et al.* 1980), but faunal exclusion studies indicate that invertebrates have a much greater influence on decomposition rates than their contribution to respiratory metabolism would suggest (Kurcheva 1960; Crossley & Witkamp 1964; Edwards & Heath 1963).

Large earthworms such as *Lumbricus terrestris* L. have an important role in woodland mull soils (Satchell 1983). *L. terrestris* is slow to colonize reclaimed peat (Curry & Cotton 1983) but when introduced experimentally it can greatly accelerate litter disappearance (Curry & Bolger 1984).

This paper presents data on *Salix* leaf fall and the influence of invertebrates on litter decomposition rates in cutover peat.

MATERIALS AND METHODS

The study site was an area of raised bog at Clonsast, Co. Offaly, which had been reclaimed following mechanical peat extraction and planted with fast-growing trees including *Alnus*, *Populus* and *Salix* species. The 50–80 cm residual mixed forest peat layer was mixed with underlying calcareous mineral subsoil by deep ploughing and disking, and appropriate applications of compound mineral fertilizers were made at planting time to facilitate establishment. The main study at Clonsast was carried out in a small plantation (50 × 50 m) of *Salix aquatica* Sm. cv. *gigantea* established in 1978. Soil characteristics were variable; typically the pH in the top 20 cm was about 7, the ash content 60–80%, N less than 1% and Ca up to 6% of dry weight. The trees were planted at a density of 4000 ha^{-1} and are coppiced at 3–4 year intervals.

Litterfall was measured between 22 September and 21 November 1981 and between 15 September and 17 November 1982. Nylon 2 mm mesh 20 cm squares were anchored with steel pins on the old litter surface and litter falling on 15 cm diameter circular subsamples was collected. There were fifteen replicates in 1981 and thirty in 1982.

A litter bag technique was used to measure litter decomposition rates. A field experiment was laid out on 12 November 1982 using two sets of nylon 20 × 15 cm

---

*Present address: Department of Zoology, University College, Belfield, Dublin 4.

bags, one with 7 mm openings to allow free access to all invertebrates and the other with 0·003 mm openings to exclude all but the microfauna. Each bag contained 7 g air-dried freshly fallen *S. aquatica* litter (90% dry wt) of typical elemental content N 1·8%, P 0·11%, K 1·63%, Ca 2·1%, Mg 0·15%, Na 0·1%.

Microplots were established at Belfield to facilitate experiments requiring frequent attention. Two wooden frames, 2·4 × 2·4 m in area, were filled to a depth of 0·6 m with soil from the field site. The bottoms and sides were lined with polythene perforated at a height of 3 cm to allow drainage. A litter layer similar to that in the field site was added and replenished periodically. A litttter bag experiment commenced on 23 November 1981 using 3 mm mesh 8 × 8 cm bags each containing 3 g air-dried *S. aquatica* litter. The bags were arranged in four groups, two of which were treated with 100 g m^{-2} naphthalene at intervals of 2 weeks to suppress faunal activity. In addition, 0·003 mm mesh bags were included for purposes of comparison with the naphthalene treatment.

Litter consumption by *L. terrestris* was measured at Belfield. Six nylon 0·5 mm mesh bags were filled to a depth of 25 cm with Clonsast soil and were buried level with the soil surface in a peat microplot. The surface area in each bag was 600 cm^2. Six worms ranging from 0·5 g to 6 g fresh wt were added to each bag and were provisioned with 22 g air-dried *S. aquatica* litter. Litter disappearance was determined at 8-week intervals and the litter was renewed. Results of litter disappearance and consumption experiments are given as means ±1 S.E.

RESULTS

Litterfall in the *Salix* plot was 441 ± 23 g dry wt m^{-2} in 1981 and 529 ± 42 in 1982. After 130 days 63·5% of the litter had disappeared from coarse mesh bags at Clonsast (Fig. 1). An anomalous increase occurred between days 130 and 151; this was probably due to throughfall from litter above the bags. After 180 days 56·5 of the original weight had been lost from coarse mesh bags compared with 37·5% from fine.

Decomposition rate in the coarse bags was virtually linear at Belfield with 43·7% of the litter disappearing within 183 days and 61·3% after 309 days (Fig. 1). The effects of faunal exclusion were quite marked, with only 24·4% of litter disappearing from naphthalene-treated bags within 183 days and 36·3% after 309 days. Results from the fine mesh bags were fairly similar and less variable: 18·1% of litter had disappeared after 155 days and 38·1% after 215 days.

Macroinvertebrates were scarce at Clonsast (Table 1), with only the litter-dwelling earthworm *Dendrobaena rubida* (Sav.) being abundant. *Lumbricus rubellus* (Hoff.), *Allolobophora chlorotica* (Sav.) and *A. caliginosa* (Sav.) were occasionally recorded. Small snails were fairly abundant, and slugs (*Deroceras* and *Arion* spp.) were occasionally seen. The macro-arthropod fauna consisted mainly of insect larva and spiders. Nematodes, Acari and Collembola were by far the most abundant groups. The dominant Collembola species were *Isotoma notabilis* Schäff, *I. viridis* Bourl., *Isotomina sphagneticola* (Linnaniemi) and *Friesea mirabilis*

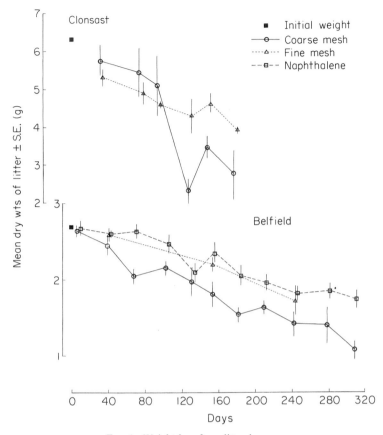

FIG. 1. Weight loss from litter bags.

TABLE 1. Numbers of invertebrates m^{-2} in the litter layer of *Salix aquatica* cvr. *gigantea*. Acari and Collembola sampled on six occasions between December 1980 and July 1892, extracted by Tullgren funnel; other fauna sampled in November 1983. Enchytraeid and nematode numbers based on four Bearmann funnel extracted samples; other taxa estimated from two composite hand-sorted samples each 1400 cm^2

Collembola	22 600	Isopoda	7
Diptera		Diplopoda	10·5
Cecidomyidae	14	Araneae	80·5
Tipulidae	31·5		
Scatopsidae	7	Opiliones	7
Muscidae	3·5		
Coleoptera		Oligochaeta	
Staphylinidae	70	Lumbricidae	53
Carabidae	21	Enchytraeidae	1 307
Curculionidae	3·5	Mollusca	364
Acari	28 500	Nematoda	238 500

TABLE 2. Litter consumption by *Lumbricus terrestris*

Period	Litter consumption per bag (g dry wt)		Daily consumption rate (mg dry wt g^{-1} fresh wt day^{-1})
	With worms	Without worms	
13 Oct–8 Dec 1983	$17 \cdot 4 \pm 0 \cdot 3$	$2 \cdot 8 \pm 0 \cdot 3$	18
9 Dec–2 Feb 1984	$8 \cdot 2 \pm 0 \cdot 9$	$1 \cdot 4 \pm 0 \cdot 3$	7
3 Feb–29 Mar 1984	$8 \cdot 2 \pm 0 \cdot 8$	$0 \cdot 6 \pm 0 \cdot 4$	$6 \cdot 8$

(Tullb.), while the most abundant Acari were *Sellnickochthonius rostratus* (Jacot), *Tectocepheus velatus* (Michael), *Olodiscus minima* (Kr), *Quadroppia quadricarinata* (Michael) and *Nanorchestes arboriger* (Berl.).

Litter consumption by *L. terrestris* was greatest during October–December when the weather was mild, and declined considerably during the colder December–March period (Table 2). Earthworm consumption accounted for 83–84% of total litter disappearance.

DISCUSSION

Litterfall values for the *Salix aquatica* plot were comparable with those for temperate deciduous woodlands on good mineral soils (Jensen 1974; Satchell 1983), but extensive sampling over a range of site conditions suggested that 100–200 g m^{-2} may be more typical for reclaimed peat.

Litter decomposition rates at Clonsast were considerably faster than 2 years previously (unpublished data), suggesting increasing site maturation, but were still considerably lower than is usual in mature woodland on mull soils where over 90% of the litter may disappear within 6 months (Heath, Arnold & Edwards 1966). The faunal exclusion studies confirm the importance of mesofaunal activity in litter decomposition, but indicate that this process is limited by the scarcity of earthworms.

Few previous studies of earthworm consumption made allowance for litter loss due to other factors, or for litter buried in the soil but not ingested. The mean ingestion rate during October–December is probably close to the maximum for this sytsem. The average for the 24-week study period (about 10 mg g^{-1} day^{-1}) agrees fairly well with estimates of litter consumption in the field over similar time periods (Curry & Bolger 1984).

It has been estimated that earthworm populations of up to 300 g m^{-2} biomass in deciduous woodlands are often capable of consuming the entire leaf fall within a few months (Satchell 1983). An earthworm biomass of 250 g m^{-2} consuming 4·5 g litter day^{-1} could consume the estimated leaf fall at Clonsast in 98–118 days under favourable conditions, but it is unlikely that the earthworm carrying capacity of *Salix* on cutover peat would ever exceed 100 g m^{-2}.

REFERENCES

Crossley, D.A. & Witkamp, M. (1964). Forest soil mites and mineral cycling. *Acarologia*, 6, 137–146.

Curry, J.P. & Cotton, D.C.F. (1983). Earthworms and land reclamation. *Earthwork Ecology* (Ed. by J.E. Satchell), pp. 215–228. Chapman and Hall, London.

Curry, J.P. & Bolger, T. (1984). Growth, reproduction and litter and soil consumption by *Lumbricus terrestris* L. in reclaimed peat. *Soil Biology and Biochemistry* (in press).

Edward C.A. & Heath, G.W. (1963). The role of soil animals in breakdown of leaf materials. *Soil Organisms* (Ed. by J. Doeksen & J. Van der Drift), pp. 76–85. North-Holland, Amsterdam.

Heath, G.W., Arnold, M.K. & Edwards, C.A. (1966). Studies in leaf litter breakdown, I. Breakdown rate of leaves of different species. *Pedobiologia*, 6, 1–12.

Jensen, V. (1974). Decomposition of angiosperm tree litter. *Biology of Plant Litter Decomposition*, Part I. (Ed. by C.H. Dickinson & G.F.J. Pugh), pp. 64–104. Academic Press, London.

Kurcheva, G.F. (1960). Role of invertebrates in the decomposition of oak litter. *Soviet Soil Science*, 4, 360–365.

Persson, T., Bååth, E., Clarholm, M., Lundkvist, H. & Söderström, B.E. (1980). Trophic structure, biomass dynamics and carbon metabolism of soil organisms in a Scots Pine forest. *Ecological Bulletins (Stockholm)*, 32, 419–459.

Satchell, J.E. (1983). Earthworm ecology in forest soil. *Earthworm Ecology* (Ed. by J.E. Satchell), pp. 161–170. Chapman and Hall, London.

Effect of introducing *Allolobophora longa* Ude on root distribution and some soil properties in New Zealand pastures

J. A. SPRINGETT

Ministry of Agriculture and Fisheries, P.O. Box 1654, Palmerston North, New Zealand

SUMMARY

1 In some New Zealand pastures, poor soil structure appears to reduce pasture production through adverse soil moisture conditions and restriction of plant roots.

2 Many soils contain only the earthworm species *Allolobophora caliginosa* and *Lumbricus rubellus* with a major zone of activity within the top 10 cm of soil.

3 Introducing a deeper burrowing species, *A. longa*, to limed soil, increased surface infiltration rates, total soil porosity at 10–20 cm, and root biomass at 15–20 cm. Pasture growth was also increased in some seasons.

4 These changes were associated with increased surface casting by *A. longa* and greater mixing of surface applied lime throughout the soil profile.

INTRODUCTION

On some sheep-grazed hill country in the East Coast of the North Island of New Zealand, pasture growth may be limited by factors related to soil structure. Soils may become very wet in winter with consequent low oxygen diffusion rates below 5–7 cm in the soil profile. Downward penetration of roots, particularly of clover is restricted. In winter, roots are seen to rot off at a depth of about 7 cm. In summer, the same soils dry out very rapidly, subjecting the shallow rooted pasture to sudden severe water stress.

These soils typically have substantial populations of *Allolobophora caliginosa* (Sav.), *A. trapezoides* (Duges) and *Lumbricus rubellus* Hoff. The top 5–7 cm of soil is well structured with no turf mat development at the surface. Penetration of worms below this level is limited to relatively few burrows formed in late spring as the worms move downwards to depths of 15–25 cm to aestivate.

This situation is typical of many New Zealand pastures which have been developed out of native bush. The indigenous earthworm fauna comprises five to nine species including litter-dwellers, top soil mixers and subsoil species (Lee 1959). When the native vegetation is cleared and grasses sown, the indigenous fauna disappears, although in some cases the subsoil species may survive for many years. Following clearing, there is a period during which the soil is exposed to direct impact by rainfall, trampling by stock, and much larger diurnal climatic variations than occurred under a tree canopy. In time, an earthworm fauna comprising a small number of accidentally introduced European species becomes established.

The changes in soil properties which occur when the native vegetation is converted to improved pasture have been reported (Stockdill & Cossens 1966). This paper reports on the changes in soil properties following the introduction of a deeper burrowing species, *A. longa* Ude., into pasture soils already colonized by *A. caliginosa*, *A. trapezoides* and *L. rubellus*.

MATERIALS AND METHODS

Study area

The trial site was situated on a sheep farm on the East Coast of the southern North Island at Porongahau. The pasture was predominantly perennial ryegrass (*Lolium perenne* L). and white clover (*Trifolium repens* L) which was rotary cultivated and resown 3 years before the first earthworm introduction. The soil was a yellow grey earth (During 1972) with a pH of 5·5. The area was grazed by sheep on a set stocked/shuffle system at 12 stock units* ha^{-1}. The climate is temperate to Mediterranean in type with moist cool winters and warm dry summers. The dry season usually lasts from December until March but there may be extended droughts or heavy summer rains.

Methods

Two separate trials were set up on the same farm in consecutive years. In the first trial, *A. longa* were introduced to plots treated with 5000 kg ha^{-1} of lime; and other limed and unlimed plots were left free of earthworms. Each of the three treatments was replicated six times. The plots were small (1 m²) and were used mainly for dry matter production measurements. Lime and earthworms were added on the same day. In the second year, larger plots (2 × 2 m) were set up with only two treatments, added *A. longa* and a control without *A. longa*, to enable destructive soil samples to be taken. Both treatments were replicated six times.

Earthworm introduction took place in August of each year. *A. longa* were collected by digging and hand-sorting from pastures known to have high populations of this species. The worms were matched for size and proportion of mature worms and were placed on the surface of the plots at a rate of 150 m^{-2}. Any worms which had not burrowed into the soil within 30 minutes were replaced. The plots were protected against birds until all worms had disappeared from the soil surface. The first trial was fenced to prevent grazing, the second trial was open. Measurements on pasture and soil properties were made over an 18 month period after earthworm introduction.

---

*A stock unit is equivalent to a 55 kg live weight Romney ewe producing 1 lamb per year and consuming 550 kg dry matter per year.

Trial 1

Pasture production was estimated by occasional cuts in October and November 1981, and April, September and October 1982. The herbage was trimmed to a height of 3 cm and subsequent growth after 3–4 weeks was measured by cutting the entire plot to a height of 3 cm. The harvested pasture on the different treatments was compared at each date of cutting.

Soil pH (2·5:1 in water) was measured after cutting soil cores into 2 cm horizons, 9 months after lime and worms had been spread.

Trial 2

Root biomass was measured in fifty cores 10 cm^2 in surface area and 20 cm deep from each treatment. The cores were subdivided into 5 cm horizons. The root biomass was measured once only in early spring (August). The cores were soaked in 10 vol hydrogen peroxide solution for 1 hour before washing and sieving to extract the root material which was then oven-dried and weighed.

Bulk density and total porosity were measured using 5 cm aluminium cores (Gradwell & Birrell 1979). The infiltration rate in the late summer was measured using a 10 cm diameter single ring, one ring on each replicate. Density and soil volumetric water content were measured in soil cores taken from beneath the infiltration rings immediately after the addition of water and a control set of cores for soil moisture determination was taken from areas not receiving artificial watering.

Destructive sampling was kept to a minimum even in the larger plots and there was no direct measurement of the earthworm population. An indirect assessment of earthworm activity was made by counting the worm casts visible on the pasture surface in May and September using 10 × 10 cm quadrats.

RESULTS

The pasture dry matter yields recorded in Trial 1 are shown in Table 1. There was a response in pasture production to the addition of lime on all five dates and an additional small significant response to the presence of earthworms in October and

TABLE 1. Pasture dry matter on Trial 1 at Porongahau

	Mean pasture dry matter production (kg ha^{-1} day^{-1} ± SD)		
	A. longa + lime	Lime	Control
Oct 81	66·7 ± 6·5*	59·3 ± 4·5	23·0 ± 17·1
Nov 81	38·3 ± 3·7*	33·1 ± 4·1	25·2 ± 6·1
April 82	20·1 ± 6·2	29·6 ± 8·9	9·5 ± 8·3
Sept 82	39·4 ± 6·5	37·3 ± 7·2	20·9 ± 14·2
Oct 82	29·3 ± 5·1*	23·1 ± 2·4	16·3 ± 13·2

* Significantly greater than lime-only treatment, $P<0.05$.

TABLE 2. The mean biomass and percentage of roots in each of four horizons in Trial 2 (mg roots per core 10 cm^2 surface area)

| | Mean root biomass (mg/core ± SD and %) | | | |
| | *A. longa* | | Control | |
Soil depth (cm)	Biomass (mg/core)	%	Biomass (mg/core)	%
0–5	814·5 ± 205·4	86·2	1278·6 ± 958·7	91·4
5–10	43·1 ± 18·7	4·4	57·2 ± 42·0	4·1
10–15	31·2 ± 11·3	3·3	33·5 ± 19·7	2·4
15–20	56·2 ± 22·0*	5·0	29·3 ± 12·7	2·1

* Significantly different $P < 0.01$.

November 1981, which disappeared in April and September of 1982 but recurred in October 1982.

The biomass of roots in the two treatments of Trial 2 is shown in Table 2. The general distribution of roots was similar in both treatments with the majority of roots in the top 5 cm of the soil profile. The variability in the 0–5 cm horizon was very large and there was no significant difference between the treatments in this horizon. However, at the lowest level sampled (15–20 cm) there was significantly more root biomass ($P < 0.01$) on the *A. longa* plots.

The effect of *A. longa* on the total porosity of the soil in Trial 2 is shown in Fig. 1. The upper 10 cm of the profile showed no significant differences but below 10 cm the *A. longa* treatment increased the total porosity ($P < 0.05$). The effect of *A. longa* on soil volumetric water content at the end of summer (March) in areas receiving no additional water and in infiltrated areas is shown in Table 3. The percentage moisture was low at this time of year but it was still significantly greater ($P < 0.05$) on the *A. longa* treatment down to 15 cm. The amount of water flowing into the soil from infiltration rings in 25 minutes is shown in Fig. 2. The *A. longa* plots were able to accept almost twice as much water as the untreated plots.

The number of worm casts on the pasture surface of Trial 1 is shown in Table 4. There were more casts in the *A. longa* plots than on the other two treatments,

TABLE 3. The natural volumetric water content of the profile and the volumetric water content of the profile under infiltration rings in soil with and without *A. longa* in March

| | Mean volumetric water content ± SD | | | |
| | Infiltrated area | | Natural area | |
Soil depth (cm)	*A. longa*	Control	*A. longa*	Control
0–5	47·2 ± 1·9*	48·0 ± 5·3	23·7 ± 3·8*	18·8 ± 2·8
5–10	45·6 ± 3·1	40·8 ± 8·7	20·2 ± 3·1*	15·8 ± 2·1
10–15	43·8 ± 5·0*	37·5 ± 5·1	18·2 ± 3·1*	14·7 ± 1·4
15–20	40·4 ± 5·0*	35·4 ± 4·9	16·5 ± 4·4	14·6 ± 1·3

* Significantly different $P < 0.10$.

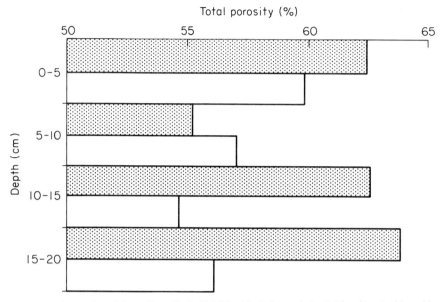

FIG. 1. The total porosity of the soil profile in Trial 2 with *A. longa* (stippled bars) and with resident worms only (open bars).

TABLE 4. The mean number of worm casts (m^{-2}) on Trial 1

	Mean number of worm casts (m^2 ± SD)		
	A. longa + lime	Lime	Control
May 82	310* ± 76·3	228 ± 47·4	152 ± 91·0
Sept 82	297† ± 94·0	181 ± 56·5	151 ± 85·7

Significantly greater than lime-only treatment: * $P < 0·05$; † $P < 0·01$.

TABLE 5. The mean pH of the soil profile at 2 cm intervals to a depth of 20 cm

	Mean pH			
Depth (cm)	*A. longa* + lime	Lime	Control	LSD (1%)
0−2	6·60	6·60	5·75	0·29
2−4	6·55	6·10	5·50	0·35
4−6	6·30	6·05	5·40	0·37
6−8	5·65	5·90	5·40	0·23
8−10	5·60	5·60	5·40	0·19
10−12	5·60	5·45	5·40	0·16
12−14	5·60	5·45	5·40	0·11
14−16	5·70	5·45	5·20	0·13
16−18	5·75	5·40	5·30	0·12
18−20	5·80	5·40	5·30	0·17

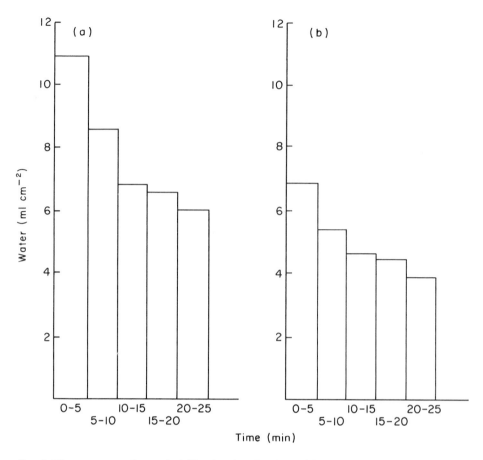

FIG. 2. The acceptance of water in infiltration rings in Trial 2: (a) *A. longa* treatment; (b) control.

although the addition of lime had also stimulated the production of casts by the resident species (*A. caliginosa*, *A. trapezoides* and *L. rubellus*). Liming increased soil pH below 10 cm only in the presence of *A. longa* (Table 5), indicating that in the presence of *A. longa* material is carried from the surface to 15–20 cm depth.

DISCUSSION

The addition of one extra earthworm species (*A. longa*) to the existing population in this New Zealand pasture soil resulted in some measurable changes in the soil structure. Soil porosity in the lower part of the profile was increased and it is suggested that it was these changes in soil structure which allowed greater root development in the 15–20 cm soil horizon. These changes occurred within 18 months of introduction. It has been shown (Kretzschmar 1983; Springett 1983) that *A. longa* transports soil vertically, taking material from the surface and depositing it at depth, and vice versa. The present work shows that when *A. longa* are active, a greater number of earthworm casts is deposited at the surface. That this also modifies the soil structure below 10 cm, and the distribution of plant roots, has been shown in the present field study and in the laboratory (Edwards & Lofty 1979).

In New Zealand pasture soils, where the earthworm fauna is dominated by the superficially active species *A. caliginosa* and *L. rubellus*, it may be important for the long-term fertility of pastures to ensure that the earthworm fauna contains a range of species occupying different ecological niches. Lavelle (1983) states that temperate grasslands have mainly (72%) anecic species. However, the category 'anecique' as described by Bouché (1977) includes *A. longa* but not *A. caliginosa*, which falls in to the 'endogées', so that most New Zealand pastures do not conform to the pattern described by Lavelle (1983). When *A. longa* is introduced, changes in soil structure occur which are sufficient to influence the growth of both roots and above-ground vegetation; suggesting that the productivity of New Zealand exotic grasslands is increased by an anecic earthworm fauna. If our aim is to understand the root environment and to manage it to improve production then detailed information on the way in which each species alters the water-holding capacity, the aeration and the nutrient availability in the root environment will be required.

ACKNOWLEDGMENTS

I would like to thank Mr J. Tully for allowing these trials on his property and Mr M. J. Roberts for technical assistance.

REFERENCES

Bouché, M.B. (1977). Strategies lombriciennes. *Ecological Bulletins (Stockholm)*, **25**, 122–132.

During, C. (1972). Fertilisers and soils in New Zealand farming. *New Zealand Department of Agriculture Bulletin*, **409**, 1–311.

Edwards, C.A. & Lofty, J.R. (1979). The influence of arthropods and earthworms upon root growth of direct drilled cereals. *Journal of Applied Ecology*, **15**, 789–795.

Gradwell, M.W. & Birrell, K.S. (1979). Methods of physical analysis of soil. *New Zealand Soil Bureau Scientific Report 10C*, 3.1–9.5.

Kretzschmar, A. (1938). Soil transport as a homeostatic mechanism for stabilising the earthworm environment. *Earthworm Ecology* (Ed. by J.E. Satchell), pp. 59–66. Chapman & Hall, London.

Lavelle, P. (1983). The structure of earthworm communities. *Earthworm Ecology* (Ed. by J.E. Satchell), pp. 449–466. Chapman & Hall, London.

Lees, K.E. (1959). The earthworm fauna of New Zealand. *New Zealand Department of Scientific and Industrial Research Bulletins*, **130**, 1–485.

Springett, J.A. (1983). Effect of five species of earthworm on some soil properties. *Journal of Applied Ecology*, **20**, 865–872.

Stockdill, S.M.J. & Cossens, G.G. (1966). The role of earthworms in pasture production and moisture conservation. *Proceedings of the New Zealand Grassland Association*, **28**, 168–183.

Soil in the ecosystem

JOHN MILES

Institute of Terrestrial Ecology, Hill of Brathens, Banchory, Kincardineshire, AB3 4BY

SUMMARY

1   The paper briefly reviews the significance of interactions between above- and below-ground biotic processes and biota in terrestrial ecosystems. It demonstrates the essential connectedness of terrestrial ecosystems, and the artificiality of separating ecosystems into soil and above-ground components.

2   The often great spatial and temporal variability of soil properties is discussed, and some of the causes analysed. Changes in labile soil propeties with time caused by *Betula pendula* are noted in more detail.

3   The other main subjects used as illustrations are: differences in death rates of buried seed populations; the significance of seed dispersal by ants; the role of rotting wood as a niche for seedling establishment; root competition; ectomycorrhizal successions associated with ageing trees; the effects of herbivores on soil processes.

4   It is concluded that the functional interactions of ecosystems are transient rather than fixed. The variability and dynamic nature of soil properties should always be borne in mind by soil biologists and ecologists.

INTRODUCTION

For this paper, I was asked to assess '. . . the significance of the biotic processes in the soil environment in relation to processes in the whole ecosystem, particularly emphasizing dynamic aspects.' Fortunately, editorial fiat solves the problem of where to strike the balance between monographic treatment and the single sentence summary, e.g. terrestrial ecosystems are always changing in time and cannot exist without soil (excluding soil-less 'micro-ecosystems', such as those of rock and bark). Accordingly, I shall briefly examine first the functional connectedness of ecosystems, and second, the variability of soil in space and time. Finally, I shall discuss a limited number of interactions between the above- and below-ground systems, choosing examples which interest me but which are poorly understood.

ECOSYSTEM CONNECTEDNESS

It is trite, but none the less true, to state that soils and soil biotic processes are an essential part of terrestrial ecosystems. Micro-organisms fix much, if not most, of the nitrogen made available to plants, and speed up the weathering of essential elements from soil minerals. Plant nutrients, critically nitrogen and phosphorus, are

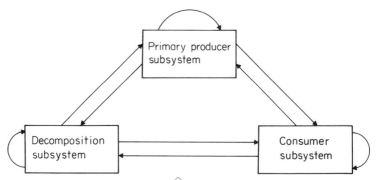

FIG. 1. Simple three-compartment model of ecosystem structure and internal links (after Swift, Heal & Anderson 1979; Andersen, MacMahon & Wolfe 1980; MacMahon 1981).

largely recirculated via decomposition of organic matter by micro-organisms. (Some plant-available, or at least ectomycorrhiza-available, soluble organic N and P may be produced from faunal feeding, and even directly from autolysis in dying tissues.) Mycorrhizas help most vascular plant taxa to tap scarce N and P. The soil fauna *inter alia* facilitates decomposition by comminuting organic remains, is preyed upon by many vertebrates, and helps to maintain soil as a well aerated rooting medium.

The simplest conceptual view of an ecosystem that is still useful, is as a three-compartment model (Fig. 1). This omits exchanges with and influences of the atmosphere, lithosphere, and other ecosystems, but otherwise neatly summarizes the internal connectedness of terrestrial ecosystems, showing the major routes of transfer of matter and/or interactions. The consumer and decomposer compartments or subsystems can be subdivided in terms of food webs; the organic detritus part of the consumer subsystems can be described as a network of molecular changes, of catabolic transformations and humus synthesis. Figure 2 is a limited expansion of this simple model. The consumer and decomposer compartments are arbitrarily divided into six and two compartments respectively. Links with the atmosphere and other ecosystems are still omitted, and also *inter alia* compartments with animal and microbial pathogens, and interactions within compartments (e.g. competition and interference between organisms, nutrient and carbon fluxes within plants, synthesis of humus polymers). The main subject material of this volume falls within five compartments: decomposers, saprovores, below-ground predators, root grazers, and plants and their associated mycorrhizas.

There are two main drawbacks to the expanded model of trophic relationships, apart from its generalization and lack of quantification. First, to avoid excessive expansion of the model, all but one of the defined compartments fall both above and below ground. One problem is that individuals of some species occur above and below ground at different stages of their life histories, e.g. many insects with winged adults have root-feeding larvae. Other species regularly commute between the soil and the surface even at the same life-history stages, e.g. the common earthworm

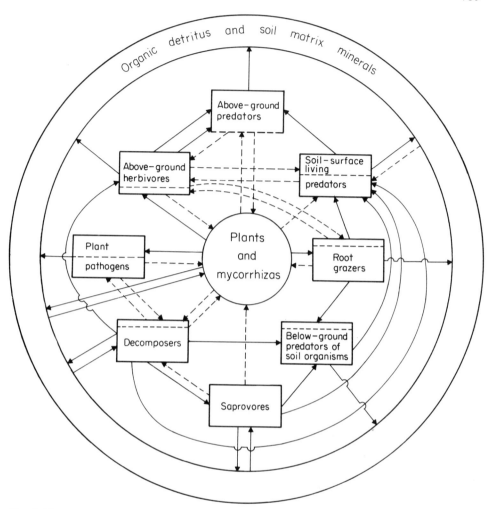

FIG. 2. Ten-compartment model of ecosystem structure. The relative importance above and below ground of organism compartments is shown by horizontal dashed lines. Dashed arrows show interactions not involving significant transfers of matter; continuous arrows show major routes of transfer of matter. Interactions coinciding with matter transfer routes are not shown separately.

Lumbricus terrestris, and most British ants. Second, many organisms have a degree of trophic flexibility such that they must be classified under more than one compartment. The honey fungus, *Armillaria mellea*, forms mycorrhizas with orchids, is a common saprophyte, but is a virulent pathogen of trees and many woody and herbaceous plants (Ramsbottom 1953; Mosse 1978). Blackbirds (*Turdus merula*) and many other birds eat a variety of invertebrates and of fruits. Badgers (*Meles meles*) depend mostly on earthworms in parts of north-west Europe, but also eat leaves, seeds, fruits, roots, fungal fruit-bodies, rodents, amphibians, rabbits (*Oryctolagus cuniculus*), insects and carrion (Kruuk & Parish 1981). They therefore feed, at least to some extent, within eight of the nine producer/consumer/decomposer boxes in Fig. 2 and, as it is likely that the occasional item of plant material infected with rust, smut or mildew will be ingested, they probably feed within all compartments! This, of course, highlights the artificiality of attempts to compress complex systems into simple models.

However, despite its limitations, Fig. 2 shows that all the predominantly below-ground compartments are linked to mainly above-ground compartments. Indeed, even at this simple, ten-compartment level, there is a web of both influences and of routes of transfer of material. This means that an impact on one compartment is likely to reverberate across the web. For example, direct addition of inorganic fertilizer influences decomposition (Swift, Heal & Anderson 1979), plant production, and the biomass and species richness of arthropod, herbivore and carnivore populations (Hurd *et al.* 1971; Hurd & Wolf 1974). No increase in small herbivorous mammals was detected in the latter study, but no other consumer groups were examined. Detailed word models can be formulated to indicate the likely extent of such repercussions. Swift *et al.* (1979) indicate how a reduction in grazing on a temperate grassland could cause a sequence of changes within the decomposer subsystem leading either to increased or reduced N availability to plants, depending on soil moisture status. Changed N availability would lead to changed N uptake by plants and possibly to eventual changes in species composition of the sward. Either could result in grazing pressures further changing with voluntary grazing. When the interactions have been quantified, mathematical models can be used to predict the degree of change (Innis 1978).

Many interactions within ecosystems do not involve transfers of significant quantities of matter, e.g. chemical interference seed and spore dispersal. Chemical interference seems to be widespread. Apart from allelopathic (plant against plant) phenomena, plants influence micro-organisms and vice versa. The ectomycorrhizas of certain conifers are inhibited by *Cladonia* spp. (Brown & Mikola 1974; Fisher 1979a) and *Calluna vulgaris* (Weatherell 1953; Handley 1963). Seeds produce an array of chemicals that inhibit various micro-organisms (Nickell 1959). This may account for the persistence of soil seed banks; without such inhibitors, most seeds might quickly rot after dispersal unless they germinated (Rice 1979; Fisher 1979b). Soil micro-organisms in turn produce chemicals that inhibit germination (Lynch 1978; Kirkpatrick & Bazzaz 1979) and subsequent plant growth (Chapman & Lynch 1983). However, micro-organisms can also promote germination under some circumstances (Abdulla 1970; Leeuwen 1981). The mechanism of this is unknown, but is presumably by the micro-organisms either producing germination-promoting chemicals, or breaking down the fruit or seed coat. There are also reports of micro-organisms producing growth-promoting auxins (Valadon & Lodge 1970; Tinker 1980), but the significance, if any, of these to plants in the field, is unknown.

Perhaps the most tangible form of internal link within most terrestrial ecosystems is not shown in Fig. 2. This is that the vascular plant components may commonly if not generally be physically linked through root grafts (Sutton 1969) and interconnections between mycorrhizal hyphae (Heap & Newman 1980; Brownlee *et al.* 1983). These phenomena and their physiological implications have not been widely and systematically investigated. However, Brownlee *et al.* (1983) and Read, Francis & Finlay (p. 193) have shown that ^{14}C-labelled assimilate is distributed via interplant mycorrhizal connections. Further, Stone (1974) demonstrated that when *Pinus resinosa* trees were girdled (which prevents transport of photosynthate to root systems), individuals with intact neighbours could survive for at least 18 years. In one instance, a whole stand was maintained by only two-thirds of the trees. Although the girdled trees did not grow as well as ungirdled ones, Stone concluded that: '. . . there are grounds for considering the entire stand root system as unitary and communal at least with respect to phloem transport.'

The ten-compartment model, or any variation or development of it, would clearly be much more useful if the compartments, transfers of material, and other interactions, could be quantified for a range of contrasting ecosystems. The compartments and transfers can be characterized by pools and flows of different elements. However, while this can now be approximated above ground for many ecosystems, there are too many gaps in the literature to permit a comprehensive approximation below-ground in any ecosystem. There are many estimates of the population size and/or biomass of particular taxonomic and functional groups of soil animals and micro-organisms, but far fewer data on turnover times (i.e. inputs to the dead organic pool) and on the functional relationships between different groups. Root grazers are one functional group whose activities are particularly poorly understood.

The model could also be characterized by an organism population approach. This would certainly be needed to quantify many of the interactions not involving substantial transfer of material, e.g. chemical inhibition or promotion of germination of or micro-organism activity. More compartments would need to be recognized, but the immediate difficulty is that for so many groups the basic problems of recognition and taxonomy are unresolved, far less those of understanding their biology and ecology.

SOIL VARIABILITY

Apart from the intrinsic complexity of soils, physical and chemical as well as biotic, their spatial and temporal variability is a further major barrier against understanding their ecology. Jenny's explanations (1941, 1980) of how soil varies with differences in state factors (the 'clorpt' equation) are well known, and provide the conceptual framework for studying soil multiformity. However, there seems to be less awareness of the scales and intricacy of soil variability.

Spatial variability

The often great variation in vegetation composition from point to point is unconcealed, if not always recognized. That soils also characteristically show great variability in space (Beckett & Webster 1971) is less evident. Variation is often on a scale so small that a single tree can be rooted in two or more quite different soils (Lyford 1974). Soil can vary abruptly in drainage and mineralogical composition.

This is particularly likely in transported materials, such as the glacial drifts commonly occurring in the British uplands (Robinson & Lloyd 1915; Kantey & Morse 1965). Differences in past land-use can also cause persistent differences in many soil properties (Van Goor 1954; Armson 1959).

Temporal variability

Many soil properties change appreciably only over long periods of time, e.g. the degree of overall mineral weathering, particular-size distribution. Other properties vary measurably during the usually shorter time-spans of secondary succession and the lives of individual plants. Different plant species can vary greatly in levels of nutrient and secondary chemical composition, and also in the extent to which they modify the chemical composition of rain dripping off their leaves and canalized as stemflow. Thus, when an individual of one species or a patch of vegetation is replaced by one different, changes in humus form and quantity, and in plant-influenced properties such as pH and exchangeable base content, must be expected. Changes in such properties occur as gradients radially from main stems and with depth (Zinke 1962; Grubb, Green & Merrifield 1969; Gersper & Holowaychuck

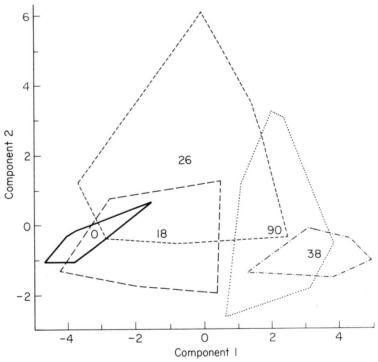

Fig. 3. Summary of changes in soil A and B horizons under a successional age sequence of *Betula pendula* stands near Advie, Morayshire: mean and range (*n* = 9) for each age-class of values of the first two components from principal component analysis of labile profile characteristics (see Fig. 4).

1970a, b). They are characteristically reversible. Page (1968, 1974) showed that surface acidity and the depth of surface organic matter accumulations varied cyclically during the life of many conifer species, the first declining and then increasing, and the second showing a contrasting trend.

I have studied, and summarized elsewhere, the changes in soil 'chemical' properties that occur when birch colonizes *Calluna vulgaris*-dominated moorland (Miles & Young 1980). These include increases in pH, exchangeable calcium, total phosphorus, and rates of nitrogen mineralization and cellulose decomposition. The appearance and structure of the soils change also. Figures 3 and 4 show the distribution and composition of the first two components from principal component analysis of labile, non-chemical properties across a site with a successional gradient from *C. vulgaris* to 91-year-old *B. pendula*. There is a clear trend in the first component, representing in particular a reduction in the number of horizons, in H horizon depth and, in the A horizon, increases in root and organic matter content, and in porosity and crumb structure. These changes, and those in colour, reflect the pronounced increase in microbial and faunal activity in the topsoil along the birch age gradient. Earthworm populations increased dramatically (Fig. 5). Figures 3 and 5 show, analogously to Page's (1968, 1974) findings under conifers, that the soil

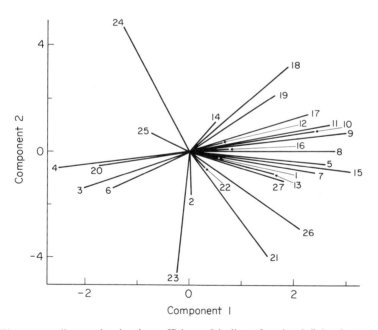

FIG. 4. Eigen vector diagram showing the coefficients of the linear function defining the two components in Fig. 3 These indicate the weighting to be placed on each soil characteristic in interpreting Fig. 3 (1–3, thickness of L, F and H horizons; 4, total number of horizons; 5–16, A horizon parameters; 5–7 colour—hue, value and chroma (Munsell scores); 8, % organic matter; 9, porosity; 10–11, structure strength and shape; 12, consistency; 13–14, sharpness; 17–26, B horizon parameters, as for parameters 5–14 in the A horizon; 27, B horizon cementation.

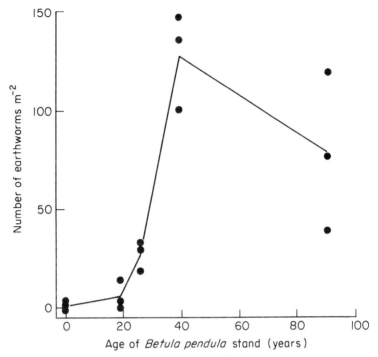

FIG. 5. Changes in total earthworm populations (number m^{-2} sampled by formaldehyde method) under the successional age sequence of *Betula pendula* stands in Fig. 5.

changes tended to reverse under very old birch, when the stand was opening up with tree senescence and death, and *C. vulgaris* was reinvading locally.

The vegetation reacts to these early soil changes, with the field layer gradually changing to one more characteristic of upland birchwoods, and increasing in species diversity. The soil changes are probably 'driven' mainly by the increased pH, with consequent effects on the vegetation. Figure 6 shows the positive correlation over all the successional stands I have studied between species density and pH.

Other plants species are associated with depressed biological activity in the soil (Miles 1981), including classically, *C. vulgaris*. Table 1 shows a decrease in growth of test plants in soil from under *C. vulgaris* in an abandoned granite quarry, compared with that from under adjacent partly colonized ground and adjacent birch stands. One reason for this depression of growth might be an accumulation of lipids in the soil. Mor humus of heathland typically has about twice the lipid content of the mull humus of broad-leaved woodlands, and these lipids are generally inhibitory to soil micro-organisms and to plant growth (Stevenson 1966: Fustec-Mathon, Righi & Jambu 1975). Another reason might be the production of phytotoxic organic acids (Jalal & Read 1983a, b). Table 1 also shows that the effectively unweathered (the quarry was abandoned in 1938), coarse-sandy granite soil was surprisingly fertile compared with old birchwood soil, and even potting compost.

TABLE 1. Mean dry weights (log$_{10}$ mg) of 10-week old plants of three species grown in soil from under three different kinds of vegetation on the floor of a disused granite quarry, from an old birchwood, and in potting compost

	Granite quarry soils from under:			Brown podzolic soils from 91-year-old *B. pendula* woodland	John Innes No. 2 potting compost	LSD at 5% level
	partly colonized ground	dominant *Calluna vulgaris*	26-year-old *Betula pendula* scrub			
Above-ground dry weights						
Luzula sylvatica (Hudson) Gaudin	1·06	0·57	1·59	0·96	1·42	0·35
Raphanus sativus L. 'French breakfast'	1·47	1·09	1·92	1·57	1·82	0·23
Rumex acetosa L.	1·31	0·43	1·46	1·44	1·64	0·67
Below-ground dry weights						
R. sativus L. 'French breakfast'	1·48	1·38	2·31	1·71	2·29	0·35

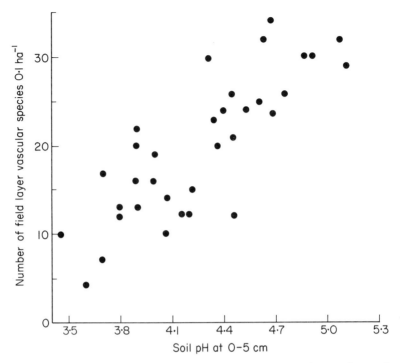

FIG. 6. Relationship between soil pH at 0–5 cm and the number of vascular plant species growing in the field layer of different aged, successional stands of *Betula pendula* and *B. pubescens* at twelve sites in Scotland and the north of England ($r = 0.891$, $P < 0.001$).

Natural granite soils in the region are now among the poorest in Scotland. This indicates how well plants may have grown as they colonized northwards after the last glaciation.

INTERACTIONS BETWEEN SOIL AND ABOVE-GROUND BIOTA

The seed bank

Persistent banks of dormant seeds (i.e. potential vegetation) seem to be ubiquitous in soil (Harper 1977; Grime 1979). However, little is known about the mechanisms controlling the size and life expectancy of buried seed populations. After dispersal, before seeds and fruits become intimately mixed into the soil, many are probably consumed by birds, small mammals and invertebrates (Janzen 1971). As they become incorporated into the soil by downwash, biological mixing, or simply being covered by litter, they are probably less readily found by predators, but quantities are probably still consumed by burrowing the soil-living animals. Earthworms, where present, ingest some seeds (Chippindale & Milton 1932; Milton 1939), and many do not survive passage through the gut (McRill & Sagar 1973; McRill 1974).

TABLE 2. 1978 seed rain, seed bank in July 1978, and estimated annual death rate k', of seeds of *Calluna vulgaris* in *C. vulgaris* dominated vegetation at varying altitudes in the Cairngorm Mountains (from Miller & Cummins 1981)

Altitude (m)	Mean annual rain of germinable seeds (10^3 m^{-2})	Numbers of buried germinable seeds (10^3 m^{-2})	Annual death rate, k'
370	420	69	0·86
580	53	68	0·44
860	0·24	14	0·02

Most buried seeds probably succumb to infection from microbial pathogens (Baker 1972). Despite the variety of likely fates, populations that are isolated from the seed rain decline in an approximately exponential manner (Roberts 1962, 1970; J. Miles, unpubl.), suggesting that the death risks are constant in time, independent of the age of the population.

Indirect evidence that the longevity of the seed bank depends on extrinsic factors is given in Table 2. It shows that the production of germinable seeds of *Calluna vulgaris* decreased dramatically with increasing altitude. At 860 m (which is near the altitudinal limit of growth of *C. vulgaris*) seed production was only 0·1% of that at 370 m. However, the seed bank (determined before the new seed crop, or at least most of it, was shed) did not diminish correspondingly. Estimated annual death rates (calculated as the loss coefficient, k', of Jenny, Gessel & Bingham 1949) were over forty times greater at 370 m than at 860 m. They suggest that one year's seed rain will disappear in 4–5 years at 370 m, but over centuries at 860 m. The lower death rates at higher altitudes are presumably mainly attributable to reduced predation and microbial infection, reflecting reduced biological activity in the soils with lower year-round temperatures. Part of the difference, however, could also be attributable to reduced catabolic rates within the seeds in a colder climate.

Seed dispersal by ants

As British ants nest in the soil (*sensu lato*), I class them here as part of the soil fauna. It has been known since the last century that ants transport seeds of many species, particularly of those that produce ant-attracting oil bodies or elaiosomes (Kerner 1891). Ant dispersal of seeds and fruits (myrmechochory) attracted considerable early attention (Uphof 1942), and Ulbrich (1939) thought it was common among woodland plants in parts of Germany. Interest in the subject waned post-war in the face of newer and more fashionable topics, but has been somewhat renewed of late. Berg (1975, 1981) has pointed out the enormous importance of myrmechochory in Australia; myrmecochorous species can constitute over half the flora in dry sclerophyll forest, and nearly half in heathland. Many dominant species of English lowland heath are collected and stored underground by *Tetramorium caespitum* (Brian, Hibble & Stradling 1965; Brian, Elmes & Kelley 1967). In temperate North America, myrmecochorous species comprised 30–44% of the herbaceous flora in West Virginian forests (Beattie, Culver & Pudlo 1979; Beattie & Culver 1981), and 30% in beech–maple forest in New York State (Handel, Fisch & Schatz 1981). In the latter case, myrmecochorous species formed 40% of the above-ground herbaceous biomass, and over half of all herbaceous stems.

Apart from the advantages of dispersal, ant transport and often burial of seeds and fruits may benefit the plant by reducing seedling competition (Handel 1978) and securing seeds from predation (Culver & Beattie 1978; O'Dowd & Hay 1980), while elaiosome or aril removal may enhance germination (Horvitz 1981).

Logs and stumps as seedling niches

Niches allowing both seedling establishment and successful subsequent growth are commonly in short supply (Miles 1979). In north temperate forests, rotting tree stumps and logs allow saplings and herbaceous species to establish (Jones 1945; Franklin & Dyrness 1973). To see this is commonplace, and sometimes these substrates are almost the only places where saplings can be found. What changes

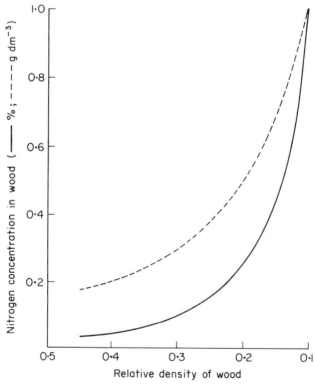

FIG. 7. Mean trend of accumulation of nitrogen in decomposing stumps of *Picea sitchensis*. Relative density is a measure of decomposition, freshly cut stumps being 0·4–0·5 g cm^{-2}, and stumps 10–20 years old, 0·1–0·15 g cm^{-2} (after Newell & Heal 1982).

TABLE 3. Density of seedlings and saplings of *Picea engelmannii* and *Abies lasiocarpa* regenerating on the trunks of fallen trees in varying states of decomposition in a *P. engelmannii – A. lasiocarpa* forest, Colorado (mean number 100 m⁻¹ of log). (After McCullough 1948; decay classes following Triska & Cromack 1980.)

			Log decay class		
	I	II	III	IV	V
Plant height class	Tree newly fallen	Early decay visible; bark intact	Bark sloughing; sapwood rotten, heartwood sound	Bark detached; heartwood rotten and crumbling	More or less totally decomposed; log outline indistinct
< 30 cm	0	7	35	724	1974
> 30 cm	0	0	0	0	66

must occur in wood before seedlings can establish? I have found no answer in the literature. Certainly, seedlings only appear in quantity when the wood is well rotted (Table 3). I suggest that this may correspond to the phase of rapid accumulation of nitrogen shown by Newell & Heal (1982) (Fig. 7). This coincides with the level of nitrogen capital needed to sustain non-nitrogen fixing plants during primary succession (Bradshaw *et al.* 1982), i.e. about 70 mg dm⁻².

Root competition

Interest in root competition, as with myrmechochory, has unaccountably languished in recent years, although many early studies demonstrated the importance of tree root competition in controlling regeneration and the composition of the field layer (Miles 1979). Figure 8 shows the dramatic flowering response of *Anemone nemorosa* when plots in a mixed *Betula pendula–B. pubescens* wood were trenched to 0·5 and refilled. The effect was short-lived but made its point with me. Previously I had only seen such displays of flowering in *A. nemorosa* on roadside verges, and had attributed this to lack of shading.

Mycorrhizas

Mycorrhizas give considerable benefits. They help most plants to tap scarce nutrients, even utilizing soluble organic forms of nitrogen (Lundeberg 1970; Stribley & Read 1974; Bowen & Smith 1981). They accelerate stem maturation (Last *et al.* 1983), while sheathing mycorrhizas at least, protect roots from pathogenic fungi (Marx 1972; Sylvia 1983). The death of mycorrhizal hyphae may account for 80% of plant return to the soil of nitrogen, phosphorus and potassium (Fogel 1970; Fogel & Hunt 1983). Recent observations of sequences of fruit-bodies with time under a mixed *Betula pendula–B. pubescens* stand (Mason *et al.* 1982; Mason *et al.* 1983), coupled with laboratory studies (Deacon, Donaldson & Last 1983; Fox 1983), suggest that mycorrhizal successions may occur, with 'late stage' fungi being unable to form mycorrhizal associations with young plants in unsterile soils. Collections of

TABLE 4. Percentage composition in September 1977 of fruit-bodies of known or probable mycorrhizal fungi under different aged stands of birch (*Betula pendula* and *B. pubescens*) growing at four sites formerly under heather moorland: (a) *B. pendula* at Tulchan, near Advie, Morayshire; (b) *B. pendula* at Kerrow, near Cannich, Inverness-shire; (c) *B. pubescens* at Craggan, Strath Avon, Banffshire; (d) *B. pubescens* at Silpho, near Scalby, North Yorkshire

(a) *B. pendula* aged (years)

	21	41	93
Laccaria proxima	17		
L. laccata	6		
Lactarius rufus	5		
Leccinum versipellis	4		
Lactarius vietus	47	15	
L. pubescens	7	29	
Russula sp.	4		
Inocybe sp.		8	
Cortinarius sp.		19	
Lactarius tabidus		15	
Paxillus involutus		4	
Xerocomus chrysenteren		4	
Cantharellus cibarius		31	29
Amanita citrina			14
Russula foetens			14

(b) *B. pendula* aged (years)

	20	29	72
Boletus impolitus	3		
Russula atropurpurea	3		
Cortinarius spp.	78	18	
Lactarius pubescens	16	5	5
Hebeloma spp.		52	
Boletus edulis		3	
Tricholoma nudum		8	
Amanita muscaria		8	32
Leccinum scabrum		8	5
Russula aeruginea		3	5
Tricholoma columbetta		3	28
Laccaria amethystina			10
Lactarius tabidus			10
Inocybe sp.			5

(c) *B. pubescens* aged (years)

	21	30	56
Cortinarius tabularis	27		
Leccinum scabrum	5		
Cortinarius armillatus	22	64	
Lactarius pubescens	15	3	
L. rufus	7	3	
Laccaria laccata	24		23
Lactarius vietus		24	10
Russula sp.		6	14
Lactarius turpis			27
Cortinarius sp.			23
Paxillus involutus			3

(d) *B. pubescens* aged (year)

	15	38	48	63	50–100
Russula sp.	29				
Lactarius rufus	29		5	4	
L. vietus	13	80	10	77	45
Amanita fulva	29	3	28	4	
Paxillus involutus		3	24		
Russula ochroleuca		3			3
Leccinum scabrum		5	33	7	45
Xerocomus subtomentosus		6			
Lactarius turpis		3			
Leccinum versipellis				4	
Russula atropurpurea				4	3

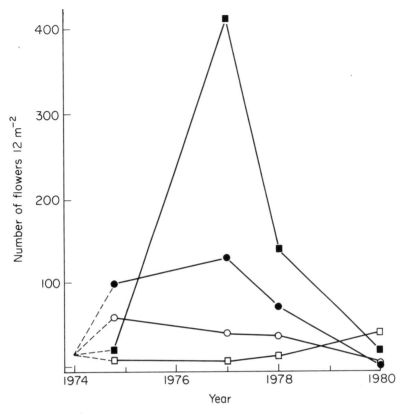

FIG. 8. Changes in flowering of *Anemone nemorosa* and *Oxalis acetosella* under a 30-year old mixed *Betula pendula–B. pubescens* wood at Hill of Brathens, Kincardineshire, after plots were root trenched to 0·5 m depth in February 1974, lined with polythene, and refilled (*A. nemorosa* trenched ■, untrenched □, *O. acetosella* trenched ●, untrenched ○).

fruit-bodies at four sites with successional sequences of *Betula* (Table 4) suggest that mycorrhizal successions, caused directly or indirectly by the ageing trees, are a widespread phenomenon. Their functional significance is, however, unknown.

Effects of herbivores

Invertebrate herbivores usually seem to consume only a small proportion of plant production. Their egesta are therefore, unlikely to be important sources of N and P. In contrast, vertebrate herbivores affect soil fertility directly by voiding faeces and urine (Wolton 1955; Peterson, Woodhouse & Lucus 1956; Lotero, Woodhouse & Peterson 1966), with consequent increases in herbage production (Wheeler 1958; Weeda 1967). This is an important mechanism promoting nitrogen and phosphorus cycling in poor upland soils in Britain (Floate 1972). During the survey of successional sequences of birch noted earlier, I gained the impression that the field layer had changed more rapidly where past grazing pressures seemed to have been greater. Although this could not be quantified, it would tend to follow if dunging and urination had speeded up the cycling of nitrogen and phosphorus.

Other effects of herbivores may be more subtle. Several studies have shown that defoliation increases primary production (thus indirectly influencing soil processes) by increasing growth rates of remaining tissues (Scott, French & Leetham 1979), by delaying senescence (Jameson 1963), and by stimulating growth through the effects of saliva (Reardon, Leinweber & Merrill 1972). In the last case, the physiologically-active chemical seems to be thiamine (Reardon, Leinweber & Merrill 1974). A similar effect has been noted in grasshoppers (Dyer & Bokhari 1976). Equally subtle is the finding that small fungus-eating mammals seem to be the primary vectors of spore dissemination of mycorrhizal fungi that produce fruit-bodies below ground (Maser, Trappe & Nussbaum 1978).

EPILOGUE

The examples given indicate how varied and important are the interactions between soil and above-ground biota and biotic processes. As a concluding example of ecosystem complexity and connectedness, Fig. 9 shows that most of the major ecosystem components directly influence three parameters, variously touched upon earlier, that directly affect plant welfare:soil pH, phytoactive chemical production, and plant nutrient supply.

It is clear that most of the functional interactions of ecosystems are transient rather than fixed. Vegetation, the predominant primary producer in terrestrial

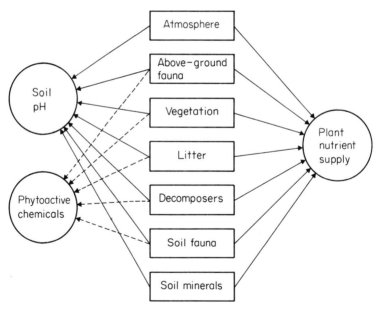

FIG. 9. Major ecosystem components influencing three parameters affecting plant welfare, soil pH, production of chemicals influencing plants, and plant nutrient supply.

ecosystems, varies continually in space and time. Even in apparently stable vegetation, episodic disturbances increasingly seem to be the rule rather than the exception. Long-term equilibria of plant-dependent soil processes are thus likely to be equally rare. As vegetation changes, the associated populations of consumers and decomposers tend to vary correspondingly, reflecting the changes in their environment in physical structure, micro-climate, plant and litter chemical composition (Burges & Raw 1967; Swift *et al.* 1979), and plant-induced soil changes (Miles 1981). An ecosystem thus has no absolute reality, other than as a concept. Any ecosystem must be defined arbitrarily, in an *ad hoc* manner. The ecosystem concept during the past two decades or so has been an invaluable peg on which to hang ecological process studies, in particular those of nutrient cycling and energy flow (Likens *et al.* 1977; Bormann & Likens 1979; Swift *et al.* 1979). However, the dynamic nature and extreme variability of soil properties must constantly be borne in mind by soil biologists and ecologists.

ACKNOWLEDGMENTS

W. F. Young and D. D. French helped in various ways, P. A. Stevens described the soil profiles, Dr P. A. Mason and K. Ingleby identified the fungal fruit-bodies, M. Randall and J. Reynolds measured the earthworm populations, and Drs G. R. Miller and D. Jenkins kindly commented on a draft of this paper.

REFERENCES

Abdulla, M.H. (1970). Preliminary study on the influence of fungal metabolites on germination of barley grains. *Mycopathologia et Mycologia Applicata*, **41**, 307–313.

Andersen, D.C., McMahon, J.H. & Wolfe, M.L. (1980). Herbivorous mammals along a montane sere: community structure and energetics. *Journal of Mammalogy*, **61**, 500–519.

Armson, K.A. (1959). *An example of the effects of past land use on fertility levels and growth of Norway spruce* (Picea abies *(L.) Karst.*). University of Toronto Faculty of Forestry Technical Report, 1.

Baker, K.F. (1972). Seed pathology. *Seed˙Biology*, Vol. II. *Germination Control, Metabolism, and Pathology* (Ed. by T.T. Kozlowski), pp. 317–416. Academic Press, New York.

Beattie, A.J. & Culver, D.C. (1981). The guild of myrmecochores in the herbaceous flora of West Virginia. *Ecology*, **62**, 107–115.

Beattie, A.J., Culver, D.C. & Pudlo, R.J. (1979). Interactions between ants and diaspores of some common spring-flowering herbs in West Virginia. *Castanea*, **44**, 177–186.

Beckett, P.H.T. & Webster, R. (1971). Soil variability: a review. *Soils and Fertilizers*, **34**, 1–15.

Berg, R.Y. (1975). Myrmecochorous plants in Australia and their dispersal by ants. *Australian Journal of Botany*, **23**, 475–508.

Berg, R.Y. (1981). The role of ants in seed dispersal in Australian lowland heathland. *Heathlands and Related Shrublands. Analytical Studies. Ecosystems of the World*, Vol. 9B (Ed. by R.L. Specht), pp. 51–59. Elsevier, Amsterdam.

Bormann, F.H. & Likens, G.E. (1979). *Pattern and Process in a Forested Ecosystem*. Springer-Verlag, New York.

Bowen, G.D. & Smith, S.E. (1981). The effects of mycorrhizas on nitrogen uptake by plants. *Terrestrial Nitrogen Cycles* (Ed. by F.E. Clark & T. Rosswall). *Ecological Bulletin (Stockholm)*, **33**, 237–247.

Bradshaw, A.D., Marrs, R.H., Roberts, R.D. & Skeffington, R.A. (1982). The creation of nitrogen cycles in derelict land. *Philosophical Transactions of the Royal Society B*, **296**, 557–561.

Brian, M.V., Elmes, G. & Kelley, A.F. (1967). Populations of the ant *Tetramorium caespitum* Latreille. *Journal of Animal Ecology*, **36**, 337–342.

Brian, M.V., Hibble, J. & Stradling, D.J. (1965). Ant pattern and density in a southern English heath. *Journal of Animal Ecology*, **34**, 545–555.

Brown, R.T. & Mikola, P. (1974). The influence of fruticose soil lichens upon the mycorrhizae and seedling growth of forest trees. *Acta Forestalia Fennica*, **141**.

Brownlee, C., Duddridge, J.A., Malibari, A. & Read, D.J. (1983). The structure and function of mycelial systems of ectomycorrhizal roots with special reference to their role in forming inter-plant connections and providing pathways for assimilate and water transport. *Plant and Soil*, **71**, 433–443.

Burges, A. & Raw, F. (1967). *Soil Biology*. Academic Press, London.

Chapman, S.J. & Lynch, J.M. (1983). The relative roles of micro-organisms and their metabolites in the phytotoxicity of decomposing plant residues. *Plant and Soil*, **74**, 457–459.

Chippindale, H.G. & Milton, W.E.J. (1932). Note on the occurrence of buried seeds in the soil. *Journal of Agricultural Science*, **22**, 451–452.

Culver, D.C. & Beattie, A.J. (1978). Myrmecochory in *Viola*: dynamics of seed-ant interactions in some West Virginia species. *Journal of Ecology*, **66**, 53–72.

Deacon, J.W., Donaldson, S.J. & Last, F.T. (1983). Sequences and interactions of mycorrhizal fungi and birch. *Plant and Soil*, **71**, 257–262.

Dyer, M.I. & Bokhari, U.G. (1976). Plant–animal interactions: studies of the effects of grasshopper grazing on blue grama grass. *Ecology*, **57**, 762–772.

Fisher, R.F. (1979a). Possible allelopathic effects of reindeer-moss (*Cladonia*) on jack pine and white spruce. *Forest Science*, **25**, 256–260.

Fisher, R.F. (1979b). Allelopathy. *Plant Disease. An Advanced Treatise*. Vol. 4. *How Pathogens Induce Disease* (Ed. by J.G. Horsfall & E.B. Cowling), pp. 313–330. Academic Press, New York.

Floate, M.J.S. (1972). Plant nutrient cycling in hill land. *Proceedings of the North of England Soils Discussion Group*, **7**, 11–27.

Fogel, R. (1980). Mycorrhizae and nutrient cycling in natural forest ecosystems. *New Phytologist*, **63**, 199–212.

Fogel, R. & Hunt, G. (1983). Contribution of mycorrhizae and soil fungi to nutrient cycling in a Douglas-fir ecosystem. *Canadian Journal of Forest Research*, **13**, 219–232.

Fox, F.M. (1983). Role of basidiospores as inocula of mycorrhizal fungi of birch. *Plant and Soil*, **71**, 269–273.

Franklin, J.F. & Dyrness, C.T. (1973). Natural vegetation of Oregon and Washington. *United States Department of Agriculture Forest Service Pacific Northwest Forest and Range Experiment Station General Technical Report*, PNW-8.

Fustec-Mathon, E., Righi, D. & Jambu, P. (1975). Influence des bitumes extraits de podzols humiques hydromorphes des Landes du Médoc sur la microflore tellurique. *Revue d'Écologie et de Biologie du Sol*, **12**, 393–404.

Gersper, P.L. & Holowaychuk, N. (1970a). Effects of stemflow water on a Miami soil under a beech tree, I. Morphological and physical properties. *Proceedings of the Soil Science Society of America*, **34**, 779–786.

Gersper, P.L. & Holowaychuk, N. (1970b). Effect of stemflow water on a Miami soil under a beech tree, II. Chemical properties. *Proceedings of the Soil Science Society of America*, **34**, 786–794.

Grime, J.P. (1979). *Plant Strategies and Vegetation Processes*. Wiley, Chichester.

Grubb, P.J., Green, H.E. & Merrifield, R.C.J. (1969). The ecology of chalk heath: its relevance to the calcicole-calcifuge and soil acidification problems. *Journal of Ecology*, **57**, 175–212.

Handel, S.N. (1978). The competitive relationship of three woodland sedges and its bearing on the evolution of ant-dispersal of *Carex pedunculata*. *Evolution*, **32**, 151–163.

Handel, S.N., Fisch, S.B. & Schatz, G.E. (1981). Ants disperse a majority of herbs in a mesic forest community in New York State. *Bulletin of the Torrey Botanical Club*, **108**, 430–437.

Handley, W.R.C. (1963). Mycorrhizal associations and *Calluna* heathland afforestation. *Bulletin of the Forestry Commission, London, 36*.

Harper, J.L. (1977). *Population Biology of Plants*. Academic Press, London.

Heap, A.J. & Newman, E.I. (1980). Links between roots by hyphae of vesicular-arbuscular mycorrhizas. *New Phytologist*, **85**, 169–171.

Horvitz, C.C. (1981). Analysis of how ant behaviours affect germination in a tropical myrmecochore *Calathea microcephala* (P. & E) Koernicke (Marantaceae): microsite selection and aril removal by neotropical ants, *Odontomachus, Pachycondyla,* and *Solenopsis* (Formicidae). *Oecologia,* **51,** 47–52.

Hurd, L.E., Mellinger, M.V., Wolf, L.L. & McNaughton, S.J. (1971). Stability and diversity at three trophic levels in terrestrial successional ecosystems. *Science,* **173,** 1134–1136.

Hurd, L.E. & Wolf, L.L. (1974). Stability in relation to nutrient enrichment in arthropod consumers of old-field successional ecosystems. *Ecological Monographs,* **44,** 465–482.

Innis, G.S. (1978). *Grassland Simulation Model.* Ecological Studies 26. Springer-Verlag, New York.

Last, F.T., Mason, P.A., Wilson, J. & Deacon, J.W. (1983). Fine roots and sheathing mycorrhizas: their formation, function and dynamics. *Plant and Soil,* **71,** 9–21.

Leeuwen, B.H. van (1981). Influence of micro-organisms on the germination of the monocarpic *Cirsium vulgare* in relation to disturbance. *Oecologia,* **48,** 112–115.

Likens, G.E., Bormann, F.H., Pierce, R.S., Eaton, J.S. & Johnson, N.M. (1977). *Biogeochemistry of a Forested Ecosystem.* Springer-Verlag, New York.

Lotero, J., Woodhouse, W.W. & Peterson, R.G. (1966). Local effect on fertility of urine voided by grazing cattle. *Agronomy Journal,* **58,** 262–265.

Lundeberg, G. (1970). Utilization of various nitrogen sources, in particular bound soil nitrogen, by mycorrhizal fungi. *Studia Forestalia Suecica,* 79.

Lyford, W.H. (1974). Narrow soils and intricate soil patterns in southern New England. *Geoderma,* **11,** 195–208.

Lynch, J.M. (1978). Microbiological problems in seedling establishment. *Microbial Ecology* (Ed. by M.W. Loutit & J.A.R. Miles), pp. 337–340. Springer-Verlag, Berlin.

MacMahon, J.A. (1981). Successional processes: comparisons among biomes with special reference to probable roles of and influences on animals. *Forest Succession, Concepts and Application* (Ed. by D.C. West, H.H. Shugart & D.B. Botkin), pp. 277–304. Springer-Verlag, New York.

Marx, D.H. (1972). Ectomycorrhizae as deterrents to pathogenic root infections. *Annual Review of Phytopathology,* **10,** 429–454.

Maser, C., Trappe, J.M. & Nussbaum, R.A. (1978). Fungal–small mammal interrelationships with emphasis on Oregon coniferous forests. *Ecology,* **59,** 799–809.

Mason, P.A., Last, F.T., Pelham, J. & Ingley, K. (1982). Ecology of some fungi associated with an ageing stand of birches (*Betula pendula* and *B. pubescens*). *Forest Ecology and Management,* **4,** 19–39.

Jalal, M.A.F. & Read, D.J. (1983a). The organic acid composition of *Calluna* heathland soil with special reference to phyto- and fungitoxicity, I. Isolation and identification of organic acids. *Plant and Soil,* **70,** 257–272.

Jalal, M.A.F. & Read, D.J. (1983b). The organic acid composition of *Calluna* heathland soil with special reference to phyto- and fungitoxicity, II. Monthly quantitative determination of the organic acid content of *Calluna* and spruce dominated soils. *Plant and Soil,* **70,** 273–286.

Jameson, D.A. (1963). Responses of individual plants to harvesting. *Botanical Reviews,* **29,** 532–594.

Janzen, D.H. (1971). Seed predation by animals. *Annual Review of Ecology and Systematics,* **2,** 465–492.

Jenny, H. (1941). *Factors of Soil Formation.* McGraw-Hill, New York.

Jenny, H. (1980). *The Soil Resource. Origin and Behaviour.* Springer-Verlag, New York.

Jenny, H, Gessel, S.P. & Bingham, F.T. (1949). Comparative study of decomposition rates of organic matter in temperate and tropical regions. *Soil Science,* **68,** 419–432.

Jones, E.W. (1945). The structure and reproduction of the virgin forests of the north temperate zone. *New Phytologist,* **44,** 130–148.

Kantey, B.A. & Morse, R.K. (1965). A modern approach to highway materials sampling. *Proceedings of the Fifth International Conference on Soil Mechanics and Foundation Engineering,* **1,** 55–58.

Kerner von Marilaun, A. (1891). *Pflanzenleben.* Vol. 2. Bibliographisches Institut, Leipzig.

Kirkpatrick, B.C. & Bazzaz, F.A. (1979). Infuence of certain fungi on seed germination and seedling survival of four colonizing animals. *Journal of Applied Ecology,* **16,** 515–527.

Kruuk, H. & Parish, T. (1981). Feeding specialization of the European badger *Meles meles* in Scotland. *Journal of Animal Ecology,* **50,** 773–788.

Mason, P.A., Wilson, J., Last, F.T. & Walker, C. (1983). The concept of succession in relation to the spread of sheathing mycorrhizal fungi on inoculated tree seedlings growing in unsterile soils. *Plant and Soil*, **71**, 247–256.

McCullough, H.A. (1948). Plant succession on fallen logs in a virgin spruce-fir forest. *Ecology*, **29**, 508–513.

McRill, M. (1974). *Some botanical aspects of earthworm activity*. Ph.D. Thesis, University College of North Wales.

McRill, M. & Sagar, G.R. (1973). Earthworms and seeds. *Nature*, **243**, 482.

Miles, J. (1979). *Vegetation Dynamics*. Chapman & Hall, London.

Miles, J. (1981). Problems in heathland and grassland dynamics. *Vegetatio*, **46**, 61–74.

Miles, J. & Young, W.F. (1980). The effects on heathland and moorland soils in Scotland and northern England following colonization by birch (*Betula* spp.). *Bulletin d'Écologie*, **11**, 233–242.

Miller, G.R. & Cummins, R.P. (1981). Population dynamics of buried seeds on mountains. *Institute of Terrestrial Ecology Annual Report*, 1980, 75–76.

Milton, W.E.J. (1939). The occurrence of buried viable seeds in soils at different elevations and on a salt marsh. *Journal of Ecology*, **27**, 149–159.

Mosse, B. (1978). Mycorrhiza and plant growth. *The Structure and Functioning of Plant Populations* (Ed. by A.H.J. Freysel & J.W. Woldendorp), pp. 269–298. North-Holland Publ. Co., Amsterdam.

Newell, K. & Heal, O.W. (1982). Stump decomposition. *Institute of Terrestrial Ecology Annual Report*, 1981, 85–86.

Nickell, L.G. (1959). Automicrobial activity of vascular plants. *Economic Botany*, **13**, 281–318.

O'Dowd, D.J. & Hay, M.E. (1980). Mutualism between harvester ants and a desert ephemeral: seed escape from rodents. *Ecology*, **61**, 531–540.

Page, G. (1968). Some effects of conifer crops on soil properties. *Commonwealth Forestry Review*, **47**, 52–62.

Page, G. (1974). Effects of forest cover on the properties of some Newfoundland forest soils and their relation to forest growth. *Canadian Forestry Service Publication*, 1332.

Peterson, R.G., Woodhouse, W.W. & Lucas, H.L. (1956). The distribution of excreta by freely grazing cattle and its effect on pasture fertility, II. Effect of returned excreta on the residual concentrations of some fertility elements. *Agronomy Journal*, **48**, 444–449.

Ramsbottom, J. (1953). *Mushrooms and Toadstools. A Study of the Activities of Fungi*. Collins, London.

Reardon, P.Q., Leinweber, C.L. & Merrill, L.B. (1972). The effect of bovine saliva on grasses. *Journal of Animal Science*, **34**, 897–898.

Reardon, P.Q., Leinweber, C.L. & Merrill, L.B. (1974). Response of sideoats grama to animal saliva and thiamine. *Journal of Range Management*, **27**, 400–401.

Rice, E.L. (1979). Allelopathy–an update. *Botanical Review*, **45**, 15–109.

Roberts, H.A. (1962). Studies on the weeds of vegetable crops, II. Effect of six years of cropping on the weed seeds in the soil. *Journal of Ecology*, **50**, 803–813.

Roberts, H.A. (1970). Viable weed seeds in cultivated soils. *Report of the National Vegetable Research Station*, 1969, 25–38.

Robinson, G.W. & Lloyd, W.E. (1915). On the probable error of sampling in soil surveys. *Journal of Agricultural Science, Cambridge*, **7**, 144–153.

Scott, J.A., French, N.R. & Leetham, J.W. (1979). Patterns of consumption in grasslands. *Perspectives in Grassland Ecology. Results and Applications of the US/IBP Grassland Biome Study* (Ed. by N.R. French), pp. 89–105. Ecological Studies 32. Springer-Verlag, New York.

Stevenson, F.J. (1966). Lipids in soil. *Journal of the American Oil Chemists' Society*, **43**, 203–210.

Stribley, D.P. & Read, D.J. (1974). The biology of mycorrhiza in the Ericaceae, IV. The effect of mycorrhizal infection on uptake of ^{15}N from labelled soil by *Vaccinium macrocarpon* Ait. *New Phytologist*, **73**, 1149–1155.

Stone, E.L. (1974). The communal root system of red pine: growth of girdled trees. *Forest Science*, **20**, 294–305.

Sutton, R.F. (1969). Form and development of conifer root systems. *Commonwealth Forestry Bureau Technical Communication*, 7.

Swift, M.J., Heal, O.W. & Anderson, J.M. (1979). *Decomposition in Terrestrial Ecosystems*. Blackwell Scientific Publications, Oxford.

Sylvia, D.M. (1983). Role of *Laccaria laccata* in protecting primary roots of Douglas-fir from root rot. *Plant and Soil*, **71**, 299–302.

Tinker, P.B. (1980). Root–soil interactions in crop plants. *Soil and Agriculture* (Ed. by P.B. Tinker), pp. 1–34. Blackwell Scientific Publications, Oxford.

Triska, F.J. & Cromack, K. (1980). The role of wood debris in forests and streams. *Forests: Fresh Perspectives from Ecosystem Analysis* (Ed. by R.H. Waring), pp. 171–190. Proceedings of the 40th Annual Biology Colloquium. Oregon State University Press, Corvallis.

Ulbrich, E. (1939). Deutsche Myrmekochoren. Beobachtungen über die Verbreitung heimischer Pflanzen durch Ameisen. *Repertorium Specierum Novarum Regni Vegetablis*, 67.

Uphof, J.C.T. (1942). Ecological relations of plants with ants and termites. *Botanical Review*, **8**, 563–598.

Valadon, L.A.G. & Lodge, E. (1970). Auxins and other compounds of *Cladosporium herbarum*. *Transactions of the British Mycological Society*, **55**, 9–15.

Van Goor, C.P. (1954). The influence of tillage on some properties of dry sandy soils in the Netherlands. *Landbouwkundig tijdschrift*, **66**, 175–181.

Weatherell, J. (1953). The checking of forest trees by heather. *Forestry*, **26**, 37–41.

Weeda, W.C. (1967). The effect of cattle dung patches on pasture growth, botanical composition, and pasture utilization. *New Zealand Journal of Agricultural Research*, **10**, 150–159.

Wheeler, J.L. (1958). The effect of sheep excreta and nitrogenous fertilizer on the botanical composition and production of a ley. *Journal of the British Grassland Society*, **13**, 196–202.

Wolton, K.M. (1955). The effect of sheep excreta and fertilizer treatments on nutrient status of a pasture. *Journal of the British Grassland Society*, **10**, 240–253.

Zinke, P.J. (1962). The pattern of influence of individual forest trees on soil properties. *Ecology*, **43**, 130–153.

Author index

Figures in italics refer to pages where full references appear; figures in bold to articles in this volume.

Abbott, L.K., 219, *223*
Abdel Wahab, A.M., 156, *157*
Abdulla, M.H., 410, *423*
Aber, J.D., 28, 30, 31, *35*: 37, 39, *42*: 175, *178*: 343, *353*: 377, *391*
Adams, R.E., 320, *331*
Addison, J.A., 303, *313*: 320, *329*
Ågren, G.I., 31, *34*: 37, *42*
Albrecht, S.L., 133, 140, *141*, *145*
Alder, J., 116, *118*
Aldrich, A.C., 131, *144*
Alexander, I.J., **37–42**: **175–179**: **267–277**: 40, *42*: 44, *64*: 175, 176, *178*: 275, 277: 386, *391*
Alexander, M., 303, *314*
Allee, W.C., 8, *18*
Allen, M.F., 320, *329*
Allen, O.N., 187, *190*
Allen, S.E., 75, *78*: 88, *104*: 269, *277*: 279, 280, *283*: 322, *329*: 333, *337*
Allmen, H. von., 246, 248, *261*
Ames, R.N., 8, *18*: 212, *215*: 227, 231
Anagnostakis, S., 309, *315*
Andersen, D.C., 408, *423*
Anderson, D.W., 6, *20*
Anderson, F., *172*, *173*
Anderson, J.M., **377–392**: 9, 15, *18*, *21*, 168, *173*: 253, 254, *261*: 268, *277*: 298, 299, 302, *313*, *314*: 303, 304, 307, 312, *317*: 328, *330*: 339, 340, *354*: 367, 368, 370, 374, *375*: 377, 378, 379, 380, 381, 382, 383, 384, 387, 389, 390, *391*, *392*: 408, 410, 423, 426
Anderson, J.P.E., 110, *118*: 185, *189*: 213, *215*: 356, *363*
Anderson, M.A., 134, 137, *143*
Anderson, R.V., 9, 16, *18*, *19*: 355, 356, *363*, *364*, *365*: 367, *375*: 378, *391*
Anderson, T., 114, 115, *119*
Andersson, F., 24, *34*
Angel, R.A., 307, *315*
Anghinoni, I., 90, *104*
App, A.A., 137, 138, *143*
Ares, J., 115, 117, *118*
Armson, K.A., 412, *423*
Arnold, M.K., 396, *397*
Arnold, P.T., **149–158**
Arnold, P.W., 73, *78*
ARS, 227, 231

Ashley, T.W., 92, 99, *104*
Athias-Binche, F., 245, 247, *261*
Atilano, R.A., 321, *329*
Atkinson, D., **43–65**: 29, *34*: 43, 44, 46, 48, 49, 50, 51, 53, 55, 57, 61, 63, *64*: 67, *70*: 88, 235, 240
Atkinson, T.G., 114, *120*
Ausmus, B.S., 30, *34*: 37, *42*
Avivi, Y., 156, *157*
Ayres, A.J., 74, *78*
Axelsson, B., 31, *34*: 37, *42*

Bååth, E., 259, *261*: 356, *364*: 367, 375: 393, *397*
Babel, U., 268, *277*
Bachelard, E.P., 25, *35*
Bailey, J.E., 326, 328, *329*
Baker, D.E., 225, *231*
Baker, G.H., 247, *261*
Baker, K.F., 186, *189*: 417, *423*
Balandreau, J.P., 130, 134, *141*
Baldani, V.L.D., 131, *142*
Baldwin, J.P., 359, *363*
Ball, D.F., 73, *78*
Balogun, R.A., 251, *265*
Baltensperger, A.A., 156, *157*
Bamforth, S.S., 13, *21*
Barber, D.A., 111, 112, *118*, *119*: 132, 133, *142*: 188, 189
Barber, L.E., 156, *157*
Barber, S.A., 28, *34*: 50, *64*: 89, 90, *104*: 359, *363*
Barea, J.M., 139, *142*
Barley, K.P., 115, *119*
Bärlocher, F., 301, *313*
Barnes, P.W., 149, *157*
Barrow, N.J., 73, *78*
Bartholomew, W.V., 8, *21*
Bates, G.H., 67, *70*
Bauer, T., 251, *261*
Baylis, G.T.S., 89, *104*
Bazilevich, N.I., 33, *36*
Bazzaz, F.A., 410, *426*
B.D.H. Chemicals Ltd., 73, *78*
Beattie, A.J., 307, 311, *316*: 418, *423*, *424*
Beck, S.M., 133, *142*
Beckel, D.K.B., 113, *119*
Becket, P.H.T., 411, *423*
Beech, D.F., 128, *143*

Behan, V.M., 253, 259, *261*: 307, *313*
Behan-Felletier, V.M., 252, *261*: 302, 303, 307, *314*
Bell, W.D., 132, *142*
Bender, E.A., 12, *18*
Benecke, P., 234, 240, *242*
Bengtsson, G., 253, *261*: 305, *314*: 367, 370, *375*
Bennett, R.A., 108, *119*
Benyon, K.I., 328, *330*
Berg, B., 172, *172*: 177, *178*: 259, *261*: 268, 276, *277*: 343, *352*: 374, *375*
Berg, R.Y., 417, *423*
Bergstrom, L., 358, *363*
Berkeley, R.C.W., 2, *18*
Berthet, P.L., 247, *261*
Bethlenfalvay, G.J., 112, *119*: 212, *215*
Bettany, J.R., 123, *125*
Bhat, K.K.S., 319, *329*
Bhattacharya, P.K. de., 133, *143*
Bhattacharya, R., 133, *143*
Bhaymik, H.D., 298, *314*
Bieleski, R.L., 319, *329*
Biermann, B., 151, *157*
Bignell, D.E., 299, *313*
Bille, S.W., 73, *79*
Billings, W.D., 26, *36*
Bingham, F.T., 417, *425*
Binkley, D., 73, *78*
Binns, E.S., 303, 305, *314*
Bird, G.W., 329, *329*
Birrell, K.S., 401, *405*
Biscoe, P.V., 26, 28, *34*
Black, I.A., 227, *232*
Black, R.L.B., 212, *217*
Blaha, M., 9, *19*: 355, *364*
Blancquaert, J.P., 246, *261*
Bleasdale, J.K.A., 290, *294*
Bledsoe, C.S., 386, 387, *391*
Bleken, E., 244, 246, *263*
Block, W.C., 256, 257, *261*
Boatman, N., 328, *329*
Bockheim, J.G., 24, *35*
Bocock, K.L., 177, *178*: 299, 300, 301, *314*, *315*
Boddey, R.H., 137, *142*
Boddy, L., 303, *314*
Bohlool, B.B., 130, 134, 138, 140, *144*, *146*: 150, 156, *158*
Bokhari, U.G., 9, *19*: 422, *424*
Bohm, W., 8, *18*: 32, *34*: 67, *70*
Bolger, T., **393–397**: 393, 396, *397*
Bolton, J., 219, *223*
Bommer, D., 32, *36*
Booth, E., 297, *316*

Booth, R.G., **279–284**: 11, *21*: 244, 245, 246, 247, 248, 254, 259, 261, 265: 281, 282, 283, 284: 321, 328, *330*, *331*: 334, *337*
Bormann, F.H., 24, *36*: 423, *423*, 425
Boström, U., 357, *364*
Bouché, M., 14, *18*
Bouché, M.B., 405, *405*
Boucher, D.H., 17, *18*
Bouton, J.H., 129, 138, 139, *146*: 156, *157*
Bowen, G.D., 108, *119*; 209, 210, *215*, *217*: 230, *231*: 342, *352*: 419, *423*
Bowring, M.F.G., 246, 251, *265*: 283, *284*
Boyer, J.S., 320, *330*
Bradshaw, A.D., 127, 140, *144*, *146*: 419, *423*
Bragg, P.L., 67, *70*
Bray, J.R., 25, 30, *34*
Brian, M.V., 417, *424*
Brian, P.W., 186, *189*
Briand, F., 13, *18*
Brill, W.J., 134, 137, *143*
Brittain, J.E., 311, *316*
Broady, P.A., 279, *284*
Brock, T.D., 1, *18*
Brookes, P.C., **123–125**: 123, 124, *124*, *125*
Brown, B.A., 299, *314*
Brown, J.C., 132, *142*
Brown, M.E., 130, 132, 133, 139, *142*: 156, *157*
Brown, M.S., 112, *119*: 212, *215*
Brown, R., **285–295**: 285, 287, 289, 290, 294, *294*
Brown, R.T., 410, *424*
Browning, V.O., 67, *71*
Brownlee, C., 195, 212, *215*: 410, 411, *424*
Bruins, E.H., 389, *391*
Buchanan, S.A., 136, *142*
Bull, A.T., 189, *191*
Bullock, H.C., 212, *216*
Burges, A., 6, 14, *18*: 298, *314*: 423, *424*
Burlingham, S.K., 130, 139, *142*
Burns, R.C., 134, *143*
Burris, R.H., 133, *145*
Butcher, J.W., 247, *264*
Butterfield, J., 351, *353*
Buwalda, J.G., 225, *231*
Buxton, P.A., 303, *314*
Buxton, R.D., 255, *261*

Cabrera, D., 137, 138, *143*
Cabrera, F., 222, *224*
Caldwell, M.M., 31, 32, *34*
Cambell, G.S., 171, *173*
Camp, L.B., 31, 32, *34*

Campbell, C.A., 274, *277*: 356, *363*
Campbell, R., 116, 117, *119*, *120*
Campion, M.K., 9, 16, *18*, *19*: 355, *364*
Cambardella, C., 8, *18*: 212, *215*
Cancela da Fonseca, J.P., 10, *18*: 245, 252, 261, *264*
Cannell, R.Q., 67, *70*
Carnahan, B., 235, *241*
Carpenter, A., **67–71**
Case, T.J., 12, *18*
Cayley, A., 95, *104*
Chabot, B.F., 88, *104*
Chakabarty, K., *130*, *145*
Chakraborty, S., 10, *18*
Chalk, P.M., 136, 137, *142*
Chalutz, E., 185, *189*
Chapman, S.J., 187, 189, *189*, *191*: 410, *424*
Charpentie, M.-J., 253, *264*: 299, 309, **316**
Chapin, III. F. S., 359, *363*
Cherrett, J.M., **67–71**
Cheshire, M.V., 9, *21*: 358, *365*
Chet, I., 260, *262*
Chiariello, N., 8, *18*: 213, *215*
Chilvers, G.A., 176, *178*
Chippindale, H.G., 214, *215*: 416, *424*
Chou, T.W., 185, *189*
Christen, A.A., 307, *314*
Christie, E.K., 88, 89, 91, 93, *104*
Christie, P., 114, 116, *119*
Chu-Chou, M., 334, *337*
Clarholm, M., **355–365**: 8, 9, *19*: 275, *277*: 297, 305, *314*, *316*: 355, 356, 357, 361, *363*, *364*: 393, *397*
Clark, F.E., 141, *142*: 172, *172*: 298, *314*: 377, *391*
Clark, W.C., 320, *330*
Clement, C.R., 128, *142*
Clowes, F.A.L., 113, *119*, *120*
Coates, D., 303, *314*
Cochrane, V.M., 163, 165, 167, *172*
Coessens, R., 246, *261*
Coggins, C.R., 162, *173*
Coldrick, G.A., 48, 57, *64*
Cole, C.V., 6, 9, 11, 13, 15, 16, *18*, *19*, *20*: 88, *104*: 159, *174*: 339, 342, *352*: 355, 356, *363*, *364*: 367, *375*: 378, *391*
Cole, D.W., 276, *277*: 386, *391*
Cole, H., 225, *231*
Cole, H. Jr., 309, *315*
Coleman, D.C., **1–21**: 5, 8, 9, 11, 12, 13, 15, 16, *18*, *19*, *20*, *21*: 32, *36*: 88, *104*: 117, *121*: 327, *331*: 339, 342, *352*: 355, 356, 357, *363*, *364*, *365*: 367, *375*: 378, *391*
Collins, V.G., 349, *353*
Colquhoun, I.A., **73–79**

Comanor, P.L., 343, *353*
Connor, E.F., 254, *261*
Contois, D.E., 164, *172*
Cornaby, B.W., 311, *314*
Cook, A.G., 259, *264*
Cook, R.J., 184, 186, *189*, *191*
Cooke, I.J., 73, 77, *78*
Cooper, J.E., 222, *224*
Cornforth, I.S., 184, *189*
Cossens, G.G., 400, *405*
Cotton, D.C.F., 255, *262*: 393, *397*
Coulson, J.C., 348, 349, 350, 351, *353*
Coupland, R.T., 109, *119*
Cousins, S.H., 339, 347, *353*
Cowley, G.T., 11, *18*
Cowling, E.B., 163, *173*
Cox, T.L., 37, *42*
Cragg, J.B., 303, *314*: 352, *353*
Crawford, D.V., 359, 361, *364*
Crawley, M.J., 293, *294*
Critchley, B.R., 259, *264*
Crocker, R.L., 6, *19*
Cromack, K. Jr., 305, *315*
Cromack, K., 419, *427*
Crossley, D.A., 298, *314*: 393, *397*
Crump, L.M., 16, *19*
Culver, D.C., 418, *423*, *424*
Cummins., R.P., 417, *426*
Curl, E.A., 305, 310, *317*
Curry, J.P., **393–397**: 255, *262*: 393, 396, *397*
Cutler, D.W., 16, *19*

Dahlman, R.C., 26, 28, 31, 34, *35*
Dalton, H., 133, *142*
Danso, S.K.A., 136, *143*
Darbyshire, J.F., *115*, *119*: 357, 361, *364*
Dart, P.J., 129, 134, 137, 138, 139, 140, 141, *142*, *143*, *147*
Darwin, C., 14, *19*
Dash, M.C., 303, *314*
Davey, C.B., 107, *120*
Davey, M.R., 139, 140, *145*
David, K.A.V., 134, *142*
Davidson, M.S., 184, 188, *190*
Davidson, R.L., 361, *364*
Davidson, S.A., 124, *125*
Davis, N.B., 283, *284*
Davis, P.R., 244, 248, 251, 254, *262*, *265*
Davis, R.C., 279, *284*: 351, *353*
Davis, R.E., 131, *144*
Davison, M.S., 131, 135, 136, 138, 140, *144*
Day, J.M. **127–147**: 129, 130, 131, 133, 134, 137, 138, 139, 141, *142*, *143*
Dazzo, F.B., 131, *146*: 156, *158*

De Angelis, D.L., 13, *19*: 233, 235, *241*
De Bertoldi, M., 328, *330*
De-Felice, J., 73, *78*
De Freitas, J.R., 136, *145*
De Goede, R.G.M., **367–376**
De-Polli, H., 132, 137, *142*
De la Cruz, A.A., 343, *353*
Deacon, J.W., 54, 63, *64*: 112, *119*: 419, *424*, *425*
Deans, J.D., 25, 27, 28, *34*: 88, *105*: 115, *119*
Demeure, Y., 13, *19*
Denman, K., 339, 347, *354*
Dennison, D.F., 251, *262*
Detling, J.K., 9, *19*
Dice, S.F., 25, *34*
Dickinson, C.H., 299, *314*: 344, *353*
Dickson, B.A., 6, *19*
Dickson, D.W., 321, *331*
Dighton, J., **339–354**: 336, *337*: 342, *353*
Dinger, B.E., 26, 30, *35*
Ditchburne, N., 175, 177, *178*
Dittmer, H.J., 115, *119*
Dobereiner, J., 129, 130, 131, 132, 133, 134, 135, 137, 139, 140, *142*, *143*, *145*, *146*
Doetsch, R.N., 116, *121*
Dommergues, Y.R., 134, *141*
Domsch, K.H., 110, 114, 115, *118*, *119*: 185, *189*: 213, *215*: 356, *363*
Donaldson, S.J., 419, *424*
Douglas, L.A., 136, *142*
Drew, E.A., 151, *157*
Drew, M.C., 82, *85*: 92, 99, *104*: 184, 185, 186, *189*
Dritschilo, W., 257, *262*
D'Sylva, B.T., 349, *353*
DuBois, J.D., 149, 150, *157*
Duddridge, J.A., 195, 200, 212, *215*, *216*: 410, 411, *424*
Dudeck, A.E., 156, *157*
Dudkiewicz, L.A., 30, *34*
Dudney, P.J., 49, *64*
Duncan, W.G., 92, *104*
Dunning, R.A., 286, *295*
Dunsworth, B.G., 32, *34*
During, C., 400, *405*
Dyer, M.I., 9, *19*: 422, *424*
Dyrness, C.T., 418, *424*

Eaglesham, A.R.J., 137, 138, *143*
Eaton, J.S., 24, *36*: 423, *425*
Edmonds, R.L., 28, 31, *34*, *36*: 37, 42: 175, *179*: 211, *217*

Edwards, C.A., 11, 14, *19*: 286, *295*: 328, *330*: 393, 396, *397*: 405, *405*
Edwards, M., 254, *265*
Edwards, N.T., 26, 27, 28, 29, 30, 33, *34*, *35*: 37, *42*: 240, *241*
Eis, S., 25, *34*
El-Nennah, M., 73, *78*
Ela, S.W., 134, 137, *143*
Eliot, E.T., 367, *375*
Elliott, E.T., 3, 9, 15, 16, *18*, *19*: 355, 357, *364*, *365*
Elliott, L.F., 171, *173*: 182, 183, 185, 186, 187, *190*, *191*
Elkins, N.Z., 311, *317*
Elmes, G.W., 245, *265*: 417, *424*
Elton, C.S., 340, *353*
Emerson, W.A., 8, *18*
Ericsson, A., 177, *178*
Erwin, T.L., 257, *262*
Eskew, D.L., 137, 138, *143*
Estey, R.H. 302, *316*: 321, *330*
Evans, E., **67–71**
Evans, G.C., 88, *104*
Evans, H.J., 156, *157*

Fager, E.W., 10, *19*
Fairbridge, R.W., 222, *224*
Fairley, R.I., **37–42**: **81–85**: 40, 42, *42*: 44, *64*: 175, *178*: 275, *277*
Farr, A., 150, *157*
Farrar, J.F., 111, *119*
Fay, P., 134, *142*
Fayle, D.C.F., 25, 27, 28, *34*
Feldman, M., 156, *157*
Fenchel, T., 361, *364*
Ferrier, R.C., **175–179**: 275, 276
Ferris, J.M., 30, *34*
Finkl, C.W. Jr., 222, *224*
Finlay, R.D., **193–217, 319–331**: 10, 253, 286, 342, 411: 253, *262*: 321, 322, *330*
Firestone, M.K., 225, 229, *231*
Firth, P., 128, *143*
Fisch, S.B., 418, *424*
Fisher, J.B., 87, *104*
Fisher, R.F., 410, *424*
Fitter, A.H., **87–106**: 87, 88, 90, 91, 94, 97, *105*, *106*: 213, *217*: 253, *265*: 286, *295*: 304, 312, *317*: 320, 321, 327, 328, *330*, *331*: 333, 336, *337*
Flegg, J.J.M., 67, *70*
Fleming, V., 211, *216*
Floyd, R.A., 113, *119*
Floate, M.J.S., 421, *424*
Flower-Ellis, J.G., 31, *43*
Flower-Ellis, J.G.K., 37, *42*

Fogel, R., **23–36**: 8: 25, 26, 28, 29, 31, 32, *34*: 37, 42, *42*: 44, *64*: 175, *178*: 211, *216*: 240, *241*: 342, *353*: 419, *424*
Ford, E.D., 25, *34*: 88, *105*: 115, *119*
Ford, J.B., **67–71**
Fortin, J.A., 302, 304, 305, 317: 320, 321, *331*
Foster, R.C., 4, *19*: 115, 116, *119*: 175, 177, *178*: 200, *216*: 336, *337*
Fowkes, N.D., 103, *105*
Fox, F.M., 419, *424*
Fox, J.A., 321, *330*
Francis, R., **193–217**: 411: 116, *119*: 203, 205, 213, *216*
Franck, O.O., 87, *106*
Frank, M.L., 368, *376*: 389, *392*
Frankland, J.C., 73, *78*: 109, *119*: 160, *173*: 300, *314*: 356, *364*
Franklin, J.F., 418, *424*
Franson, M.A., 227, *231*
Frazer, D., 260, *262*
Freckman, D.W., 13, *19*: 311, *317*
French, N.R., 422, *426*
Fried, M., 135, 136, *143*, *145*
Frissel, M.J., 159, 160, *173*, *174*: 389, *392*
Frizzel, M.J., 378, *391*
Frogatt, P.J., 140, *147*
Führer, E., 10, *19*
Fustec-Mathon, E., 414, *424*

Gadgil, P.D., 177, *178*: 214, *216*: 268, 274, 276, *277*
Gadgil, R.L., 177, *178*: 214, *216*: 268, 274, 276, *277*
Galpin, M.F.J., 162, *173*
Gams, W., 114, 115, *119*: 374
Ganley, J., 255, *262*
Garay, I., 259, *262*
Gardner, R.H., **233–242**: 233, 235, *241*
Garrett, S.D., 297, *314*
Garwood, E.A., 67, *70*: 88, *105*
Gaskins, M.H., 129, 131, 138, 139, 140, *141*, *143*, *146*: 156, 157, *158*
Gaur, A.C., 138, *145*
Gerdemann, J.W., 225, *232*: 319, 320, *330*
Gerson, U., 260, *262*
Gersper, P.L., 412, *424*
Gessel, S.P., 24, *35*: 417, *425*
Ghilarov, M.S., 254, *262*
Gibson, D.J., **73–79**
Gilbert, O.J., 177, *178*
Giller, K.E., **127–147**: 8: 137, 138, *143*
Gilmore, S.K., 307, *314*
Gilmour, C.M., 133, *142*: 187, *190*
Gilpin, M.E., 12, *18*

Giovannetti, M., 219, *224*: 328, *330*
Gisin, H., 255, *262*
Glassman, C.A., *392*
Glover, S.U., 329, *329*
Goodlass, G., 135, *143*: 184, *190*
Gordon, J.K., 130, *143*
Gore, A.J.P., 349, 350, *353*
Gorham, E., 127, 129, *143*
Goss, M.J., 9, *20*: 358, *364*
Gosz, J.R., 268, 275, 276, *277*: 377, *392*
Goto, H.E., 334, *337*
Gottsche, D., 235, *241*
Govi, G., 67, *70*
Grable, A.R., 137, *145*
Grace, L.J., 334, *337*
Gradwell, M.W., 401, *405*
Graham, J.H., 212, *216*
Granhall, U., 140, *143*, *145*
Gray, L.E., 319, *330*
Gray, T.R.G., 24, *34*: 241, *242*
Greaves, M.P., 115, *119*: 357, 361, *364*
Green, H.E., 412, *424*
Green, R.E., 290, *295*
Greenham, D.W.P., 58, *64*
Greenland, D.J., 128, *143*
Greenslade, P.J.M., 248, 257, 260, *262*, *263*: 339, 345, *353*
Grégoire-Wibo, C., 244, 246, *262*
Gregory, P.J., 26, 28, *34*: 359, 361, *364*
Greig-Smith, P., **73–79**: 254, *262*
Grier, C.C., 28, *36*: 37, *42*: 175, 177, *178*, *179*: 211, *217*: 268, 275, 276, *277*
Grier, C.G., 31, *34*
Gries, C.C., 377, *392*
Griffin, D.M., 299, 301, *314*
Griffiths, E., 4, *19*: 187, *190*
Grime, J.P., 260, *262*: 339, 345, *353*: 416, *424*
Grime, P.G., 214, 215, *216*
Grimnes, K.A., 245, *262*
Grimshaw, H.M., 75, *78*: 269, *277*: 280, *283*
Griselli, M., 328, *330*
Groenwold, J., 67, *71*
Grubb, P.J., 412, *424*
Gunn, K.B., 111, *118*: 185, 186, *190*
Gunnarsson, T., 370, *375*
Gupta, P.L., 73, *78*
Gurung, H.P., 49, 50, *64*
Gussin, E.J., 183, 185, 187, 188, *190*

Habte, M., 303, *314*
Hackney, C.T., 343, *353*
Haeck, J., 248, *262*
Hagvar, S., 255, *262*: 303, *315*
Haines, B.L., 386, *391*

Hale, W.E., 336, *337*
Hale, W.G., 248, 255, 256, 257, *262*
Hall, I.R., 219, 221, *224*: 320, *330*
Hallé, F., 87, *105*
Halm, B.J., 124, *125*
Halstead, R.L., 124, *125*
Hancock, J.G., 10, *20*: 139, *146*
Handel, S.N., 418, *424*
Handley, W.R.C., 410, *425*
Hanlon, R.D.G., 168, *173*: 299, 304, 312, *314*: 328, *330*: 367, 375
Hansen, G.K., 111, *119*
Harding, D.J.L., 67, *70*, 302, *315*
Hardy, R.W.F., 134, 138, 139, 140, *143*, *144*, *145*
Harley, J.L., 162, *173*: 177, *178*: 208, 210, 211, 212, *216*: 286, *295*: 333, *337*
Harmer, R., **267–277**: 127, *144*: 277
Harper, J.L., 293, *295*: 416, *424*
Harper, S.H.T., 183, 184, 186, 188, 189, *190*, *191*
Harris, D., 134, 141, *142*
Harris, J.R.W., 244, 248, 251, 252, 254, *262*, *265*
Harris, W.F., 24, 26, 27, 28, 29, 30, 31, 33, *34*, *35*: 37, *42*: 234, 240, *241*, *242*
Harrison, A.T., 149, 157
Hartenstein, R., 303, *315*
Harvey, A.E., 175, *178*
Hassall, M., 168, *173*: 311, *315*
Hassell, M.P., 250, *263*
Havill, D.C., 73, *78*
Hay, M.E., 418, *426*
Hayes, M.H.B., 381, *391*
Hayman, D.S., 221, *224*: 227, *231*: 320, 328, 329, *329*, *330*
Head, G.C., 29, *35*: 44, 45, 50, 51, 55, *64*, *65*: 67, *70*: 113, *119*
Head, G.S., 67, *71*
Heal, O.W., **339–354**: 236, *241*: 268, 277: 298, 299, 300, 301, 302, *315*, *317*: 339, 340, 343, 345, 349, 350, *353*, *354*: 377, 379, 384, *391*, *392*: 408, 410, 418, 419, 423, *426*
Healey, I.N., 246, 253, *261*, *262*: 302, 303, *313*
Heap, A.J., 109, *120*: 213, *216*: 410, *424*
Heath, G.W., 393, 396, *397*
Hedley, M.J., 123, *125*
Heil, R.D., 6, *18*
Heilman, P., 24, *35*
Heitkamp, D., 29, *35*
Helal, H.M., 132, *143*
Hellman, O., 103, *105*
Hengeveld, R., 248, *262*

Henis, Y., 137, 138, *143*, *145*: 156, 157
Henry, C.M., 112, *119*
Hering, T.F., 301, *315*
Hermann, K., 37, *42*
Herrmann, R.K., 25, 26, 28, 29, *35*, *36*
Hermosilla, W., 256, *262*
Hesse, P.R., 227, *231*
Heytler, P.G., 138, 139, *145*
Hibble, J., 417, *424*
Hibbs, D.E., 87, *104*
Hickman, J.C., 8, *18*: 213, *215*
Hicks, D.J., *88*, *104*
Higgins, R.C., 251, *262*
Hilbert, D.W., 9, *19*
Hill, S., 187, *191*
Hill, S.B., 252, 253, 259, *261*: 302, 303, 307, *313*, *314*
Hislop, J., 73, 77, *78*
Hislop, R.G., 251, *263*
Hodgson, J., 194, *217*: 219, 220, *224*
Hodkinson, I.D., 251, *262*
Hoffman, G., 67, *70*
Holben, F.J., 128, *146*
Holden, J., 112, *119*
Holding, A.J., 222, *224*
Hole, F.D., 14, *19*
Holling, C.S., 252, *263*
Holm, E., 235, 241, *241*
Holowaychuk, N., 412, *424*
Holsten, R.D., 134, *143*
Honda, H., 87, *104*
Horn, H., 87, *105*
Horton, K., 15, 16, *19*
Horvitz, C.C., 418, *425*
Houchins, J.P., 133, *145*
Howard, A.J., 117, *119*
Howson, G., 343, *353*
Hozumi, K., 293, *295*
Hubbell, D.H., 131, 139, 140, *143*, *146*, 156, *157*, *158*
Huck, M.G., 44, *64*: 67, *70*: 88, *105*
Hughes, M.K., *241*
Huhta, V., 244, 256, 257, 259, *263*
Huish, S.A., **377–392**: 9, *18*: 299, *313*: 367, 368, 370, 374, *375*, 379, 381, 382, 383, 384, *391*
Humphreys, W.F., 236, *241*
Hunt, B., 26, 28, *34*
Hunt, G., 25, 26, 28, 29, 31, *34*: 37, 42, *42*: 175, *178*: 211, *216*: 419, *424*
Hunt, H.W., 8, 9, 13, 16, *18*, *19*, *21*: 159, 160, 163, 169, *173*, *174*: 355, *364*: 367, *375*: 389, *391*
Hunt, R., 342, *353*
Hunter, P.J., 69, *70*, *71*

Huntjens, J.L.M., 358, *364*
Hurd, L.E., 410, *425*
Hussey, R.S., 10, *19*: 321, *330*
Hutchinson, G.E., 6, *19*
Hutchinson, K.J., 257, *263*
Hutchinson, S.A., 309, *315*
Hutchinson, T.C., 214, *216*
Hutson, B.R., 244, 246, 248, 256, *263*
Huţu, M., 255, *263*
Huxley, P.A., 67, *70*

Ichimura, T., 252, *264*: 302, *317*
Ignaciuk, R., 127, *144*
Ikonen, E., 256, *263*
Inbal, E.,. 156, *157*
Ineson, P., **377–392**: 9, *18*: 299, *313*: 339,
    345, *353*: 367, 368, 370, 374, *375*: 378,
    379, 380, 381, 382, 383, 384, 387, 389,
    390, *391*
Ingham, E.R., 11, *19*: 164, 165, 166, 167,
    168, *173*: 227, *231*
Ingham, R.E., 3, 11, 15, 16, *18*, *19*
Ingold, C.T., 306, *315*
Innis, G.S., 410, *425*
Ito, O., 137, 138, *143*

Jackson, G.D., 75, *78*
Jackson, R.M., 130, *142*
Jacot, A.P., 307, *315*
Jager, G., 389, *391*
Jaggard, K.W., 289, 290, 291, *295*
Jaiyebo, E.O., 128, *143*
Jalal, M.A.F., 414, *425*
Jambu, P., 414, *424*
James, S., 17, *18*
Jameson, D.A., 422, *425*
Janos, D., 194, *216*
Jansen, E., 168, 171, *173*
Jansson, S.L., 356, *364*
Janzen, D.H., 416, *425*
Jenkinson, D.S., **123–125**: 123, 124, *124*,
    125: 128, 129, *143*: 358, *364*: 389, *391*
Jennings, D.H., 162, *173*
Jenny, H., 5, *19*: 411, 417, *425*
Jensen, C.R., 111, *119*
Jensen, H.L., 133, *143*: 188, *190*
Jensen, V., 133, *143*: 235, 241, *241*: 396,
    397
Jepson, P.C., 290, *295*
Johnen, B.G., 8, *20*: 111, *119*: 151, *157*
Johnson, D.D., 274, *277*
Johnson, N.M., 423, *425*
Johnson, R.A., 15, *21*
Jones, D., 187, *190*
Jones, E.W., 418, *425*

Jones, F.G.W., 286, *295*
Jones, H.E., 349, 350, *353*
Jongerius, A., 10, *19*
Joosse, E.N.G., 244, 247, 252, *263*
Jordan, C.F., 175, *178*
Jow, W., 117, *120*
Juma, N.G., 172, *173*
Juniper, B.E., 113, *120*
Jurgensen, M.F., 175, *175*

Kaczmarek, M., 248, *263*
Kalanchova, L., 113, *121*
Kamel, M., 309, *315*
Kantey, B.A., 412, *425*
Kapulnik, Y., 138, 139, *143*, *146*: 156, *157*
Kapustka, L.A., **149–158**: 149, 150, 151,
    157
Karg, W., 251, *263*
Karppinen, E., 257, *263*
Kascht, L.J., 87, *106*
Kattoulas, M., 245, *264*
Katznelson, H., 108, 114, *120*: 130, *144*
Keay, P.J., 140, *147*
Keeler, K.H., 17, *18*: 149, *157*
Keeney, D.R., 14, *21*
Kelley, A.F., 417, *424*
Kelly, M., **393–397**
Kempf, J.S., 87, *105*
Kerner von Marilaun, A., 417, *425*
Kevan, D.K.McE., 259, 261: 298, *315*
Keyes, M.R., 31, *34*: 175, 177, *178*, *179*
Khanna, P.K., 234, 240, *242*
Kiffer, E., 252, *261*
Kigel, J., 138, 139, *143*: 156, *157*
Kilbertus, G., 10, *19*, *21*
Kilham, K., **225–232**
Killham, K.S., 225, 229, *231*
Kinerson, R.S. Jr., 26, 27, 28, 33, *34*: 240,
    241
King, K.L., 257, *263*
Kinloch, R.A., 321, *331*
Kira, T., 24, 26, *35*: 293, *295*
Kirkby, E.A., 387, *391*
Kirkpatrick, B.C., 410, *425*
Kjøller, A., 233, 234, 236, *241*
Kjondal, B.R., 303, *315*
Klein, D.A., 9, 13, *19*: 110, *120*: 164, 168,
    173
Klepper, B., 67, *71*: 88, *105*
Klingmuller, W., 140, *144*
Kloepper, J.W., 139, *144*
Klucas, R.V., 130, *145*
Knapp, E.B., 171, *173*
Knight, C.B., 307, *315*
Knuth, D.E., 96, *105*

Koelling, M.R., 31, *35*
Kolesnikov, V.A., 27, 28, *35*
Koob, K., 306, *316*
Koopman, B.O., 103, *105*
Kormanik, P.P., 101, *104*
Kosslak, R.H., 130, *144*
Kouchecki, H.K., 194, *217*: 219, 220, *224*
Kramer, P.J., 212, *216*
Krause, D., 117, *120*
Krebs, J.R., 283, *284*
Kretzschmar, A., 404, *405*
Krieg, N.R., 131, *146*
Kronauer, R.E., 87, *105*
Kroontje, W., 257, *265*
Kruuk, H., 409, *425*
Kubiena, W., 4, *20*
Kubiena, W.L., 299, 301, *315*
Kucera, C.L., 31, *35*
Kucey, R.M.N., 112, *120*
Kudeyarov, V.M., 378, *391*
Kühnelt, W., 10, *20*
Kumar, R., 299, 300, *315*
Kummerow, J., 117, *120*
Kumi, J.W., 32, *34*
Kunc, F., 110, *121*: 356, *365*
Kunin, R., 73, *78*
Kurcheva, G.F., 393, *397*

Lacy, R.C., 253, *263*: 336, *337*
Ladd, J.N., 123, 124, *125*: 389, *391, 392*
Lambers, H., 111, *120*
Lambert, D.H., 225, *231*
Landsberg, J.J., 103, *105*
Larsen, M.K., 175, *105*
Larsen, R.I., 114, *120*
Larson, R.L., 131, *144*
Last, F.T., 54, 63, *64*: 419, *424, 425*
Latter, P.M., 343, 349, *353*
Lattimore P.T., **149–158**
Laurenroth, W.K., 8, *21*
Lavelle, P., 14, *20*: 342, 344, *353*: 405, *405*
Lawley, R.A., 109, 116, 117, *120*
Lawrence, D.B., 29, *35*
Lawton, J.H., 250, *263*
Le Tacon, F., 221, *224*
Lee, J.A., 127, *144*
Lees, K.E., 399, *405*
Leetham, J.W., 422, *426*
Leeuwen, B.H. van., 410, *425*
Lefroy, R.D.B., 88, 89, 92, 93, *105*
Leinaas, H.P., 244, 246, *263*
Leinweber, C.L., 422, *426*
Leonard, M.A., **377–392**: 367, *375*: 381, 383, 384, *391*
Leopold, L.B., 87, *105*

Lethbridge, G., 131, 135, 136, 138, 140, *144*: 184, 188, *190*
Levi, M.P., 163, *173*
Leyden, R.F., 229, *232*
Liebermann, M., 185, *189*
Likens, G.E., 24, 33, *36*: 423, *423, 425*
Lin, W., 139, 140, *144*
Linberg, T., 140, *143*
Lindberg, T., 177, *178*: 268, 276, *277*
Lindeman, R.L., 339, *353*
Linder, S., 31, *34*: 37, *42*
Lindermann, R.G., 151, *157*: 212, *216*
Lister, A., 283, *284*
Littell, R.C., 129, 138, 139, *146*: 156, *157*
Lloyd, J.E., 299, 300, *315*
Lloyd, W.E., 412, *426*
Lodge, E., 410, *427*
Lofs-Holmin, A., 357, *364*
Lofty, J.R., 14, *19*: 405, *405*
Lohm, U., 236, *242*: 259, *261*: 367, *375*
Longstaff, B.C., 244, 248, 250, 251, 252, 254, *263, 265*
Loring, S.J., 257, *263*
Lotero, J., 421, *425*
Lotka, A.J., 17, *20*
Lousier, J.D., 13, *21*, 177, *178*
Lowe, W.E., 109, *120*: 36, *365*
Lowry, O.H., 150, *157*
Lubchenco, J., 305, 306, *315*
Lucas, H.L., 421, *426*
Lucas, J.A., 139, 140, *145*
Luff, M.L., 248, 256, *263*
Lundeberg, G., 419, *425*
Lundgren, B., 109, *120*: 259, *261*: 367, *375*
Lundkvist, H., 259, *261*: 297, *316*: 356, *364*: 393, *397*
Lussenhop, J., 299, 300, *315*
Luther, H.A., 235, *241*
Luxton, M., 10, *20*: 233, 236, *241, 242,* 297, 298, 303, *315, 316*: 352, *354*
Lyford, W.H., 32, *35*: 115, *120*: 411, *425*
Lyles, G.L., 75, *79*
Lynch, J.M., **181–191**: 16: 2, *18*: 108, 111, 112, 113, *118, 119, 121*: 132, 133, 140, 142, *146*: 181, 182, 183, 184, 185, 186, 187, 188, 189, *189, 190, 191*: 410, *424, 425*
Lyr, H., 67, *70*

MacArthur, R.H., 339, 345, *353*
McBride, J., 129, *144*
McCalla, T.M., 186, *190, 191*
McClaugherty, C.A., 28, 30, 31, *35*: 37, 39, 42: 175, *178*: 349, *353*
McClellan, J.F., 11, *18*

McClung, C.R., 131, 132, 140, *144*, *146*
McCoy, E.D., 254, *261*
McCready, C.C., 210, 211, *216*
McCullough, H.A., 419, *426*
McDonald, L., 8, *21*
MacDonald, N., 94, *105*
MacDonald, R.M., 131, *144*: 328, *330*
Macfadyen, R.M., 9, 14, *18*, *20*: 352, *353*
McGowan, M., 26, 28, *34*: 359, 361, *364*
McGill, W.B., 159, 160, 163, 169, *173*, *174*: 389, *391*
McGinty, D., 24, 31, *35*
McGrath, S.P., 124, *124*
McGraw, A.C., 101, *105*
McIlveen, W.D., 39, *315*
McIlwaine, R.S., 14, *120*
McKercher, R.B., 6, *21*
McLean, S.F. Jr., 236, *241*
MacLean, S.F., 350, *353*
MacMahon, J.A., 408, *425*
McMahon, J.H., 408, *423*
McMahon, T.A., 87, *105*
McMillan, J.H., 253, *263*: 303, 307, *315*: 320, *330*: 336, *337*
McMinn, R.G., 25, 27, 32, *35*
MacNamara, C., 307, *315*
McNaughton, S.J., 410, *425*
Macrea, C., 73, *78*
McRill, M., 416, *426*
Madwick, H.A.I., 25, *35*
Magnusson, C., 30, *35*
Mahmoud, H., 214, *216*
Malajczuk, N., 305, *315*
Malibari, A., 195, 200, 212, *215*, *216*: 410, 411, *424*
Malloch, D.W., 17, *20*
Malone, J.P., 114, *120*
Margaris, N.S., 181, *191*: 245, *264*
Marks, G.C., 175, 177, *178*: 336, *337*
Marks, P.L., 24, *36*
Marriel, I.E., 130, *143*
Marrs, R.H., 127, *144*: 419, *423*
Marsh, B., 227, *231*
Marshall, K.C., 2, *20*
Martin, J.K., 111, 112, *119*, *120*: 132, *142*
Martin, J.P., 187, *191*
Martin, J.R., 133, *144*
Marx, D.H., 320, *330*: 419, *425*
Maser, C., 422, *425*
Mason, P.A., 54, 63, *64*: 336, *337*: 342, *353*: 419, *425*
Matson, P., 73, *78*
Matson, P.A., 274, *277*
Matsui, E., 132, 137, *142*, *144*, *146*

Matthews, S.W., 131, *141*
May, R.M., 12, *20*
Mayer, R., 234, 240, *242*
Meier, C.E., 175, 177, *179*: 211, *217*
Meier, R.L., 37, *42*
Melillo, J.M., 28, 30, 31, *35*: 37, 39, *42*: 175, *178*: 268, 275, 276, *277*: 343, *353*: 377, *391*, *392*
Melin, E., 209, *216*
Melling, J., 2, *18*
Mellinger, M.V., 410, *425*
Menge, J.A., 212, *216*: 321, *329*
Merrifield, R.C.J., 412, *424*
Merrill, L.B., 422, *426*
Merrill, W., 163, *173*
Mertens, J., 246, *261*
Meshram, S.U., 138, *144*
Metz, L.J., 311, *315*
Mikkonen, M., 244, *263*
Mikola, P., 410, *424*
Milam, J.R., 129, 138, 139, *146*
Miles, J., **407–427**: 413, 414, 418, 419, 423, 426
Miller, G.R., 417, *426*
Miller, H.G., 377, *392*
Miller, P.M., 309, *315*
Miller, R.D., 274, *277*
Miller, T.D., 139, *144*
Millier, C.R., 134, *141*
Mills, J.T., 303, *315*
Milton, W.E.J., 416, *424*, *426*
Mitchell, M.J., 244, 246, 247, 252, *263*: 299, 303, *314*, *315*: 320, *330*
Mitchell, R.L., 225, *231*
Moir, W.H., 25, *35*
Møller, C.Mar., 235, *242*
Mooney, H.A., 8, *18*: 213, *215*
Moorby, J., 88, 89, 91, 93, *104*
Moore, A.W., 128, *143*, *144*
Moore, E.E., 175, *179*
Moore, J.C., 15, 16, *19*
Morse, R.K., 412, *425*
Mosse, B., 212, *216*: 219, 21, *224*: 230, *231*: 328, *329*: 333, *337*: 409, *426*
Müller, D., 235, *242*
Mundie, C.M., 9, *21*
Mundie, M., 358, *365*
Muraleedharan, V., 253, *263*
Murphy, J., 101, *105*
Myers, R.J., 73, *78*
Myers, R.J.K., 274, *277*

Nagelkerke, C.J., 375, *376*
Nair, P.K.R., 73, 77, *78*
Nakas, J.P., 110, *120*

Nannier, G., 10, *19*
Narayanan, A., 88, 89, *105*
Nataf, L., 259, *262*
Naylor, D., 48, 57, *64*
Neal, J.L., 114, *120*: 131, *144*
Nelsen, C.E., 320, *330*
Nery, M., 130, *143*
Newbold, P.J., 23, 24, *35*
Newell, K., 10, *20*: 303, 304, 312, *315*: 320, *330*: 336, *337*: 418, 419, *426*
Newman, E.I., **107–121**: 8, 132: 63, *64*: 108, 109, 111, 116, 117, *119*, *120*, *121*: 213, *216*: 410, *424*
Nicholson, T.H., 187, *191*: 193, 212, *216*
Nicholson, P.B., 299, 300, 301, *315*
Nickell, L.G., 410, *426*
Nielsen, J., 235, *242*
Nihlgard, B., 24, *35*
Nilsson, H., 209, *216*
Nohrstedt, H.O., 134, *144*, *145*
Nommik, H., 278, *392*
Norman, A.G., 130, *146*: 298, 299, *316*, *317*: 361, *365*
Norstadt, F.A., 186, *191*
Northover, M.J., 279, *283*
Nosek, J., 333, *337*
Nur, I., 138, 139, *143*, *146*: 156, *157*
Nurminen, M., 257, *263*
Nussbaum, R.A., 422, *425*
Nye, P.H., 24, 26, *35*: 88, 93, *105*: 128, *143*: 319, *329*: 359, *364*
Nykvist, N., 211, *216*

Oades, J.M., 4, *21*: 213, *217*
Ocampo, J.A., 329, *330*
O'Dowd, D.J., 418, *426*
Odum, E.P., 23, *35*
Odum, H.T., 11, *20*
Ogawa, H., 24, 26, *35*: 293, *295*
O'Hara, G.W., 139, 140, *145*
Ohlrogge, A.J., 92, *104*: 113, *119*
Okon, Y., 130, 133, 138, 139, 140, *143*, *144*, *145*, *146*: 156, *157*
Old, K.M., 10, *18*: 187, *191*
Oldeman, R.A.A., 87, *105*
O'Neill, R.V., **233–242**: 233, 234, *242*
Orlov, A.Ya., 177, *178*
Overgaard Nielsen, C., 302, *315*
Overton, W.S., 26, 28, *36*: 37, *42*
Ovington, J.D., 25, 29, *35*: 73, *78*: 128, *145*
Ovington, J.E., 128, *145*

Pacovsky, R.S., 112, *119*: 212, *215*
Page, G., 413, *426*
Page, M.B., 73, *78*

Paget, D., 328, *329*
Parish, T., 409, *425*
Parker, C.A., 128, *145*
Parker, G.G., 276, *277*
Parker, L.W., 311, *317*
Parkinson, D., 9, *20*: 115, *121*: 168, *173*: 177, *178*: 241, *242*: 248, 252, *263*: 298, 299, 303, 304, 309, 310, 311, *313*, *314*, *315*, *316*, *317*: 320, *329*, *330*: 336, *337*: 345, *365*: 367, *376*
Parkinson, J.A., 75, *78*: 269, *277*: 280, *283*
Parle, J.N., 299, *316*
Parmalee, R., 311, *317*
Palmerley, S.M., 212, *217*
Panting, L.M., 186, *190*
Papavizas, G.C., 107, *120*
Park, O., 8, *18*
Park, T., 8, *18*
Parnas, H., 159, *173*
Parr, J.F., 298, *316*
Parr, T.W., 250, 256, 257, 258, *264*, *265*
Parton, W.J., 6, *20*: 172, *173*
Pastor, J., 24, *35*
Patriquin, D., 131, 132, *144*
Patriquin, D.G., 131, *145*
Paul, E.A., 109, 112, *120*, *121*: 132, 140, *145*, *146*: 172, *173*: 188, *191*: 356, *365*
Paustian, K., 159–174: 8: 358, *364*: 389, *391*
Pavlychenko, T.K., 115, *120*
Payne, T.M.B., 108, 114, *120*: 130, *144*
Pearson, V., 212, *216*
Pederson, W.L., 130, *145*
Pelham, J., 419, *425*
Penn, D.J., 183, 185, 187, 188, *199*
Perfect, T.J., 259, *264*
Persson, H., 25, 27, 28, 29, 31, *34*, *35*: 37, 40, *42*: 88, *105*: 115, 117, *120*: 175, 176, 177, *178*: 276, *277*
Persson, J., 172, *173*
Persson, T., 236, *242*: 297, *316*: 356, *364*: 368, *375*: 379, *392*: 393, *397*
Petersen, E.J., 133, *143*
Petersen, H., **233–242**: 10, *20*, 233, 236, *242*: 244, 248, 252, *264*: 297, 298, 303, *316*: 352, *354*
Peterson, R.G., 421, *425*, *426*
Pherson, D.A., 307, 311, *316*
Phillips, H.L., 113, *120*
Phillips, J., 311, *317*
Phillips, J.M., 227, *231*
Phillips, M.J., 15, *20*
Phillips, R., 336, *337*
Pianka, E.R., 310, *316*
Pickett, S.T.A., 87, *105*

Piearce, T.G., 15, *20*
Pierce, R.S., 423, *425*
Pigeon, R.F., 1, *20*
Pimm, S.L., 12, *20*
Piper, S.R., 28, *36*
Pirozynski, K.A., 17, *20*
Pitcher, R.S., 67, *70*
Platt, T., 339, 347, *354*
Poinsot-Balaguer, N., 10, *18*: 252, *261*
Pomeroy, L.R., 348, *354*
Ponge, J.-F., 253, *264*: 299, 309, *316*
Poole, T.B., 254, *264*: 303, 307, *316*
Popović, B., 275, 276, *277*
Porter, L.K., 8, *18*: 137, *145*: 212, *215*
Postgate, J.R., 133, *142*: 187, *191*
Powlson, D.S., **123–125**: 123, 124, *124*, *125*
Prabhoo, N.R., 253, *263*
Pratt, P.F., 77, *78*
Prenzel, J., 177, *178*: 359, *364*
Prikryl, Z., 110, 111, 113, *120*, *121*
Primrose, S.B., 185, *191*
Prokopy, R.J., 251, *263*
Prosser, J.I., 160, *173*
Pruden, G., 123, *125*
Pryn, S.J., 186, *191*
Puckridge, D.W., 133, *144*
Pudlo, R.J., 418, *423*
Pugh, G.J.F., 339, 344, 345, *353*, *354*

Quarmby, C., 75, *78*
Quarmby, J., 269, *277*: 280, *283*
Quesenbury, K.H., 129, 138, 139, *146*: 156, *157*

Raffensperger, E.M., 307, *314*
Rai, S.N., 138, *145*
Rajski, A., 255, *264*
Rambelli, A., 328, *330*
Ramsbottom, J., 409, *426*
Randall, R.J., 150, *157*
Rao, A.V., 138, *146*
Rao, V.R., 132, *145*
Rapp, M., 276, *277*
Raven, J.A., 387, *392*
Raven, P.H., 17, *20*
Raw, F., 298, *314*: 390, *392*: 423, *424*
Rawes, M., 350, *354*
Rayner, A.D.M., 303, *314*
Read, D.J., **193–217**: 8: 411: 116, *119*: 194, 195, 200, 203, 205, 211, 212, 213, *215*, *216*, *217*: 219, 220, *224*: 410, 411, 414, 419, *424*, *425*, *426*
Reardon, P.Q., 422, *426*
Redlin, M.J., 175, *179*
Reddy, K.B., 88, 89, *105*

Reichle, D.E., 26, 30, *34*, *35*: 297, *316*
Reid, C.P.P., 8, 11, 13, 15, *18*, *20*: 88, *104*: 212, *215*, *217*: 227, 230, *231*: 339, 342, *352*
Reid, J.B., 9, *20*: 358, *364*
Reiners, W.A., 33, *35*: 127, 129, *143*, *145*: 268, 275, 276, *277*: 377, *392*
Reith, J.W.S., 225, *231*, *232*
Rennie, D.A., 135, 136, *145*
Rennie, R.J., 135, 136, *145*: 149, 150, *157*
Restall, S.W.F., 184, *191*
Reuss, J.O., 159, 160, 163, 169, *173*: 389, *391*
Reyes, V.G., 299, *316*
Reynders, L., 130, 132, 138, 139, *145*, *146*
Reynolds, E.R.C., 115, *120*
Rhodes, L.H., 225, *232*: 302, *316*: 319, *330*
Rice, E.L., 181, *191*: 410, *426*
Rice, W.A., 188, *191*
Rich, J.R., 329, *329*
Richer, A.C., 128, *146*
Richter, K.O., 305, *316*
Riedel, R.M., 302, *316*
Riffle, J.W., 302, *316*: 320, *330*
Righelato, R.C., 164, *173*
Righi, D., 414, *424*
Riley, J.P., 101, *174*
Roberts, H.A., 417, *426*
Roberts, J., 25, *36*: 115, *120*
Roberts, R.D., 127, *144*: 419, *423*
Robertson, G.P., 386, *392*
Robertson, L.S., 257, *263*
Robertson, P.O., 184, *191*
Robinson, D., 88, 89, 92, 93, *105*
Robinson, G.W., 412, *426*
Robson, A.D., 219, *223*
Rodin, L.E., 33, *36*
Rodrigues, N.S., 137, *144*
Rogers, W.S., 44, 45, 46, 47, 50, 59, *64*, *65*: 67, 70, *71*
Rolhf, F.J., 291, *295*
Romell, L.G., 211, *217*: 268, *275*, *277*
Roncadori, R.W., 10, *19*: 321, *330*
Rorison, I.H., 73, *78*: 88, 89, 92, 93, *105*: 222, *224*
Rosenbough, N.J., 150, *157*
Rosswall, T., 140, *145*: 259, *261*: 275, *277*: 357, 358, *364*: 367, *375*: 377, *391*
Roth, L.M., 306, *316*
Rouatt, J.W., 108, 114, *120*: 130, *144*
Rovira, A.D., 91, *105*: 107, 108, 115, 116, *119*, *121*: 185, *191*: 361, *364*
Rundgren, S., 252, *261*: 305, *314*: 367, 370, *375*
Runge, M., 378, *393*

Runham, N.W., 69, *71*
Ruschel, A.P., 136, 137, *144*, *145*, *146*
Rusek, J., 10, *20*: 251, *264*
Russell, Sir E.J., 2, 17, *20*
Russell, E.W., 185, *191*
Russell, R.S., 24, 29, 30, 33, *36*: 184, *191*
Russel-Smith, A., 259, *264*
Rutter, P.R., 2, *18*

Safir, G.R., 320, 326, 328, *329*, *330*
Sagar, G.R., 244, *264*: 416, *426*
Saggar, S., 123, *125*
Saker, L.R., 82, *85*: 92, 99, *104*
Salati, E., 132, 137, *142*, *145*, *146*
Salawu, E.O., 302, *316*: 321, *330*
Salisbury, E.J., 214, *217*
Sanders, F.E., 212, *217*: 319, *331*
Sandon, J., 16, *19*
Santantonio, D., 24, 26, 28, 29, 31, *35*, *36*:
  37, 39, 40, *42*: 175, *178*
Sapata, F., 136, *143*
Saretsalo, L., 103, *105*
Sarig, S., 139, *146*: 156, *157*
Satchell, J.E., 379, 390, *392*: 393, 396, *397*
Sator, C., 32, *36*
Sauerbeck, D.R., 8, *20*: 111, *119*: 132, *143*
Savage, H.M., 97, *106*
Schaefer, B., 9, *19*
Schaefer, R., 14, *20*: 342, *353*
Schank, S.C., 129, 131, 138, 139, *144*, *146*:
  156, 157
Schank, S.H., 156, *157*
Schatz, G.E., 418, *424*
Schende, S.T., 138, *144*
Schenk, N.C., 321, *331*
Schidt, K.P., 8, *18*
Schildknecht, H., 306, *316*
Schippers, B., 108, *121*: 356, *365*
Schnürer, J., 357, *364*
Schramm, J.R., 209, *217*
Schroth, M.N., 10, *20*: 139, *144*, *146*
Scott, J.A., 422, *426*
Seastedt, T.R., 9, *20*: 297, 311, *316*: 378,
  392
Selm, A.J. van., 246, *265*
Seniczak, S., 246, 247, *264*
Shaefer, B., 355, *364*
Shafer, S.R., 302, *316*
Shamoot, S., 8, *21*
Sharpley, A.N., 14, *21*
Shaver, G.R., 26, *36*
Shaw, P.J.A., **333–337**: 253: 321, *331*
Shaw, R.H., 26, 28, *36*
Shaw, T.C., 73, *78*

Sheikh, N.A., 319, *331*
Shen, S.M., 123, *125*
Shields, J.A., 109, *121*: 356, *365*
Shugart, H.H., 233, 235, *241*
Sibbesen, E., 73, 74, 77, *78*
Siccama, T.G., 24, *36*
Sims, J.R., 75, *78*
Sims, P.L., 31, *36*
Sinclair, J., 9, *19*: 355, *364*
Singh, J.S., 8, *21*: 31, 32, *36*: 117, *121*
Sinha, R.N., 303, *315*
Sivakumar, M.V.K., 26, 28, *36*
Skeffington, R.A., 127, 140, *144*, *146*: 419,
  423
Skinner, M.F., 209, *217*
Slater, J.H., 189, *191*
Sloger, C., 131, 134, 138, 140, *144*, *146*
Smalley, J.L., 73, *78*
Smart, J.S., 94, 95, 96, *106*
Smith, A.M., 184, *191*
Smith, F.A., 387, 386, *392*
Smith, F.E., 12, *21*
Smith, J.L., 116, *121*
Smith, K.A., 135, *143*: 184, *190*, *191*
Smith, O.L., 159, *173*
Smith, R.L., 129, 131, 138, 139, *144*, *146*:
  156, *157*
Smith, S.E., 177, *178*: 208, *216*: 419, *423*
Smith, V.R., 73, 77, *78*
Smith, W.H., 29, 30, 33, *36*: 342, *354*: 386,
  392
Snider, R.J., 247, 257, *263*, *264*
Snider, R.M., 245, *262*
Söderström, B.E., 9, *21*: 109, *121*: 163, 164,
  165, *173*: 176, *178*: 259, 261: 269, 275,
  277: 297, *316*: 356, *364*: 367, 374, *375*:
  393, *397*
Sohlenius, B., 30, *35*: 169, *173*: 275, *277*:
  297, *316*, 356, *364*: 367, 375
Sokal, R.R., 291, *295*
Sollins, P., 26, 30, *35*, *392*
Southwood, T.R.E., 339, 345, *354*
Sow, B., 342, *353*
Sparkes, K.E., 11, *21*: 244, 248, 249, 254,
  259, *264*, *265*: 321, *331*
Sparling, G.P., 9, *21*: 131, 135, 140, *144*:
  220, 221, *224*: 358, *365*
Spasoff, L., 321, *330*
Splatt, P.R., 377–392
Spokes, J.R., 328, *330*
Springett, B.P., 311, *316*
Springett, J.A., **399–405**: 14, *21*: 311, *316*:
  404, *405*
Spycher, G., *392*
St John, T.V., 15, *20*: 194, *217*: 327, *331*

Staaf, H., 31, *34*: 172, *172*: 275, 277: 374, *375*
Staffeldt, E.E., 343, *353*
Stamou, G.P., 245, *264*
Standen, V., 303, *316*
Stanford, G., 358, *365*
Stansell, J.R., 67, *71*
Stanton, N.L., 254, *264*
Stefaniak, O., 246, 247, *264*
Steingraeber, D.E., 87, *106*
Steingröver, E., 111, *120*
Stelter, H., 244, *264*
Stevens, W., 1, *21*
Stevenson, F.J., 127, *146*: 414, *426*
Stewart, J.W.B., 6, *20*, *21*: 123, 124, *125*
Stewart, W.D.P., 156, *157*
Stockdill, S.M.J., 400, *405*
Stone, E.L., 411, *426*
Stone, L.B., 103, *106*
Stotzky, G., 130, *146*: 299, *317*: 361, *365*
Stout, J.D., 13, *21*: 302, *317*
Straalen, N.M. Van., 244, 247, *264*
Stradling, D.J., 417, *424*
Strahler, A.N., 94, *106*
Stribley, D.P., **219–224**: 221, *224*: 225, *231*: 419, *426*
Struwe, S., 233, 234, 236, *241*
Stuttard, R.A., 302, *315*
Sudhaus, W., 256, *264*
Sutherland, J.B., 184, *191*
Sutherland, J.R., 302, 304, 305, *317*: 320, 321, *331*
Suthipradit, S., 128, *143*
Sutton, R.F., 410, *426*
Swaby, R.J., 188, *190*
Swain, T., 17, *21*
Swaminathan, K., 225, *232*
Swank, W.T., 377, *392*
Swift, B.L., 299, *314*
Swift, D.M., 8, *21*
Swift, J.J., 10, *21*
Swift, M.J., 268, *277*: 298, 300, 301, 310, *317*: 339, 340, 344, 345, 346, 348, *354*: 377, 379, 384, *391*, *392*: 408, 410, 423, *426*
Swift, R.S., 381, *391*
Syers, J.K., 14, *21*
Sylvia, D.M., 419, *426*

Takeda, H., 246, 252, *264*: 302, *317*
Talbot, P.H.B., 307, *317*
Talibudeen, O., 73, 77, *78*: 222, *224*
Tang, C.S., 185, *191*
Tarrand, J.J., 131, *146*
Tate, C.M., 311, *316*

Tate, K.R., 124, *125*
Tavares, M., 221, *224*
Taylor, A.A., 73, *78*
Taylor, G.S., 115, *121*
Taylor, H.M., 26, 28, *36*: 44, *64*: 67, *70*, *71*: 88, *105*
Taylor, L.R., 280, *284*
Taylor, P.J., 85, *85*
Tennant, D., 82, *85*
Thaxter, R., 221, *224*
Theodorou, C., 210, *215*
Thilipoca, H., 128, *143*
Thomas, M., **67–71**
Thompson, A.R., 328, *330*
Thornton, J.D., 162, *173*
Thornton,R.H., 114, *121*
Thrower, L.B., 116, *121*
Thrower, S.L., 116, *121*
Tiedje, J.M., 299, *316*
Tien,T.H., 139, 140, *146*
Tien, T.M., 156, *157*
Tilbrook, P.J., 279, *284*
Tilman, D., 73, 77, *79*
Timmer, L.W., 229, *232*
Timonin, M.I., 306, *317*
Tinker, P.B., **219–224**; 24, 26, *35*: 212, *216*, 217: 220, 221, *224*: 225, *231*, *232*: 319, *331*: 333, *337*: 410, *427*
Tinuer, P.B., 359, *364*
Tisdall, J.M., 4, *21*: 213, *217*
Tomlin, A.D., 87, *105*
Tomlinson, T.B., 87, *105*
Torrey, J.G., 113, *120*
Tosi, L., 251, *264*
Touchot, F., 10, *21*
Trappe, J., 194, *217*
Trappe, J.M., 422, *425*
Trinick, M.J., 220, *224*
Triska, F.J., 419, *427*
Troeng, E., 31, *34*: 37, *42*
Trofymow, J.A., 3, 11, 15, 16, *18*, *19*
Truog, E., 187, *190*
Tucker, S., 311, *317*
Turin, H., 248, *262*
Turk, A., 67, *70*
Turner, S.M., 108, 109, 114, *121*
Twinn, D.C., 236, *242*: 302, *317*
Tyler, M.E., 129, 138, 139, *146*: 156, *157*

Ulber, B., 253, 254, *264*: 286, *295*
Ulbrich, E., 417, *427*
Ulrich, B., 234, 240, *242*
Umali-Garcia, M., 131, *146*: 156, *158*
Uphof, J.C.T., 417, *427*

Usher, M.B., **243–265**: **279–284**: 11, *21*: 73, *79*: 213, *217*: 243, 244, 246, 248, 251, 252, 253, 254, 256, 259, *262*, *264*, *265*: 281, 282, 283, *284*: 286, *295*: 304, 312, *317*: 321, 327, 328, *331*: 333, 334, 336, *337*

Vaituzis, Z., 116, *121*
Valadon, L.A.G., 410, *427*
Valpas, A., 246, 257, *263*, *265*
van Berkhum, P., 131, 134, 138, 140, 141, *142*, *144*, *146*: 150, 156, *158*
van den Driessche, R., 177, *178*
van der Drift, J., 168, 171, *173*: 309, *317*
van Goor, C.P., 412, *427*
van Gundy, S.D., 13, *19*: 321, *329*
van Selm, A.J., 375, *376*
van Straalen, N.M., 375, *376*
van Veen, J.A., 159, 160, 172, *173*, *174*: 378, 389, *391*, *392*
van Vuurde, J.W.L., 108, *121*, 356, *365*
Vancura, V., 110, 111, 113, *120*, *121*: 356, *365*
Vannier, G., 10, *21*
Vartiovaara, U., 188, *191*
Veal, D.A., 189, *191*
Vegter, J.J., 248, *265*
Venkateswarlu, B., 138, *146*
Veresoglou, D.S., 90, *106*
Verhoef, H.A., **367–376**: 15: 246, 265: 374, 375, *376*
Verma, B.C., 225, *232*
Vescio, L.S., 149, *157*
Victoria, R.L., 137, *142*
Vidaver, A.K., 130, *145*
Vilkamaa, P., 256, *263*
Vincent, B., 2, *18*
Visser, S., **297–317**: 10: 9, *20*: 168, *173*: 252, *263*: 303, 304, 309, 310, 311, *315*, *317*: 320, *331*: 336, *337*: 367, *376*
Vitousek, P.H., 127, 129, *143*
Vitousek, P.M., 268, 274, 275, 276, *277*: 377, *392*
Vlassak, K., 130, 132, 138, 139, *145*, *146*
Vogt, D.J., 175, *175*
Vogt, K.A., 28, 31, *34*, *36*: 37, *42*: 175, 177, *179*: 211, *217*
Vokou, D., 181, *191*
von Bulow, J.F.W., 135, *142*
Vos, J., 67, *71*
Vose, P.B., 136, 137, *144*, *145*, *146*

Waid, J.S., 114, 115, *121*
Waide, J.B., 377, *392*
Waiss, A., 186, *190*

Waiss, A.C., 185, *191*
Waldorf, E., 245, *265*
Walker, C., **219–224**: 221, *224*: 419, *425*
Walker, T.W., 6, *21*
Walkley, A., 227, *232*
Wallace, J.M., 185, *191*
Wallwork, J.A., 298, 303, *317*
Wang, G.M., **219–224**
Wani, S.P., 137, 138, *143*
Warcup, J.H., 10, *18*: 301, 307, 311, *317*
Wareing, P.F., 156, *157*
Warembourg, F.R., 132, *146*
Warnock, A.J., 213, *217*: 253, *265*: 286, *295*: 304, 312, *317*: 321, 327, 328, *331*: 333, 336, *337*
Watanabe, I., 137, 138, *143*
Waterhouse, P.L., 73, *79*
Watson, A., 108, *120*
Watt, K.E.F., 14, *21*
Weatherell, J., 410, *427*
Webb, D.P., 298, *317*
Webb, N.R., 245, *265*
Weber, N.A., 306, *317*
Webster, R., 411, *423*
Weeda, W.C., 421, *427*
Weier, K.L., 274, *277*
Weil, R.R., 257, *265*
Weiser, G.C., 139, *146*
Welbank, P.J., 85, *85*
Werner, C., 95, 96, *106*
West, P.W., 75, *79*
Wetselaar, R., 128, *143*
Weydad, F., 251, *264*
Wheeler, J.L., 421, *427*
Whelan, J., 256, *265*
Whipps, J.M., 111, 112, 113, 114, *121*: 132, 140, *146*
White, G.C., 50, *64*
White, J., 293, *295*
White, J.W., 128, *146*
Whitehead, D.C., 85, *85*
Whitford, W.G., 311, *317*
Whitney, G.G., 87, *106*
Whitt, D.M., 128, *147*
Whittaker, J., 9, *20*
Whittaker, J.B., 252, *263*: 304, 310, *315*: 320, *331*: 336, *337*: 348, 349, 350, *353*: 367, *376*
Whittaker, J.W., 303, *317*
Whittaker, R.H., 24, 30, 33, *36*: 255, *265*
Whittingham, J., 213, *217*
Wicklow, D.T., 252, 253, *265*: 299, 300, *315*, *317*
Wiggins, E.A., 305, 310, *317*
Wilkes, J.O., 235, *241*

Williams, B.L., 275, *277*
Williams, E.C., 30, *34*
Williams, E.D., 85, *85*
Williams, J.T., 73, *79*
Williams, S.T., 24, *34*: 241, *242*
Williams, T.E., 128, *142*
Williams, W.M., 73, *78*
Wilson, D.S., 17, *21*
Wilson, E.D., 339, 345, *353*
Wilson, J., 54, 63, *64*: 419, *425*
Wilson, J.M., 220, *224*
Wilson, S.A., 67, *70*
Winston, P.W., 256, *265*
Wiren, A., 275, *277*: 367, *375*
Witen, A., 259, *261*
Witkamp, M., 307, 309, *317*: 368, *376*: 389, *392*: 393, *397*
Witteveen, J., 374, *376*
Witty, J.F., 134, 136, 140, *147*
Woldendorp, J.W., 360, 361, 362, *365*
Wolf, L.L., 410, *425*
Wolfe, M.L., 408, *423*
Wolton, K.M., 421, *427*
Wood, M., 222, *224*
Wood, T.G., 15, *21*

Woodhouse, W.W., 421, *425*, *426*
Woodmansee, R.G., 149, *158*: 389, *391*
Woodmansee, R.W., 159, 160, 163, 169, *173*
Woods, F.W., 212, *217*
Woods, L., 16, *18*: 355, *365*
Woolston, R.E., 113, *121*
Wynne-Edwards, V.C., 17, *21*

Yang, S.F., 185, *189*
Yeates, G.W., 13, 14, 15, *21*: 236, *242*
Yocom, D.H., 252, *65*: 299, 300, *317*
Yoda, K., 293, *295*
Yodzis, P., 13, *21*
Yoneyama, T., 137, 138, *147*
Yoshida, T., 137, 138, *147*
Young, W.F., 413, *426*

Zachariae, G., 10, *21*
Zaidi, Z., 14, *20*
Zak, B., 320, *331*: 333, *337*
Zar, J.H., 227, *232*
Zettel, J., 246, 248, *261*
Zinke, P.J., 412, *427*
Zuberer, D.A., 139, *146*

Subject index

Figures in italics refer to Tables or Figures.

Acari (mites), 13
  in Antarctic moss-turf, 283
  and no-till agriculture, 15–16
  oribatid mites, beech litter, 236–9
  on reclaimed peat, 395–5
Acetylene reduction methods, nitrogen
    fixation, 134–5
Acid rain pollution, monitoring of soil
    sulphate, 77
Aggregation
  soil arthropods, 282
  soil structure, macroaggregates (crumb), 4
    microaggregates, *4*
Agro-ecosystem, mathematical model, 6, 7
  minimum or zero tillage, 15–16
Agropyron, nitrogen fixation, 149–51
Agrostis, drought resistance, 90–1
Agrostis–Festuca grasslands, 350
Allium mycorrhiza–fauna interactions,
    322–7
Allolobophora
  New Zealand pastures, 399–405
  on peat, 394
  see also Earthworms
Alnus, on reclaimed peat, 393
Aluminium, pH and winter oats, 220, 222–3
  inhibition of mycorrhiza, 223
Amanita, slug deterrent alkaloids, 305
Ammonium, losses through nitrification,
    381–2
  animal and root effects, in microcosms,
    385
Amoeba, in soils, *3*, *357*, 359–62
  see also Protozoa
Anaerobic soils, effects of saprophytes, 182
Antarctic moss-turf habitat, collembola,
    279–82
Anthoxanthum specific root length, 90
Antibiosis, microbial saprophytes, 182
Ants, and seed dispersal, 417–18
Aphelenchus, feeding on mycorrhiza, 320
Apple
  root production and distribution, 45
  growth, root observation laboratory,
    51–8

Bacteria *see* Microbes: Nitrogen fixation
Baermann funnels, extraction of soil fauna,
    395
Beech forest
  biomass model, *236*

faunal food source allocation, *237*
fine root production rate, 238–40
microbial biomass production, 240–1
root and microflora productivity, 233,
    235–7
'Benomyl' fungicide, phosphate uptake by
    mycorrhizal plants, 323, 326
Beta, herbivory by soil arthropods,
    285–94
  consequences of damage, intraspecific
    competition, 293
  destructive growth analysis, 286–7
  lethal effects, 288–90
    sub-lethal effects, 290–2
  pesticide treatments, 286
Betula
  colonizing *Calluna* moorland, 413–16
  interspecific mycorrhiza with *Pinus*,
    199–201
Biocides, specificity and pertubation
    experiments, 12–13
Biomass estimates
  root production, 8
    balancing transfers, 41–2
    calculation of flux, 37–40
    decision matrix, 40
    standing crop, various micro-organisms,
    357
  see also Earthworms
Biotechnology, soil inoculation, prospects,
    188–9
Blaniulus, in sugar beet growth, 285
Blanket peat
  transitions, 350–2
  trophic structure, *349*
Botrytis, pore size and fungal sporulation,
    299
Bouteloua nitrogen fixation, 151, *153*, *154*,
    155
Briza specific root length, 90

Calcium
  changes, deciduous woodland, *385–7*
  measurement by ion exchange, 75–7
Calluna moorland
  colonized by *Betula*, 413–16
  depressed growth of test plants, *415*
Carbofuran, effect on root-feeding
    arthropods, 286, 292
Carbon-14 methods
  active roots, criteria, 32

ectomycorrhiza, 195–200
    field experiments, 200
    observation chambers, 195–200
    rhizodeposition, 110–14
C$_4$ plants, nitrogen fixation potential, 133
Carbon flows, structural, metabolic and soil,
        7
    in fungal growth, 160–3
Carbon: nitrogen ratio, soil model, 7, 129
    carbon sources in rhizosphere, 107,
        110–15
    in fungal growth, 162–3
    grazing effects, 168–71
    measurements, 369
    nitrogen dynamics, winter, 370–1
Carlo-Erba Elemental Analyser 1106,
        (nitrogen and carbon), 369
Cellulose
    degradation, 107, 115–15
    and ligno-cellulose, 183
Chamaecyparis mycorrhizal transfer, 200–1
Chlamydospore recovery, 219
Chlorfenvinphos insecticide, and collembola,
        323, 326
Chlorpropham, 286
Citrus, effect of fungal-feeding nematodes,
        10
Coal mining, restoration of soil, 81–5
Cobalt deficiency, 226, 229–31
    uptake enhancement, *228*
    VAM inoculation, 227
Coleoptera, litter decomposition on
        reclaimed peat, 395
Collembola
    Antarctic species, distribution, 279–80
    beech litter, simulations, 237–40
    detoxification of phenols and clay
        ingestion, 10
    earthworm burrows, 69
    effects of grazing intensity, 372
        field study, 368–9, 370–1
        laboratory study, 369, 371–3
        nitrogen, coniferous forest, 367–75
        dynamics, winter, 370–1
    fungal spores as food, 307
    fungus interactions, 372–4
    grazing and fungal growth, 9
    gut content analysis, 324
    mycorrhizal associations, 325
        and shoot yield, 325–7
    on peat, 394
    pine forests, 15
    population dynamics, 243–7, 249–59
        Brillouin index, *258*
    sugar beet attack, 285

Colonizers, of new resources, high inoculum
        potential, 310–11
Comminution of plant debris and fungal
        colonization, 298–301
Computer modelling, ecosystem data
        synthesis, 233
    changes, microbial production, 240–1
    root production, 238–40
    generalized equation, 234
    trophic classification and energy flows,
        236
Copper deficiency, 226, 229–31
    uptake enhancement, *228*
    VAM inoculation, 227
Coprophilous fungi, rabbit dung, and
        dipteran larvae, 299
Core break method, in root measurement,
        82
Cortex, root, early disintegration, 112–113
Cynodon, nitrogen gain, *128*

Dactylis
    nitrogen gains, 128
    specific root length, 90
    VAM and trace nutrients, 225–9
Dendrobaena on peat, 394
    see also Earthworms
Denmark, beech forest, 233–40
Deschampsia
    shade tolerance and mycorrhiza, 214
    specific root length, 90
Detritus accumulation, beech forest, 235–40
Diapause in soil arthropods, 246
Diazotrophs *see* Nitrogen-fixation
Diptera
    larval density, and fungi, *300*, 302
    litter decomposition on reclaimed peat,
        395
Dispersal
    of microbial inoculum, by insects, 306
        by Collembola, 307–10
    of soil fauna, rate, 247–8
Douglas fir *see Pseudotsuga*
Dune grassland soil, nutrient supply, 74–7

Earthworms
    beech litter, simulations, 238–9
    burrows, utilization by slugs, *69*
    increases with tree felling, 259
    litter consumption, 394
    litter decomposition, reclaimed peat,
        393–6
    and nematode populations, 14–15
    New Zealand, effect of introduction,
        399–405

standing crop values, 357
symbiosis with microflora, 342
vertical transport, 404–5
Ecosystems
function and soil interaction, 10
lack of reality, 423
properties and development, 5
simplest concept, *408, 422*
ten compartment model, 409–11
see also Computer modelling
Eigen vector diagram, soil characteristics, *413*
Enchytraeidae, 302, 303, 395
and earthworm burrows, 69
Endomycorrhiza *see* Mycorrhizas, vesicular-arbuscular
Ethylene production, acetylene reduction assay, 134
Evolutionary perspective, mycorrhizas, 17
Exclusion experiments, collembolas, *Pinus* stand, 368–75
Exudate losses, roots, 29–30, 342
see also Rhizodeposition

Faeces, microarthropod origin, 298–301
Fagus, root and microflora productivity, 235–41
see also Beech
Faunal exclusion techniques *see* Exclusion experiments: Litter bag techniques: Lysimeters
Fenuron, 286
Festuca, VAM and trace nutrients, 225–30
Folsomia (collembola)
feeding experiments, 321
ingestion of clay, 10
rate of development, 244–7
Formation of soil, equation *5–6*
Fragaria (strawberry), root production, 45, 59–60
Frankia (bacteria), evolution, 17
Fungal pathogens of sugar beet, 286
Fungi
feeding preferences, collembola, 335
growth and nitrogen mineralization, 159
beech forest model, 235–41
grazing effects, simulations, 168–71
model, nitrogen dynamics, 161–3
simulations and growth experiments, 163–8
parasitic, and sugar beet, 286
in rhizosphere, 107–10
species in pine forest, 420
specific biocides, 12–13
successional sequences, 300–1

toxicity and grazing arthropods, 336
wood decaying species, 336
see also Microbes: Mycorrhizas

Gamasina, beech litter, 237
Geophagy, soil ingestion by micro and macro fauna, 15
Glomeris (millipede) mineral N fluxes, 379–80
effects of feeding on nitrogen leaching, *380*
used in modelling effect, litter feeders on nitrogen mineralization, 382–5
Glomus
Collembolan densities and mycorrhizal food, 322–4
and shoot yield, 325–7
rate of growth with Collembola, 304
veracity of identity, 221
E3 mycorrhiza, 194, 219, 225–31
Glucose, to mimic root-derived C, 359
Gossypium, effect of fungal-feeding nematodes, 10
Gram-negative bacteria, 356
Graminae, roots, % in primary production, 26, 27, *31*
Grass root production following opencast mining, 81–5
Grassland
developed from bush clearance, New Zealand, 399–405
effect of introducing non-native earthworms, 400–5
dune, nutrient supply rates, 73–7
nitrogen gains, 128
Grazing
high input systems, 85
microbial–faunal interactions, *9*
microbial food sources, 302–6
and mineralization of N, *12*
effect on root production, 84
see also Herbivory: names of fauna

Harvest method
root biomass determination, 24–6, 31
root production: mass ratio, 33
Herbivores, effects on plant production, invertebrates, vertebrates, 421–2
Herbivory losses, root production, 30, 33
Hestehave model, Danish beech forest, 233–40, *236*
Historical aspects, 2
Holcus, specific root length, 89
Hordeum (barley)
N-budget, fertilized soil, 357–8

specific root length, 90, *93*
Horizons, surface organic, 268
Humus formation
  beech forest, 237–9
  sitka spruce forest, 267, 274
Hyphae
  cytoplasmic v. evacuated, 160, 170–1
  'vessel' hyphae, 200
  see also Mycorrhizas

Ilex mycorrhizal transfer, 200–1
Infiltration rings, water measurement,
    401–2, *404*
Inocybe in pine stands, 335, 420
Insects *see* names of orders; genera
Invertebrates, soil, and microbial
    communities, 297–313
  comminution, channelling, mixing,
      298–301
  grazing, 302
    dispersal, 306
    microbial communities, 303, 310–13
    microbial food sources, 302–3
    see also specific names
Ion-exchange resins, measurement of soil
    nutrient status, 73–7
Isotope dilution methods, nitrogen fixation,
    135–7

Juncus peat bog, 350

Laccaria, 335, 420
Lactarius, 335, 420
Leaching, nutrients, 183–4
Lichens, and nitrogen fixation, 140, 156
Lime, below surface transport by
    earthworms, 400–5
Liming experiments *see*
    Vesicular-arbuscular mycorrhiza
Liriodendron
  root losses, 28
    photosynthate loss to nematodes, 30
  root production: mass ratio, 33
  seasonal standing crop variation, 27
Litter bag techniques, 393–4
Litter-feeding animals, effect on nitrogen
    metabolism, ANREG regression
    model, 382–5
  experimental approach, 382–4
Lolium
  fungal colonization, 109, 114
  nitrogen gains, 128, *154, 155*
  S-23, 81–5
  split-root experiment, 92
  VAM and trace nutrients (Cu, Co, Mg),
    225–30

Lumbricids on blanket bog, 350
  see also Earthworms
Lysimeters
  deciduous woodland, 387–9, 390
  fine root persistence, 175
  leachate measurement, 268–9

Magnesium deficiency, 226, 229–31
  measurement by ion exchange, 75–7
  uptake enhancement, *228*
  VAM inoculation, 227
Malus (apple), root production, 45, 51–8
Manganese, pH and winter oats, 220, 222–3
Marasmius in *Pinus* stands, 335
Mass flow, in soil solution, 359
Merulius lachrymans, ingestion of spores by
    woodlice, 307
Microaggregates, orders of magnitude, 4
Microbes (microbial)
  bacteria and fungi, abundance, 107–10
  biomass in soil, 123–4, 235–41
    see also Beech forest
  growth rate, 108
  mineralization *see* Nitrogen
  motility, 118
  populations
    in blanket peat, *349*
    in macroaggregates, *3, 4*
    primary decomposers, 6
  saprophytes, 181–8
  see also Bacteria: Fungi: Nitrogen
      fixation: Protozoa
Microcosm experiments
  animal and root effects, lysimeters, 387–9
    microcosms, 384–7
  interactive effects, N cycling, 381
  wheat and nitrogen uptake, *362*
Millipedes, and sugar beet growth, 285
  see also Glomus
Mineralization
  of organic N and P, *11, 12*
  in soil formation, 6
  see also Nitrogen mineralization
Mites *see* Acari
Mollusca, litter decomposition, reclaimed
    peat, 395
Moor House Nature Reserve
  litter decomposition, 341
  succession on eroded peat, 257
  trophic structure, summary, *349*
Muhlenbergia, nitrogen fixation, 151, 153
Mutualism, and group selection, 17
Mycorrhizas
  and *Betula* stands, 419, 420
  'bridges', 8, 116

early land plants, 193
ectomycorrhizas, definition, 194
   direct nutrient transfer, 213–15
   experimental chambers, 195
   inter-plant connection, *196*, 410
   toxic substances, 305
external mycelial phase, 208–15
   interplant carbon transfer, 214–15
   phosphorus transfer, 213
   respiration, 211
-fauna interactions, 319–21
   collembola extraction, 322
fine root persistence, 175–7
mixed-species experiments, 116–17
nitrogen fixation, 151
nutrient cycling, 193
production estimates, 28, 31–2
root classification, 25–6
specificity, 194
standing crop estimates, 26
trace element uptake, 225–31
vesicular-arbuscular (VAM), 10–12, 15,
   91
   definition, 194
   direct transfer, nutrients, 212–15
   experimental methods, 202–8
   nematodes, 320–21
   water gradient, 200
see also Fungi
Nebraska Sand Hills grassland, nitrogen
   fixation, 149–56
Nematodes, on reclaimed peat, 394, 395
   bacterial feeding, 15–16
   beech litter, simulations, 238–9
   food preferences, 302
   fungal feeding, 9
   trophic interactions, 13
   in water films, 14
New Zealand, effect of *Allolobophora* on
   root distribution, 399–405
Nitrate
   application in opencast mine soils,
      81–5
   enhanced rate of decomposition on peat,
      352
   measurement by ion exchange, 75–7
Nitrogen dynamics, coniferous forest,
   367–75
Nitrogen fixation, 127–41
   annual net gains, *128*, 129, 140–1
   asymbiotic heterotrophs, 127–9
   bacteria, 130–2
      localization, 156
   energy sources, 132–3
   inoculation, growth responses, 138–9

measurement, 134–8
   acetylene reduction assay, 134, 150
   bioassay, 150, 152
   isotope dilution, 135–8
   overestimation, 140
response to isolates, 152–6
specific associations, 131
Nitrogen flows, structural, metabolic and
   soil, *7*
in fungal growth, 160–3
   simulations and growth experiments,
      163–8
mineralization, and grazers, *12*
   simulations, 168–71
Nitrogen losses, oak woodland, lysimeters,
   388
Nitrogen mineralization
   deciduous forest, 377–90
      animal and root effects, in lysimeters,
         387–9
      in microcosms, 384–7
      animals, direct and indirect effects, 378
      litter animals, general effects, 390
      net mineralization, 377
      process variables, nutrient flux rates,
         379
         ammonium losses and nitrification,
            381
         organisms, 382
   N-budget, barley crop, 357, *358*
   root surface interactions, *360*
      experimental evidence, 361
   effects of roots, 358
      glucose additions, 359
      root derived N, 362
   soil organic matter, 356
Nitrogen transformations, spruce humus,
   267, 274–7
No-till agriculture, 15–16
Nutrients
   cycling *see* names of minerals
   immobilization, 183
      and prevention of leaching, 183
   turnover rate and pool size, 6

Omnivory, and trophic level, 13
Onychiurus
   fungi associated with, 307–9
   grazing preferences, 309–10
   on *Pythium*, parasitic on *Beta*, 285–6
Opencast mining, restoration of soil, 81–5

Parthenogenesis in collembola, 244
Paspalum and nitrogen fixation, 129, 131
Peat, reclaimed cutover, role of

invertebrates in decomposition, 393–6
Pectin, degradation, 114
Ped, definition, 5
Pennisetum (millet), inoculation with N-fixing bacteria, 139
Perturbation experiments, 11–12
pH, and lime additions, *403*
Phenmedipham, post-emergence treatment, 286
Phleum sward
and apple root production, 45
root growth, 61–3
specific root length, 90
Phosphate
inorganic and organic, flows, *6*
measurement by ion exchange, 75–7
microbial, and protozoan biomass, 16
and no-till plots, 16
mineralization, with bacterial-feeding nematodes, *11*
with soil-eating termites, 15
reduced rate of decomposition, 352
uptake in mycorrhizal plants, 319, 323–8
Picea (Sitka spruce)
competitive saprophytic colonization, 304
immobilization of nitrogen, 276
interspecific mycorrhiza with *Pinus*, 199–201
mycorrhiza, collembolan food preference, 333–6
root biomass, 25, 27, 28
fine roots, annual production, 54
seasonal fluctuations, *41*
Pinus spp.
interspecies mycorrhizas, 199
mycorrhiza, collembolan food preferences, 333–6
nitrogen dynamics, collembolan grazing, 367–75
nitrogen gains, *128*
root biomass, 25
current annual increment, 27
fungal feeding nematodes, 30
root losses, 28
root size, *28*
Plant spacing, multinomial theorem, 290
Poa, specific root length, 90
Podsol, nitrogen dynamics, 367–75
Polytrichum cover, Antarctic moss turf, 279–83
Population dynamics, 243
natural communities, 254
adversity-selection, 260
changes, 256

management, 257
perturbation experiments, 259
site type, 255
species diversity, 254
single species, developmental rate, 246
fecundity, 244–6
immigration and emigration, 247
mortality, 247
two or three species, herbivory, 252
interspecific competition, 248–50
predation, 250
third level predators, 253
Populus, on reclaimed peat, 393
Porosity, effect of earthworms, 402–5
Potassium changes, deciduous woodland, 385–7
Potassium, measurement by ion exchange, 75–7
Prairies, net primary production, 31
Pre-emergence herbicides, 286
Primary production
contribution of roots, different ecosystems, *31*
sitka spruce, seasonal fluctuations, *41*
see also Biomass estimates
Propham, 286
Protozoa
biomass, and phosphorus, 16
ingestion by earthworms, 15
role in supplying nitrogen to plants, 355, 357, 359
Pseudotsuga (Douglas fir)
mycorrhizal losses, 29
net primary production, 31–33
root length and age, 25, 27, *28*
root, production: mass ratio, 33
Pythium
on *Beta*, 285
pore size and sporulation, 299

Quercus (oak) detoxification of phenols by collembolans, 10

Rhizobium, evolution, 17
Rhizodeposition, 107
carbon-14 methods, 110–14
Rhizopogon and Ca oxalate production, 305
Rhizosphere, bacterial and fungi, relative abundance, 107–10
interactions and translocations, 115–18
see also Nitrogen fixation
Rhizotron *see* Root observation laboratory
Root competition, *Betula* and *Anemone*, 419, *421*

Root distribution, effect of *Allolobophora*, 399–405
Root exclusion experiments, 267–76
　discussion, 274–6
　FDA stained hyphae, 273
　lysimeter installation, 268
　nitrogen measurement, exchangeable and net nitrogen, 271
　　throughfall, 270
　organic matter, 274
　sampling and analysis of leachate, 269
Root grafts, *Pinus* girdling experiments, 410–11
Root morphology
　specific root length (SRL), 87
　　genetic variation, 89
　　interaction with other organisms, 100–2, 115–17
　　mechanisms, 93
　　mycorrhizas, 91, 100–2
　　pathogens, 91
　　quantitative description, 94–9
　　root diameter, 93, 97–8
　　search theory, 102–3
　　soil conditions, 89
　　spatial and temporal heterogeneity, 91–2
　　topology, 94, 99
Root observation laboratory, 43–63
　apple root growth, 51–8
　　browning, 51–3, 55
　　decomposition, 53
　　mineral reserves, 57–8
　　periodicity, 51–2
　　woody roots, 55–7
　grass root growth (*Phleum*), 61–3
　　density, 63
　　production and decomposition, 61–2
　materials and methods, 45–6
　　periodicity, 49
　　quantifying data, 50
　　root density, 47
　　root distribution, 50
　　sample position and size, 48–50
　root growth and invertebrate activity, 67–70
　　design, 67–9
　　results and discussion, 69
　strawberry root growth, 59–60
　　browning, 59–60
　　production and decomposition, 60
　　effect of tree competition, 59
　use in characterizing periodicity of growth, 45

Root production, grass, following opencast mining, 81–5
　effect of grazing, 84
Roots
　active and inactive, criteria, techniques, 32
　fine root production, calculation, 37–40
　　decision matrix, 40
　　example, 40–2
　　population dynamics, 37
　　seasonal fluctuations, spruce, *41*
　　simulation, beech forest, 233–40
　primary production, 8, 23–33
　　biomass changes, 24–9
　　exudate losses, 29–30
　　fine roots, various ecosystems, *28*
　　growth rings, volume increments, 27
　　herbivory losses, 30–1
　　methods of calculation, 23–4
　　microbe interactions, 8
　　net primary production, 31–3
　　persistence of excised roots, 175–7
　　root cap tissue, 113
　see also species names
Rumex and watering rates, 97–9
Russula, in pine stands, 335, 420
Rusts and smuts, insect dispersal of spores, 306

Salix, litter decomposition on reclaimed peat, 393–6
Saprophytes, microbial, on straw, 181–8
　beneficial and detrimental effects, 181–2
　infections of seeds, 186–7
　nutrient cycling, 287–8
　nutrient immobilization, 183
　phytotoxin production, acetic acid, 185
　　ethylene, 184
　　patulin, 186
　soil stabilizing micro-organisms, 187–8
Scaptocoris, production of fungicidal substances, 307
Schizachyrium, nitrogen-fixation, 151, *154*, *155*
Scotland, North East, mineral deficiency, 225–31
Scutigerella in sugar beet growth, 285–6
Seed banks, longevity, 416–17
Seed dispersal by ants, 417–18
Seedling niches, decomposing logs, *418*
Self thinning in beet, and grazing, 293–4
Sequential harvesting, error in production measurement, 8
Signy Island, Antarctica

moss bank ecology, 279–82
  trophic structure, *351*
Size of microfauna, and role in soil
    community, 14
Slugs, 394
  utilization of earthworm burrows, 69
Soil ingestion, by micro- and macrofauna, 15
Soil–pest complex and sugar beet, 285
Soil variability, 411–16
  Eigen vector diagram, 413
  spatial, 411
  temporal, 412
South Orkney, moss-turf collembolans,
    279–82
Soybean, root biomass, 26
Springtails *see* Collembolans
Spruce, sitka *see Picea*
Standing crop biomass, various
    micro-organisms, *357*
Starch, degradation, 114
Straw decomposition, 181–8
Stubble burning, 7
Stubble-mulch plots, 16
Sugar-beet *see Beta*
Swedish arable lands project, 7
Swift's model, fundamental niches, *344–5*,
    350
System-level manipulation, reduction of
    decomposers by gamma rays, 11
  perturbation experiments, 11–12

Timothy grass *see Phleum*
Tomocerus (Collembola)
  effects on nitrogen, coniferous forest, 367
  ingestion of clay, 10
Transport, vertical, by earth worms, 404–5
Trenching *see* Root exclusion
Trifolium and mycorrhiza, *101, 102*
  interaction with *Lolium*, 116
  mycorrhiza–fauna interactions, 322–7
Trophic relationships, surface horizon, *3*
  length of food chain, 12
  model, prairie, *13*

Trophic structure, resource quality, 339–51
  below-ground, 341
    exploitation phase, 345
    fundamental microbial niches, *344*
    interaction phase, 346
    root-litter: weight loss, *343*
    Swift's model, 344–5
  physical restraints, 347
    quantitative approach, 348, *349*
Tullgren funnels, extraction of soil fauna,
    369, 395
Turnover rate of nutrients, and pool size, 6

Vertical transport, by earthworms, 404–5
Verticillium (fungus), and *Tomocerus*,
    367–75
Vesicular-arbuscular mycorrhizae (VAM),
    10–12, 15, 91
  'arterial' hyphae, *204, 207, 209*
  C-14 experimental methods, 202–8
  definition, 194
  direct nutrient transfer, 212–15
  liming experiments and VA mycorrhizas,
    219–23
    aluminium and manganese, changes,
    222
    pH and endophyte ratio, 220–1
    sudden change of pH 221–2
  trace and minor element uptake, 225–31
  see also Mycorrhiza
Vital staining technique, active roots, 32

Weed diversity, and soil-pest grazing, 293–4
Wheat growth, nitrogen uptake, *362*
  root biomass, 26
Woodland, nitrogen gains, 128
  see also Beech forest: *Betula: Picea:
    Pinus: Pseudotsuga*

Zero-tillage, dry-land wheat, 15